Primary Productivity of Grass Ecosystems of the Tropics and Sub-tropics

Primary Productivity of Grass Ecosystems of the Tropics and Sub-tropics

Edited by

STEPHEN P. LONG,

Professor of Biology,
University of Essex

MICHAEL B. JONES,

Lecturer and Fellow,
Trinity College, Dublin

MICHAEL J. ROBERTS,

Department of the Environment,
London

CHAPMAN & HALL

London · New York · Tokyo · Melbourne · Madras

Published by Chapman & Hall, 2–6 Boundary Row, London SE1 8HN

Chapman & Hall, 2–6 Boundary Row, London SE1 8HN, UK

Chapman & Hall, 29 West 35th Street, New York NY10001, USA

Chapman & Hall Japan, Thomson Publishing Japan, Hirakawacho Nemoto Building, 7F, 1-7-11 Hirakawa-cho, Chiyoda-ku, Tokyo 102, Japan

Chapman & Hall Australia, Thomas Nelson Australia, 102 Dodds Street, South Melbourne, Victoria 3205, Australia

Chapman & Hall India, R. Seshadri, 32 Second Main Road, CIT East, Madras 600 035, India

First edition 1992

Typeset in 10/12 pt Palatino by Columns Ltd, Reading, Berkshire
Printed in Great Britain by T.J. Press (Padstow) Ltd, Padstow, Cornwall

ISBN 0 412 41020 6

A catalogue record for this book is available from the British Library

Library of Congress Cataloging-in-Publication Data

Primary productivity of grass ecosystems of the tropics and sub-tropics / edited by Stephen P. Long, Michael B. Jones, and Michael J. Roberts.
p. cm.
Includes bibliographical references and index.
ISBN 0–412–41020–6
1. Grasslands—Tropics. 2. Grassland ecology—Tropics. 3. Primary productivity (Biology)—Tropics. I. Long, Stephen P. (Stephen Patrick), 1950– . II. Jones, M. B. (Michael B.), 1946– . III. Roberts, Michael J. (Michael John), 1953– .
QK938.P7P75 1992 91–21499
581.5′2643′0913—dc20 CIP

Contents

Contributors

J.A.N. DE MELLO
Instituto Nacional de Pesquisas da Amazônia (INPA), Cx.Postal 478, 69083 Manaus, AM, Brazil.

J.P. EVENSON
Thai-Australian Project, Department of Botany, University of Queensland, St. Lucia, Brisbane, 4067 Australia.

A.-X. GAO
Sub-Tropical Forest Institute, The Chinese Academy of Forestry, Fuyang, Zhejiang Province, China.

E. GARCIA-MOYA
Centro de Botanica, Colegio de Postgraduados, Chapingo, Mexico 56230, Mexico.

D.O. HALL
Department of Biosphere Sciences, King's College London, University of London, Campden Hill Road, London, W8 7AH, UK.

Q.-M. HUANG
Sub-Tropical Forest Institute, The Chinese Academy of Forestry, Fuyang, Zhejiang Province, China.

S.K. IMBAMBA
Department of Botany, University of Nairobi, PO Box 30197, Nairobi, Kenya.

M.B. JONES
School of Botany, Trinity College, University of Dublin, Dublin 2, Ireland.

W.J. JUNK
Max-Planck-Institut fur Limnologie, AG Tropenökologie, Postfach 165, D-2320 Plön, Germany.

Contributors

APINAN KAMNALRUT
Faculty of Natural Resources, Prince of Songkhla University, Hat Yai, Thailand 90110.

J.I. KINYAMARIO
Department of Botany, University of Nairobi, PO Box 30197, Nairobi, Kenya.

D.-Y. LI
Shanghai Institute of Plant Physiology, Academia Sinica, 300 Fenglin Road, Shanghai 200032, China.

S.P. LONG
Department of Biology, University of Essex, Colchester, CO4 3SQ, UK.

P. MONTANEZ CASTRO
Centro de Botanica, Colegio de Postgraduados, Chapingo, Mexico 56230, Mexico.

R.J. OLEMBO
Deputy Assistant Executive Director, United Nations Environment Programme, PO Box 30552, Nairobi, Kenya.

M.T.F. PIEDADE
Instituto Nacional de Pesquisas da Amazônia (INPA), Cx.Postal 478, 69083 Manaus, AM, Brazil.

G.-X. QIU
Shanghai Institute of Plant Physiology, Academia Sinica, 300 Fenglin Road, Shanghai 200032, China.

M.J. ROBERTS
Department of the Environment, Romney House, 43 Marsham Street, London, SW1T 3PY, UK.

J.M.O. SCURLOCK
Department of Biosphere Sciences, King's College London, University of London, Campden Hill Road, London, W8 7AH, UK.

Y.-K. SHEN
Shanghai Institute of Plant Physiology, Academia Sinica, 300 Fenglin Road, Shanghai 200032, China.

Z.-W. WANG
Shanghai Institute of Plant Physiology, Academia Sinica, 300 Fenglin Road, Shanghai 200032, China.

D.-D. YANG
Sub-Tropical Forest Institute, The Chinese Academy of Forestry, Fuyang, Zhejiang Province, China.

Project co-ordinator's preface

This book arises from the research conducted under United Nations Environment Programme Project FP1305-83-01(2405)(new) The Primary Productivity and Photosynthesis of Semi-Natural Ecosystems of the Tropics and Sub-tropics. This study commenced in 1984, and forms part of an ongoing project through continued support from the UNEP and many other sources. At the time this work was being planned information on the production, photosynthesis and carbon flow of the world's largest terrestrial biome (Hammond, 1990), tropical grasslands, was sparse in comparison to other biomes. Further, research on temperate grasslands had suggested that the methods used for grasslands in the International Biological Programme (IBP) would have underestimated production and that this error might be of even greater significance in tropical environments (Beadle *et al.*, 1985). This was of particular significance, since the IBP studies provided the key source of information on ecosystems within for the tropics. At the same time new developments and advances in understanding of field measurement of photosynthesis and remote sensing of production provided opportunities which were not available to the earlier IBP studies. Thus a major gap in our understanding of current production and C-flows in the biosphere was identified. In 1984, this gap was recognized to be of particular significance given the growing concern about possible changes in climate and the need to provide sound baseline data and appropriate methods which would allow long-term monitoring of response to change.

The long-term objective of the project was to develop expertise on grassland primary production within developing countries of the tropics and sub-tropics, and to establish a network of regional centres, supported by the co-ordinating centre at King's College and technical, research and database management through a Technical Support Unit established at the University of Essex. Achieving these objectives has involved five major developments: (1) Identification of centres; (2) Training; (3) Establishment of an Information Centre; (4) Identification and documentation of methods; (5) Establishment of a central database and data analysis. Five regional centres were approached and established. Their success is clearly demonstrated by the contributions that each has made in the form of an individual chapter to this book and publications elsewhere. However, these are not a collection of isolated studies, a uniform approach was agreed prior to commencing studies and has been used at all sites as exemplified by the use of a common methods and aims section in Chapter 1. This book also contains the methods developed between the centres during this project.

Training has been provided through a series of international hands-on practical courses, and a new training book arising from these courses is also published by Chapman & Hall (Hall *et al.*, 1992). During this period over 200 researchers from 25 countries have participated in our 3 week courses. An information centre has been established at King's College, whilst the University of Essex has recently published the 3rd edition of its tabulations and comparisons of equipment for field productivity and environmental physiology research (Bingham and Long, 1990) which has been made freely available to over 4000 research scientists, mostly in developing countries. This information will also be incorporated in the new training book (Hall *et al.*, 1992). The success of the project in providing new and more detailed information on carbon flow, photosynthesis, productivity, and remote sensing in tropical grasslands, is underlined by the many new findings and detailed data presented in this book. A measure of our success in developing expertise is the fact that four of the senior authors of the chapters were students of the early training courses of this project, who are now regularly invited to act as experts and teachers on training courses themselves and to contribute to international scientific conferences. The results presented (Chapters 2–8) provide some of the most detailed data available to date on the role of tropical and sub-tropical grasslands as both sources and sinks for carbon in the global C-cycle. Information from the database developed by the project is already being incorporated into some of the major models currently being developed by international groups for predicting ecosystem change in response to global climate change.

Although this project has been supported by many international and national agencies, a co-ordinated programme was only possible through the support provided by the UNEP. Professor R.J. Olembo of the UNEP foresaw the need to improve knowledge of photosynthesis, production and carbon flow in tropical grasslands, was instrumental in establishing the project and has maintained an active interest in this work throughout. The new information in this book on the potential of tropical grasslands as a major sink of CO_2, at a time when rising atmospheric CO_2 levels have become such a major environmental concern shows just how adroit this foresight was.

D.O. Hall,
King's College, London.

Hammond, A.L. (ed.) (1990) **World Resources 1990-1**. World Resources Institute/UNEP/UNDP/Oxford University Press, Oxford.
Beadle, C.L., Long, S.P., Imbamba, S.K., Hall, D.O. and Olembo, R.J. (1985) **Photosynthesis in Relation to Plant Production in Terrestrial Environments.** UNEP/Tycooly International, Oxford.
Hall, D.O., Scurlock, J.M.O., Bolhár-Nordenkampf, H.R., Leegood, R.C. and Long, S.P. (eds) (1992) **Photosynthesis and Production in a Changing Environment: A field and laboratory manual.** Chapman & Hall, London.
Bingham, M.J. and Long, S.P. (1990) **Equipment for Crop and Environmental Physiology: Specifications, sources and costs.** 3rd edn UNEP/University of Essex, Colchester.

Abbreviations and Symbols

A photosynthetic rate of CO_2 uptake per unit surface area (μ-mol m^{-2} s^{-1})
A_{sat} light saturated A (μ-mol m^{-2} s^{-1})
A_{max} maximum net photosynthesis (μ-mol m^{-2} s^{-1})
AP_n above-ground net primary production (g m^{-2} yr^{-1})
B dry weight of live plant material – biomass (g m^{-2})
BP_n below-ground net primary production (g m^{-2} yr^{-1})
C carbon
C_3 photosynthetic pathway where three carbon compounds are the first product
C_4 photosynthetic pathway where four carbon compounds are the first product
D dry weight of dead plant material (g m^{-2})
E_c light energy conversion efficiency (measure of efficiency with which S_{abs} is used to produce dry matter
E_i efficiency with which vegetation absorbs light
GCMs Global Circulation Models
IBP International Biological Programme
L leaf area index (leaf area per unit ground area; dimensionless)
l plant losses through death (g m^{-2} yr^{-1})
LAD leaf area duration
NDVI normalised difference vegetation index
PAR photosynthetically active radiation (400–700 nm; W m^{-2})
PFD photon flux density (μ-mol m^{-2} s^{-1})
*P*n net primary productivity (g m^{-2} $month^{-1}$; g m^{-2} yr^{-1})
r relative rate of decomposition of plant material(g g^{-1} $month^{-1}$)
RDI Radioactive dryness index
RuBP Ribulose bis-phosphate
R/NIR red to near infra-red reflectivity ratio
SAI stem area index
S_a incident solar radiation received above the canopy (MJ m^{-2})
S_{abs} solar radiation intercepted by plant canopy (MJ m^{-2})
S_b incident solar radiation penetrating to the base of the canopy (MJ m^{-2})
SOM soil organic matter
UNEP United Nations Environment Programme
W_d absolute rate of decomposition of plant material (g m^{-2} $month^{-1}$)
ε_c light energy conversion 'efficiency' (measure of the 'efficiency' with which S_{abs} is used to produce dry matter; g MJ^{-1})
ε_i efficiency with which vegetation absorbs light (dimensionless)
Ψ leaf water potential

1

Introduction, aims, goals and general methods

S.P. LONG and M.B. JONES

1.1 INTRODUCTION

Whilst the productivity and standing carbon stocks of tropical forests have attracted considerable attention, particularly in respect of implications for the 'global greenhouse effect', less attention has been given to other natural and semi-natural communities of the developing world. This is despite their recognized importance in the global carbon budget (Hall, 1989). For example, it has been estimated that while tropical forests store only about 19% of the total carbon sequestered by terrestrial communities each year, tropical grasslands are suggested to store about 26% of the total (Gifford, 1980; Gates, 1985). Sub-tropical bamboo forests form another major community type which has been neglected in respect of production ecology. In surveying studies of primary production and carbon turnover in forests world-wide, Vogt *et al.* (1986) identified 270 studies – not one concerned bamboos. This book concerns both tropical grasslands and sub-tropical bamboo forests (Figure 1.1). These communities have two features in common. First, the dominant or co-dominant species are graminaceous. Second, both are principally communities of the developing world. Understanding of the precise quantitative roles of vegetation in exchange and removal of carbon in the global carbon cycle requires reliable estimates of both standing stocks of plant biomass, including that below-ground, and reliable estimates of productivity. Such information is now of increased importance in providing a reference and in aiding understanding of how exchanges of carbon in these communities may respond to the predicted climatic changes and their potential for ameliorating the 'Greenhouse effect' by sequestering carbon (Houghton *et al.*, 1990). The surveys of biomass and productivity provided by the International Biological Programme (IBP), stimulated further and more detailed studies within temperate

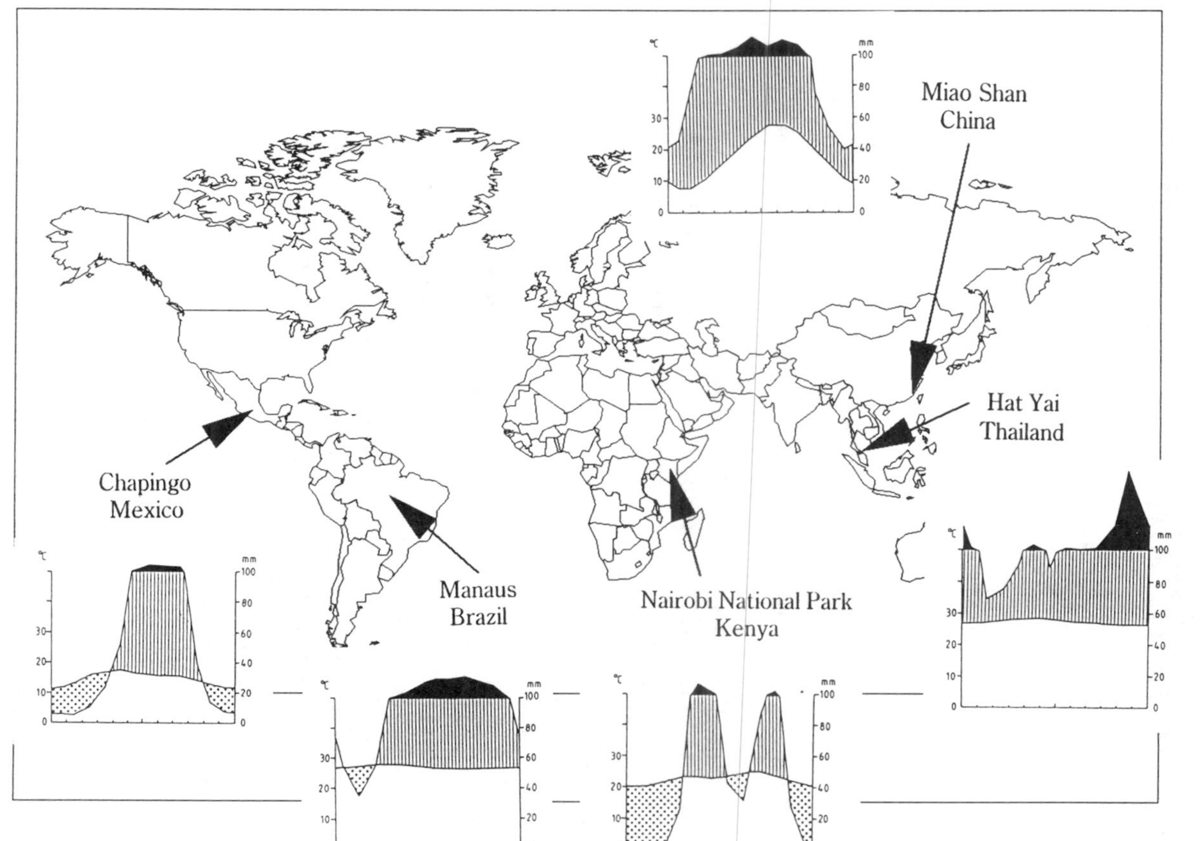

Figure 1.1. The location of the study sites and climate diagrams representative of the sites, redrawn from Walter *et al.* (1975). The horizontal axis on the climate diagrams represents January to December and the vertical axis is mean monthly temperature (left) and monthly precipitation (right). Dotted areas represent periods of relative drought (temperature is the upper line, precipitation the lower in these portions of the diagram) for the regions concerned and, vertical shading represents relatively humid periods when the vegetation is unlikely to be subjected to drought. Black shading represents periods with mean monthly precipitation in excess of 100 mm; here the vertical scale is reduced to 1/10 of that below 100 mm precipitation.

and cold climate ecosystems. These have demonstrated the importance of taking full account of turnover of vegetation in estimating productivity, of below-ground biomass, and of variation between years (Linthurst and Reimold, 1978; Long and Mason, 1983; Vogt *et al.*, 1986). To stimulate similar work, investigating these same points within ecosystems of developing countries, the United Nations Environment Programme began support in 1984 of five studies of productivity in different locations. The objectives of this work were not to duplicate the already extensive data of IBP (Cooper, 1975; Coupland, 1979), but to promote intensive investigations of the significance of factors which received little attention in previous studies – in particular, below-ground biomass, measurement of turnover rates, short-term dynamics of above- and below-ground mass, rates of decomposition, and year-to-year variation in relation to climatic fluctuations. To maximize comparability of the results obtained techniques were common, as far as practicable, for all vegetation types. Work was co-ordinated through a co-ordinating centre and a technical support unit in the UK, and through annual meetings between the centres and an international panel of advisers.

This chapter explains why we consider the existing data on these communities to be deficient, the objectives of the work undertaken and the common methods used.

1.2 SCOPE, DEFINITION AND STATUS OF TROPICAL GRASSLANDS

Estimates of the area occupied by tropical grasslands vary from 15.0 to 24.6 million square kilometres of the Earth's surface (Lieth, 1978; World Resources Institute, 1990). In terms of area occupied this is now the largest terrestrial biome (World Resources Institute, 1990) and in terms of biomass it is second only to the tropical forest biome (Lieth, 1978). In addition, periodically inundated grasslands, a form of wetland ecosystem, occupy 0.7–1.0 million square kilometres in South America alone, of which one-third is in the Amazon basin (W.J. Junk, personal communication).

For the purposes of this book, tropical grasslands are defined as any ecosystem within the tropics in which graminaceous species are a dominant or co-dominant character of the vegetation. Thus, this consideration includes not only purely herbaceous communities, but also the mixed grass and tree communities of the savanna and savanna forests. Tropical grasslands typically occur between the equatorial forests and the hot deserts. These grasslands are therefore not only climatically distinct from temperate grasslands, but are often biogeographically separated by desert regions. Nevertheless these two groups of grasslands do share a few taxa in common. For example, C_4 grasses of the genus *Andropogon* (bluestems) are important components of both tropical and temperate grasslands (Hopkins, 1974). Moving from the tropical rainforests towards the deserts, grass-dominated communities occur naturally where soil moisture becomes insufficient to support a closed forest canopy. The point at which this transition occurs will depend on precipitation, humidity, soil, frequency of fire and grazing pressure. In general though, the transition will occur when the dry season extends to 5–7 months, and rainfall is in the region of 800–2000 mm. In progressively drier areas the proportion of trees to grass diminishes (Hopkins, 1974). However, in contrast to the vast treeless expanses of the temperate prairie and steppe grasslands, much tropical grassland consists of a continuous grass cover with isolated small trees which have developed under the influence of fire and grazing, i.e. the savanna regions of Africa, South America and tropical Australia. Typical savanna grassland is represented in this work by a study site in Nairobi National Park, Kenya (Chapter 2). Eastern and southern Africa is the principal region of the tropics where a diverse and dense fauna of native large ungulates is preserved in comparatively natural and unmanaged ecological surroundings. McNaughton (1985) has suggested that it is only here that the adaptive traits of grasses which have evolved under grazing by dense native ungulate populations have been retained. In South-East Asia despite high annual precipitation, the seasonality of the monsoonal climate produces dry periods of sufficient duration to allow regular fires and prevent formation of a closed forest canopy. Here these openings in the forest allow the development of savanna-like grasslands (Kayll, 1974). This community is represented in the present work by a study site at Hat Yai in southern Thailand (Chapter 3). Trees may only be absent from tropical grasslands where edaphic factors, other than availability of water, prevent their establishment. This is likely to include areas of high salinity and areas which are regularly flooded. Examples of these are provided in the present work by a saline grassland in the volcanic valley of Mexico (Chapter 4) and grassland on the floodplain in the central Amazon (Chapter 5).

Although some areas of tropical grassland clearly represent a natural climax for the prevailing climatic conditions, many tropical grasslands are anthropogenic in origin. The long dry season of these areas coupled with high production in the preceding wet season makes them highly susceptible to fire. Lamotte (1967) at Lamto on the Ivory

Coast and Charter and Keay (1960) in Nigeria demonstrated that protection of these areas of savanna from fire allowed quite rapid reversion to closed secondary forest. Fires within tropical grasslands do occur naturally but in many areas a long tradition of burning the vegetation exists to improve the grazing value of these lands. Fire may play a key part in limiting the diversity and extent of the tree cover and allowing persistence of the grasses (Hopkins, 1974; Phillips, 1974). Fire will also extend tropical grassland into tropical forest zones by eliminating the fire-sensitive trees of these areas. Some of these extensions of tropical grasslands may be ancient in origin, others very recent. For example, Kayll (1974) considers that firing has for centuries maintained the savanna-like grasslands interspersed with the forests of South-East Asia. There is historical evidence of annual burning of savanna in West Africa for up to 2500 years (Hopkins, 1974). Indeed over much of the continent of Africa, fire is important in controlling the density of the tree population, and many grasslands and much of the open savanna have developed under the influence of frequent burning (Kowal and Kassam, 1978). Today, many cleared areas of the Amazon rain forest are now maintained as semi-natural grazing lands. Whilst much of this land will apparently revert to secondary forest when abandoned, a proportion does not (Saldarriaga, 1986; Scott, 1986). Whilst human activity may be causing expansion of tropical grasslands at the expense of forest, it is also causing loss of grassland to the gain of the desert. On the dry edges of the tropical grassland overgrazing and possible increased incidence of drought is resulting in loss of grass cover and desertification. A further threat to the extent of tropical grassland is presented by the predicted pattern of climatic change in response to increased 'greenhouse gases'. A compilation of the predictions of the major global circulation models (GCMs) suggests that the doubling of the 'greenhouse effect' due to rising levels of CO_2, CH_4, N_2O and other hetero-atomic gas molecule concentrations in the atmosphere, expected by the middle of the next century, will result in major climatic changes. These include a global rise in temperature and a decrease in soil moisture of between 2 and 4 cm in many of the present savanna areas (extrapolated from Houghton *et al.*, 1990). Such a change may be expected to accelerate the rate of replacement of savanna by desert on the dry margins of the biome. However, as others have pointed out, elevation of CO_2 will increase plant water use efficiency which may counteract the effect of decreased soil moisture levels (Idso, 1989). Current management of savannas may already be contributing to the 'greenhouse effect'. Savanna fires in Africa have increased in the last two decades to the extent that 75% may now be burned annually. Directly, this will add *ca* 1 million tonnes of carbon

(= 1 Pg) to the atmosphere (Hao *et al.*, 1990), although much may be recently fixed carbon. Indirectly, by increasing the frequency with which shoots must regrow, below-ground carbon reserves in savannas could be depleted and the potential for below-ground storage removed. A further effect of this burning is to increase atmospheric ozone concentrations over North and West Africa to 40 – 70 μmol mol^{-1}, as a result of increased atmospheric levels of hydrocarbons and nitrogen oxides following savanna fires. This tropospheric ozone, besides being directly phytotoxic will also contribute to the 'greenhouse effect' (Andrae, 1990; World Resources Institute, 1990).

Although the largest areas of tropical grassland are found in Africa, both north, east and south of the West African tropical forest, extensive areas of tropical grasslands also occur in South America, both north and south of the Amazon, in India, South-East Asia and in northern Australia (Figure 1.1). Through most of this area the grasslands are of similar basic character; a continuous area of grass with isolated trees all adapted to survival of periodic burning (McNaughton, 1985). Few of the major species are common to more than one of these geographic regions. However, the grasses of all of these sites are almost entirely C_4 whilst the trees and shrubs are C_3 (Mitsui and Black, 1982). Tropical grassland is the only biome in which plants utilizing C_4 photosynthesis constitute the major primary producers, in all others C_3 species predominate. This gives increased significance to this biome when considering ecosystem reactions to elevated CO_2 given the well-known physiological differences between these two photosynthetic groups in their capacity to respond to increased CO_2. As a consequence of these physiological differences it is predicted that the productivity of these C_4 species will respond less positively to the fertilization effect of increasing atmospheric CO_2 concentrations (Lemon, 1983).

1.3 IMPORTANCE OF TROPICAL GRASSLANDS

For the countries in which they occur, tropical grasslands have two immediate economic values: (1) as tourist attractions, particularly in the African savanna where these ecosystems support large and diverse populations of large herbivores and carnivores; and (2) for agriculture, where they are grazed under varying intensity by domesticated livestock but are being increasingly used for crop production.

The recognized importance of the African savannas to the conservation of a wide range of large mammals has resulted in an extensive system of game reserves, providing large expanses of natural vegetation. However, outside of many of these reserves the land is under ever-increasing pressure for increased agricultural production to support growing human populations. The result is that many reserves are in danger of becoming biogeographic islands. Indeed, McNaughton (1985) considers the Serengeti region to be the Earth's last vast unmanaged grazing ecosystem which occupies a clearly defined ecological region.

Destruction of rain forest for agricultural expansion has been widely recognized as a global problem. However, the areas with the most pressing needs for increased food production are not the tropical rain forest zones, but the semi-arid tropics where grasslands form the natural vegetation. Of the original area of African savanna and grassland, estimated at ca 7×10^6 km^2, just 40% (i.e. 2.8×10^6 km^2) remain, although this figure is partially compensated for by grassland created through the burning of dry forest (World Resources Institute, 1990). In the wetter zones, cultivated land obtained from savanna can support a wide range of both perennial and annual crops (Kowal and Kassam, 1978). Sorghum and maize now form the major food crops of these cultivated lands, although there is a wide range of cash crops, ranging from pineapple to cotton. Average grain yields for individual countries vary, but for sorghum range from 420 kg ha^{-1} in Senegal to 750 kg ha^{-1} in Ghana, however yields of up to 4500 kg ha^{-1} have been obtained on experimental stations. Maize yields vary from 500 kg ha^{-1} in Dahomey to 1100 kg ha^{-1} in Ghana, but yields of 4000 and 8000 kg ha^{-1} have been obtained on experimental stations in the savanna in Sudan and Guinea, respectively (Kowal and Kassam, 1978). Dry matter yields for maize under good management of up to 18 600 kg ha^{-1} have been recorded for maize in the savanna at Samaru in northern Nigeria (Kowal and Kassam, 1973). This compares to dry matter yields for the annually burned grassland of just 2950 kg ha^{-1} at this location (Kowal and Kassam, 1978).

1.3.1 Conversion to arable cropping and effects of carbon reserves

Conversion to arable cropping would not necessarily result in increased plant productivity. McNaughton (1985) has shown that the trophic web based on grazing may be sustained with no net loss to the plant. This is because of compensatory growth which occurs under moderate levels of grazing.

Terrestrial ecosystems are major repositories of carbon containing about 2.1–2.3 $\times 10^{18}$ g (Oechel and Strain, 1985). This is an amount which is some three times the current atmospheric pool of carbon.

Changes in the amount of carbon in terrestrial biomass and dead plant remains, brought about by conversion to arable cropping, could have profound effects on the flux of carbon to or from the atmosphere. Arable crops which may leave large areas of surface bare for significant periods of the year increase vulnerability to surface erosion during periods of heavy rain, and removal of soil organic matter and stored carbon.

If there is an increase in primary production following the rise in atmospheric CO_2 levels, terrestrial ecosystems could be a major sink for CO_2 through accumulation of dead organic matter. This is because soil organic matter has a mean residence time of many decades. However, climatic changes which accompany CO_2 rise could have a major impact on decomposition of the stored carbon. The direction of these changes in tropical grasslands though is by no means certain. If it is assumed that climate change will result in decreased soil moisture levels and increased temperature (UNEP, 1987), then it may be expected that tropical grassland soils will be increasingly exposed and their organic matter more vulnerable to oxidation. On the other hand, growth of plants in elevated CO_2 increases the C/N ratio of the resulting plant material and this change in quality may depress microbial decomposition rates (Long and Hutchin, 1990).

1.3.2 Overgrazing

The maintenance of grasslands by grazing animals is determined by the intensity of grazing. Grazing involves reduction in leaf area with its effects on tiller development and shoot and root growth. But there are also changes in microclimate and other factors such as trampling and return of dung and urine. Overgrazing can have a variety of effects, including modification of microclimate by reducing the insulating effect of vegetation, changed moisture infiltration and runoff, and increased vulnerability to surface erosion, and hence soil organic matter and carbon loss. Under intense grazing pressure the more palatable species are lost and less desirable species predominate. Increases in unpalatable shrubs often make the continued utilization of the more arid and semi-arid lands for livestock production possible only by repeated burning (Hopkins, 1974; Kowal and Kassam, 1978).

Agricultural overgrazing of semi-natural tropical grasslands may increase the requirement for fertilizers as nutrient reserves are depleted. On the other hand, some grazing will promote nutrient cycling and may stimulate plant growth by maintaining the pool of labile nutrients (Kowal and Kassam, 1978; Coupland, 1979; McNaughton *et al.*, 1983).

1.4 PREVIOUS PRODUCTIVITY STUDIES OF TROPICAL GRASSLANDS: THE CASE FOR NEW INTENSIVE STUDIES

Cursory examination of productivity figures for arable and natural grassland systems of the tropics might suggest that conversion of land to arable would greatly increase total dry matter productivity (Lieth, 1978; Buringh, 1980). However, knowledge of these natural ecosystems is based mainly on the IBP studies. As we outline below, the methodology used in these studies may have led to a serious and variable underestimation of production and turnover of plant biomass in these communities.

Net primary production (P_n) is the total photosynthetic gain, less respiratory losses, of plant matter by vegetation occupying a unit area of ground (Milner and Hughes, 1968; Cooper, 1975; Lieth, 1978; Linthurst and Reimold, 1978; Coupland, 1979; Jordan, 1981; Roberts *et al.*, 1985). Over any one period, this must equal the change in plant mass (ΔW) plus any losses through death (L), both above- and below-ground (Roberts *et al.*, 1985):

$$P_n = \Delta W + L \qquad [1]$$

Essentially, P_n is the measure of the amount of plant matter available to consumer organisms.

Previously Bourlière and Hadley (1970) reported net above-ground production for 22 tropical grasslands based on peak standing dry matter, alone. Below-ground production was not considered. The most extensive data on the productivity of tropical grasslands are provided by the IBP studies, reviewed by Singh and Joshi (1979). They reviewed 21 studies (eight published) of net primary production in tropical grasslands of India and Africa, a few also include estimates of below-ground production. In these studies below-ground net primary production (BP_n) was estimated by 'trough–peak' analysis of the biomass (Singh and Joshi, 1979). That is, where biomass (W) per unit area was found to increase between harvest intervals, BP_n over the interval was considered to equal the increase, but where W decreases or remains unchanged over an interval between harvests, BP_n is assumed to be zero. The annual BP_n is then obtained by summing the estimates for each harvest interval, i.e. the positive increments are summed. This was the recommended IBP method for grasslands (Milner and Hughes, 1968; Coupland, 1979), termed here the 'IBP standard method'. This method assumes that production and death are discrete processes, any overlap of the two processes will lead to underestimation of BP_n. Above-ground net primary production (AP_n) of these tropical grasslands was apparently estimated by 'trough–peak' analysis of live and standing dead material (Singh and Joshi, 1979). In this case AP_n is determined via a decision matrix (Singh *et al.*, 1975). Positive increments in biomass are summed, as in the 'IBP standard method', but where positive increments in the amount of standing dead coincide with positive increments in biomass, these are also added to the total. Thus, this procedure would correct for amounts of material lost through death during periods of biomass increase, if no decomposition occurs. Since this is unlikely the method will again lead to underestimation. Three assumptions which could seriously influence the estimate of production therefore underlie the procedures used in the IBP studies of tropical grasslands.

1. That during any interval between harvests death of material does not occur if production is occurring and vice versa, i.e. formation of new organs or parts of organs does not overlap with loss through death of organs or parts of organs. In mixed grasslands this procedure will also fail to take account of the different cycles of growth and death in different species, thus a species which is most productive at the time when total biomass is declining would be ignored.
2. That P_n cannot be negative, either above or below ground, since only positive increments are considered. Again this is clearly an over-simplification. Since these grasslands are dominated by perennials with underground or surface storage organs significant transfers of assimilate between above- and below-ground components will be expected. For example, as dry or cold seasons approach, translocation of matter from the shoots to below-ground storage organs can occur. During this period the above-ground biomass will decline and the below-ground biomass rise. Using the IBP computational procedures this will give a positive BP_n and zero AP_n, and therefore a positive total plant net production, even though in reality no new material is being formed, but existing material reallocated. Similarly, when favourable growth conditions return, production of new shoots from material stored in underground organs would be recorded as net production of the whole plant.
3. That all increases in biomass represent P_n. Variability in biomass over the study sites will mean that the average biomass recorded on different dates may vary simply because of random fluctuations in the samples obtained. An apparent positive increment in biomass between two dates could result simply by random chance and although over the year these random fluctuations may be self-cancelling, selection of only positive increments in calculating P_n

will mean that the greater the variability between the samples the greater the overestimation of production (Singh *et al.* 1975). Singh *et al.* (1975), in their analyses of the United States–IBP grassland productivity studies, suggest that this error may be compensated for by only including positive increments where the biomass on one sampling date is significantly greater ($P<0.1$) than on the previous sampling date. However, this will mean that small, but real increases in biomass may go undetected. In a relatively uniform natural coastal grassland, Hussey and Long (1982) found that harvests of 40 quadrats, of the optimum sampling size, placed by a randomized design were necessary to provide a mean estimate of biomass with a standard error within 10% of the mean. Thus, even at this level of sampling a 25% increase in biomass between two dates would not be statistically significant. In the IBP tropical grassland studies sample numbers were considerably lower, e.g. 10 (Singh and Yadava, 1974).

Assumption 1 would therefore lead to underestimation of P_n, assumption 2 to overestimation, and assumption 3 without statistical restraints would lead to overestimation but with statistical restraints could lead to substantial underestimation. Despite these theoretical objections, Singh and Joshi (1979), in an analysis of some 30 methods of computing P_n as applied to the data gathered for United States grasslands in the IBP studies, concluded that the 'trough–peak' analyses used in calculating P_n in the tropical grassland studies and many other IBP grassland sites were among the best procedures.

Determination of actual net primary production would require correction of change in biomass for losses due to death, as defined in equation (1). Losses (L) may be determined by correcting the change in dead vegetation mass between sampling intervals by the amount lost due to decomposition measured by use of litter bags or labelling of dead material (Roberts *et al.*, 1985) or by direct observation of organ death (Jackson *et al.*, 1986). Calculations based on simultaneous measurement of ΔW and L of equation (1) make none of the assumptions of the IBP standard method and other 'trough–peak' analyses. No such method was included in the comparisons made by Singh *et al.* (1975) who only considered methods in which it was assumed that production could be estimated from changes in vegetation mass alone. Methods were evaluated on their ability to discriminate between different grasslands and although 'trough–peak' analysis methods were favoured, it was noted that all methods were closely correlated and that a constant correction factor could be used to relate different methods. For example, multiplication of P_n estimated from peak biomass by 1.41 would give P_n as estimated by the 'IBP standard method' of summing the positive increments in plant mass.

Linthurst and Reimold (1978) compared the production estimated for three United States coastal grasslands obtained from maximum biomass, as used by Bourlière and Hadley (1970), and from the 'trough–peak' analysis, as used by Singh and Joshi (1979), with a method in which P_n was calculated from change in biomass corrected for simultaneous losses through death. Two important conclusions can be reached from this work: (1) that P_n estimated by both use of maximum standing crop and 'trough–peak' analysis seriously underestimated P_n, when account was taken of simultaneous losses of material, by 50%–85% and by 10%–70%, respectively; and (2) that the degree of underestimation varied markedly with location and with species composition and that there was no constant conversion factor. This conclusion is in sharp contrast to that of Singh *et al.* (1975), but unsurprising in view of the fact that the errors accruing from the three assumptions underlying the IBP methods will vary independently. Similar conclusions to those drawn from the work of Linthurst and Reimold (1978) can be drawn from other comparisons of methods made in temperate grasslands (Bradbury and Hofstra, 1976; Long and Mason, 1983). These findings bring the IBP estimates of primary production in tropical grasslands into question. Not only may these be serious underestimates, but they may even fail to rank grasslands correctly in terms of their actual productivity. Although the potential errors in the methods used in most of the IBP studies of grassland P_n were well recognized (Cooper, 1975; Coupland, 1979) the degree of underestimation was not quantified. This would require application of the IBP standard methods and methods correcting for losses on the same vegetation. Underestimation of P_n further affects understanding of ecosystem functioning in tropical grasslands, since the estimates of P_n provided the basis for calculation of vegetation death and decomposition in the IBP syntheses of tropical grasslands (Singh *et al.*, 1975).

1.5 REMOTE SENSING AND SOLAR RADIATION CONVERSION EFFICIENCY

The preceding section has emphasized the necessity for intensive study of biomass and litter dynamics, coupled with measurement of decomposition and other losses needed to gain a more accurate assessment of primary production in these ecosystems. Given the heterogeneity of tropical grasslands, destructive determination of biomass and dead vegetation necessitates the collection of a large number of samples if the changes in mean quantities per unit area with time are to be determined with any precision. Non-destructive techniques, including remote sensing methods, therefore have considerable potential in allowing the same sites to be studied with respect to time and in allowing larger numbers of samples to be examined. Studies of arable and tree crops, both in temperate and tropical climates, have shown a linear relationship of dry matter accumulation against the cumulative quantity of solar radiation intercepted by the crop. For a large number of C_3 crops the slope of this relationship has been shown to be *ca* 1.5 g (dry wt) MJ^{-1}, whilst for C_4 plants maxima are *ca* 2.3 g MJ^{-1} (Beadle *et al.*, 1985). If a similar constancy occurs in natural vegetation then it will provide a relatively simple non-destructive method of estimating productivity. Solar radiation interception can be measured by the use of tube solarimeters mounted above and below the canopy (Nobel and Long, 1986). It is assumed that this relationship reflects a fairly constant efficiency of energy use in photosynthesis by closed canopies. In a healthy arable crop there is little loss of material through death prior to maturity, and thus the biomass present will be a close reflection of the net production. In perennial grasslands however, as outlined above, biomass will be a poor reflection of production. Little attention has been given to the relationship between P_n and solar radiation interception in natural vegetation. This series of studies of tropical grasslands involving intensive measurements of biomass dynamics and estimation of P_n provided an ideal opportunity for examining the possibility that measurement of solar radiation interception may provide a non-destructive method of measuring production in tropical grasslands.

Remote sensing, using the change in the near-infra-red to red reflectivity ratio (R/NIR) has been widely used in the estimation of standing biomass (Chapter 7). However, the relationship will vary with canopy structure and leaf area as a proportion of total shoot biomass (Monteith and Unsworth, 1990), thus in natural vegetation this relationship will change with time of year and community composition. Monteith and Unsworth (1990) show from theory that R/NIR is directly related to the capacity of a canopy to intercept radiation. Thus, if P_n is linearly related to the quantity of solar radiation intercepted, then the R/NIR must provide a non-destructive measure of P_n. Indeed, this has been shown to be an effective method of determining sugar-beet production from the air (Steven *et al.*, 1983). As Warrick (1986) has noted, this technique therefore has much apparent potential for the remote sensing of primary production in natural vegetation over large areas, however as yet there have been no extended evaluations of the method with detailed ground truth studies in natural vegetation.

1.6 AIMS AND GOALS

Thus, while there are many estimates of the productivity and standing biomass of tropical grasslands, great uncertainty surrounds the extent to which these underestimate the true production and the significance of variability in time and space. A key need then is for detailed study of the dynamics and turnover of live and dead vegetation. Such studies of biomass dynamics and turnover require frequent destructive harvests followed by sorting of material. If sufficient replicates are to be taken to provide statistically meaningful results, then such studies will be highly labour intensive.

Despite this, the primary aim of this work was to provide more detailed information of the productivity of tropical grasslands which would take full account of turnover of plant biomass. A secondary aim was to examine the potential of measurement of reflected red/far-red ratios in estimating the primary production of tropical grasslands. This is because the ratio of red to far-red light reflected from vegetation has been mechanistically linked to net primary production (Monteith and Unsworth, 1990). The specific goals of this work are given below.

1. To develop an agreed set of practical procedures for measurement of vegetation dynamics, above and below ground, for the determination of production. This common methodology would be used, as far as possible, at all sites to maximize comparability.
2. Establishment of a network of centres conducting these studies to provide baseline data from which the impact of future environmental change could be assessed.
3. To determine the magnitude of underestimation of production and carbon sequestration of the IBP and earlier studies.
4. To record and analyse seasonal, year-to-year, and site-to-site variation in production.
5. To determine relationships between climatic variables and changes in biomass and production.
6. To examine leaf and canopy photosynthetic CO_2 assimilation to provide a basis for understanding seasonal changes in production.
7. To examine the feasibility of non-destructive procedures of light interception measurement and remote sensing of reflected radiation, as non-destructive probes for measuring production and biomass.

Goal 1 forms the subject of the final section of this chapter. The remote sensing aspects of goal 7 are examined in detail in Chapter 7. The remaining goals are addressed in Chapters 2–5.

Achievement of goals 2–5 necessitated intensive studies, i.e. a high replicate number of samples taken at regular intervals. Given the time-consuming nature of harvesting and careful sorting of vegetation the decision was taken to concentrate on a single study-site within each region, rather than diffusing effort onto a larger number of sites. Sites were therefore chosen to represent major grassland types contrasting in form and edaphic character (Table 1.1).

1.7 BAMBOO FORESTS

In common with tropical grasslands, bamboo forests and scrub are not only dominated by graminaceous species, but are also confined predominantly to the developing world. However, bamboos form woody stems and for the larger species methods for the measurement of production in trees rather than grasses are the more appropriate. World-wide there are some 1300 species of bamboos, about 1000 of these from Asia, and of those 300 from China (Willis, 1973). One of these species, *Phyllostachys pubescens*, forms large monotypic stands and forms nearly 70% of the area of bamboo forest in China. Since many bamboos, including *P. pubescens* are of economic importance for their woody stems, a considerable body of information on their economic yield has developed. However, ecological productivity requires a knowledge not only of wood yield, but also of carbon gain in leaves, roots and rhizomes. Like other woody plants they represent a major reservoir of terrestrial carbon and systems with potential to ameliorate rates of increase in atmospheric CO_2 concentration. Chapter 6 provides a detailed study of the biomass dynamics and primary production of a *P. pubescens* forest of sub-tropical China.

1.8 STUDY SITES

Detailed information of the five study sites are given in the individual chapters. Figure 1.1 indicates the location of the five sites chosen for this United Nations Environment Programme project and Table 1.1 summarizes the conditions of each site. The conditions at each site are briefly outlined below.

Table 1.1. Site descriptions

Terrestrial grasslands	*Saline grassland* *Montecillos Chapingo Mexico*	*Dry savanna grassland* *National Park Nairobi Kenya*	*Sub-humid savanna* *Klong Hoi Kong Hat Yai Thailand*	*Floodplain grassland* *Marchantaria Manaus Brazil*	*Bamboo stand* *Miao Shan Fuyang China*
Location	19°N 98°W	1°S 36°E	6°N 100°E	3°S 60°W	30°N 120°E
Elevation (m)	2220	1500	100	20	1500
Precipitation (mm y^{-1})	700	950	2100	3130	950
Average monthly temperatures (°C):					
Minimum	12 (January)	12 (July)	27 (December)	26 (February)	3 (January)
Maximum	19 (May)	19 (February)	29 (June)	28 (August)	27 (July)
Solar radiation (MJ m^{-2} y^{-1})	6810	6500	5970	6410	3259
Soil type	Solonet	Dark 'cotton' deposits	Humic gley	Alluvial red earth	Acidic
Grassland type	Saline	Dry Savanna	Humid 'Savanna'	Emergent macrophyte	Tall bamboo
Dominant species	*Distichlis spicata*	*Pennisetum menzianum* *Themeda triandra*	*Eulalia trispicata* *Lophopogon intermedius*	*Echinochloa polystachya*	*Phyllostachys pubescens*
Agricultural use of adjacent areas	Pasture	Sorghum Pineapple	Sorghum Rice	Panicums	Tea Rice

1.8.1 'Dry' grassland sites

Kenya

This site was typical dry savanna grassland within the Nairobi National Park. Rainfall is bimodal (Figure 1.1) with most rains falling in May and a lesser peak of rainfall in October. Growing seasons are therefore normally May–August and October–December. However, in 1984 (immediately preceding this study) the May rains failed. The site was protected by an exclosure, but would normally be lightly grazed by wildebeest and other game animals. Similar land in Kenya is currently being ploughed for sorghum, cotton and pineapple production.

Thailand

Measurements were made at the Prince of Songkhla University field site at Klong Hoi Khong, about 40 km from Hat Yai. The climate here is wet monsoonal, with maximum rainfall in October–December and a brief drier season in February/March (Figure 1.1). The site is a semi-natural humid grassland with occasional stands of dipterocarp trees, typical of grassland found throughout southern Thailand. A grass community is maintained by periodic burning. To avoid complications in determining productivity, the site was protected by a fire-break for the duration of this study. This site is now a relict of the original grassland, the surrounding land having been cleared for production of rice, sorghum and a variety of other arable crops.

Mexico

The site is a saline grassland at the Colegio de Postgraduados field site, Montecillos (about 50 km from Mexico City). The climate is dry sub-humid, with maximum rainfall in July/August and a dry season from December to February (Figure 1.1). The study area contained an almost pure stand of a halophytic grass. The grass community is maintained by periodic burning.

1.8.2 Floodplain grassland

Brazil

The study site is a stand of emergent macrophyte vegetation on the island of Marchantaria in the Amazon floodplain, about 40 km from Manaus. This has a typical tropical rain forest climate, uniformly warm, with heavy rainfall except during a short drier season from

August to September (Figure 1.1). Seasonality of vegetation growth is determined predominantly more by the river level than by climate, the growing season coinciding with the period of inundation. Many of these sites are now burnt in the dry season and planted with arable crops which can be harvested before the sites are flooded during the next wet season.

1.8.3 Bamboo forest

China

This forest is almost entirely a monotypic stand of *Phyllostachys pubescens*, located in Fuyang County and is 30 km south-west of Hangzhou City in eastern China. The climate is humid/warm temperate, with hot and humid summers and with winter temperatures sufficiently low to allow snow to fall and settle on the canopy (Figure 1.1). The seasonality of growth is determined by climate.

1.9 METHODS

The first goal of this study was to establish a common methodology for the measurement of productivity and to ensure direct comparability of results. The demography of the plants of the inundation grassland site in the Amazon (Chapter 5) and the bamboo forest site in China (Chapter 6) demanded specific and separate methods for much of the determination of primary production and these are described in the appropriate chapters. However, for the three tropical grassland sites, identical methods were possible. Prior to the commencement of field measurements in 1984, an initial workshop was held to determine a procedure that would meet the study objective within the manpower resources available. General procedures for measuring the production of natural herbaceous communities have been described in detail by Roberts *et al.* (1985). This section describes the specific approach chosen for the studies reported here.

At each site, a 1 ha area of vegetation considered typical of the site was chosen. This was fenced and fire-breaks were constructed to allow control of grazing and burning. Net primary production (P_n) over a year will be manifest as the increase in live plant material, i.e. plant biomass (ΔB), both above ground and below ground, plus losses through death to other levels in the ecosystem (l); equation (1). In these sites dead material will remain *in situ* until it is completely decomposed by micro-organisms or removed by invertebrates whilst large grazers and fire are excluded. Thus l will equal the increase in the quantity of dead plant material (ΔD) corrected for the amount decomposed over the year (A).

$$l = \Delta D + A \qquad [2]$$

Substituting for L in equation (1), we obtain the following expression for determining P_n:

$$P_n = \Delta B + \Delta D + A \qquad [3]$$

To determine P_n following the definition of equation (3), changes in both live and dead vegetation were measured at monthly intervals. Dry weight of both live and dead vegetation present at each site was determined monthly by clipping to ground level sets of 20 quadrats, 0.25 m × 1.0 m, located by a randomized block design (Roberts *et al.*, 1985). This quadrat size was determined as the most 'cost-efficient' following the procedures of Wiegert (1962). Soil cores of 5 cm in diameter were removed from the centre of 5–20 of these quadrats, to a depth of 15 cm, and organic material extracted by washing over a sieve of 2 mm mesh. Use of 1 and 0.5 mm mesh sizes did not recover significantly (t-test, $P>0.05$) greater quantities of live root, but greatly increased the time required for the sorting of samples. For dead material, ability to pass through a 2 mm mesh sieve is used throughout as the arbitrary division between recognizable dead vegetation (D, of equation 3) and particulate organic matter. Preliminary studies suggested that extraction of soil cores to 15 cm depth was adequate to remove more than 90% of the root system by weight.

Above-ground material was sub-sampled to approximately 100 g fresh weight before sorting (Roberts *et al.*, 1985). Below-ground material was divided into fine roots (*ca* <1 mm diam.) and coarser material. Fine root matter was sub-sampled to 1.0 g, and the coarser material sorted entirely. Leaves were sorted into live and dead matter on the basis of tissue appearance, dead parts being removed from otherwise green leaves. Stems were sorted likewise, taking care to remove dead sheaths from living stems. Roots were divided on a similar basis, using vital staining (tetrazolium salts) where visual discrimination was not otherwise possible. Tetrazolium staining was cross-checked by microscopic examination of cut root tissues where the results of surface staining were ambiguous. The sorted plant material was then thoroughly washed and dried to constant weight at *ca* 105°C.

Decomposition losses (W_d) were determined monthly by litter bags. A portion of dead material obtained at random from each harvest (*ca* 2 g) was placed in each of 20 litter bags of 2 mm nylon mesh, 8.0 cm ×

6.0 cm, and recovered after 1 month from the field. Contents were washed over a sieve of 2 mm mesh and dried to constant weight. Controls in which the litter bags were filled, taken to the field and then immediately returned and processed showed no significant loss of material. The loss of material from the bags is a measure of the rate at which a random sample of the dead vegetation at the site at the start of each interval between harvests would decompose over the month. Decomposition of dead shoot material was measured at the ground surface, whilst for roots and rhizomes the litter bags were inserted 5 cm below ground, with the soil carefully replaced above the bag to minimize disturbance.

Relative rates of decomposition (r) were expressed as the proportion of initial dry weight lost within the month:

$$r = \ln(w_i/w_{i+1})/(t_{i+1} - t_i) \quad [4]$$

Where t_i and t_{i+1} are the dates on which litter bags were placed in the field and removed from the field, respectively. On average the elapsed period in this study was 30 days. w_i and w_{i+1} are the dry weights of dead plant material placed in and recovered from the litter bags at times t_i and t_{i+1}, respectively.

In the context of this study decomposition represents loss of dead material due to leaching, breakdown to small particles (< 2 mm diam.) and consumption by invertebrates. Total decomposition (A) was estimated as the products of the relative decomposition rate and the mean quantity of dead material over a given month.

$$A = r_i[(D_i - D_{i+1})/2] \quad [5]$$

Where r_i is the relative rate of decomposition over the period t_i to t_{i+1}, typically 30 days. D_i and D_{i+1} are the mean quantities of dead vegetation either above or below ground at the start and end, respectively, of the period t_i to t_{i+1}.

The integrated quantity of incident solar radiation (S_{abs}) intercepted by the vegetation with respect to time at each site was determined with tube solarimeters (Types TSL and TSLM, Delta-T Devices, Burwell, UK). Solarimeters, connected to millivolt integrators (Type MVI, Delta-T Devices), were placed at the base and above the canopy. The quantity of solar radiation intercepted over a given interval was then given by the difference between the incident radiation and the quantity reaching the base of the canopy (Nobel and Long, 1986):

$$S_{abs} = S_o - S_b \quad [6]$$

Where S_o is the quantity of solar radiation received above the canopy

(MJ m^{-2}). S_b is the quantity of incident solar radiation which penetrates to the base of the canopy (MJ m^{-2}).

1.10 SUMMARY

1. Tropical grasslands and savannas form the world's largest terrestrial biome, and a major site of potential sequestration of carbon from the global carbon cycle.
2. Estimates of the productivity of tropical grasslands are based largely on the International Biological Programme (IBP) studies. Methods used in IBP for grasslands are known to underestimate production, and it is argued that this underestimation is both highly variable and may be particularly significant in tropical grasslands. Very few studies of tropical grasslands have taken any account of either turnover of biomass or below-ground biomass and production.
3. Bamboo forests represent another community of the developed world which may form a significant sink for rising global CO_2 levels. The ecological net primary productivity of bamboos has received little attention.
4. The aims and goals of the overall project are developed and explained.
5. The common methods developed and employed for the project are described.

REFERENCES

Andrae, M.O. (1990) Biomass burning in the tropics: impact on environmental quality and global climate. *Population and Development Review*, in press.

Beadle, C.L., Long, S.P., Imbamba, S.K.*et al.* (1985) *Photo synthesis in Relation to Plant Production in Terrestrial Environments*. UNEP/Tycooly, Oxford.

Bourlière, F. and Hadley, M. (1970) The ecology of tropical savannas. *Annual Reviews of Ecology and Systematics*, **1**, 25–152.

Bradbury, I.K. and Hofstra, G. (1976). Vegetation death and its importance in primary production measurements. *Ecology*, **57**, 209–11.

Buringh, P. (1980) Limits to the productive capacity of the biosphere, in *Future Sources of Organic Raw Materials* (eds L.E. St.Piere and G.R. Brown), Pergamon Press, Oxford, pp. 325–83.

Charter, J.R. and Keay, R.W.J. (1960). Assessment of the Olokemeji fire control experiment 28 years after institution. *Nigerian Forestry Information Bulletin*, **3**, 1–24.

Cooper, J.P. (ed.) (1975) *Photosynthesis and Productivity in Different Environments*, IBP Vol. 3, Cambridge University Press, Cambridge.

Coupland, R.T. (ed.) (1979) *Grassland Ecosystems of the World*, IBP Vol. 18, Cambridge University Press, Cambridge.

Gates, D.M. (1985) Global biospheric response to increasing atmospheric carbon dioxide concentration, in *Direct Effects of Increasing Carbon Dioxide on Vegetation* (eds B.R. Strain and J.D. Cure), Department of Energy, Washington, pp. 171–84.

Gifford, R.M. (1980) Carbon storage by the biosphere, in *Carbon Dioxide and Climate: Australian Research* (ed. G.I. Pearman), Australian Academy of Sciences, Canberra, pp. 167–81.

Hall, D.O. (1989) Carbon flows in the biosphere: present and future. *Journal of the Geological Society, London*, **146**, 175–81.

Hao, W.M., Liu, M.H. and Crutzen, P.J. (1990) Estimates of annual and regional releases of CO_2 and other trace gases into the atmosphere from fires in the tropics, in *Proceedings of the Third International Symposium on Fire Ecology*, Springer- Verlag, New York, in press.

Hopkins, B. (1974) *Forest and Savanna*, 2nd edn, Heinemann, London, 154 pp.

Houghton, J.T., Jenkins, G.J. and Ephraums, J.J. (1990) *Climate Change: The IPCC Scientific Assessment*, Cambridge University Press, Cambridge, 364 pp.

Hussey, A. and Long, S.P. (1982) Seasonal changes in weight of above- and below-ground vegetation in a salt marsh at Colne Point, Essex. *Journal of Ecology*, **70**, 757–71.

Idso, S.B. (1989) *Carbon Dioxide and Global Change: Earth in Transition*, Tempe, IBR Press.

Jackson, D.J., Long, S.P. and Mason, C.F. (1986) Net primary production, decomposition and export of *Spartina anglica* on a Suffolk salt marsh. *Journal of Ecology*, **74**, 647–62.

Jordan, C.F. (ed.)(1981) *Tropical Ecology*. Hutchinson Ross, Stroudsberg.

Kayll, A.J. (1974) Use of fire in land management, in *Fire and Ecosystems* (eds T.T. Kozlowski and C.E. Ahlgren), Academic Press, New York, pp. 483–511.

Kowal, J.M. and Kassam, A.H. (1973) Water use, energy balance and growth of maize at Samaru, northern Nigeria. *Agricultural Meteorology*, **12**, 391–406.

Kowal, J.M. and Kassam, A.H. (1978) *Agricultural Ecology of Savanna*, Clarendon , Oxford.

Lamotte, M. (1967) Recherches ecologiques dans la savane de Lamto

(Côte d'Ivoire): présentation du milieu et programme de travail. *La Terre et la Vie*, **21**, 197–329.

Lemon, E.R. (ed.) (1983) *The Response of Plants to Rising Levels of Atmospheric Carbon Dioxide*. West View Press, Boulder.

Lieth, H.F. (ed.)(1978) *Patterns of Primary Productivity in the Biosphere*, Hutchinson Ross, Stroudsberg.

Linthurst, R. and Reimold, R.J. (1978) An evaluation of methods for estimating the net aerial primary production of estuarine angiosperms. *Journal of Applied Ecology*, **15**, 919–31.

Long, S.P. and Hutchin, P. (1990) Primary production in grasslands and coniferous forests in relation to climate change: an overview. *Ecological Applications*, **1**, 139–56.

Long, S.P. and Mason, C.F. (1983) *Saltmarsh Ecology*, Blackie, Glasgow.

McNaughton, S.J. (1985) Ecology of a grazing ecosystem: the Serengeti. *Ecological Monographs*, **55**, 259–94.

McNaughton, S.J., Wallace, L.L. and Coughenour, M.B. (1983) Plant adaptation in an ecosystem context: effects of defoliation, nitrogen, and water on growth of an African C4 sedge. *Ecology*, **64**, 307–18.

Milner, C. and Hughes, R.E. (1968) *Primary Production of Grassland*. IBP Handbook 6, Blackwell, Oxford.

Mitsui, A. and Black, C.C. (eds)(1982) *Handbook of Biosolar Resources*, Vol.1, CRC Press, Boca Raton.

Monteith, J.L. and Unsworth, M.H. (1990) *Principles of Environmental Physics*, 2nd edn, Arnold, London.

Nobel, P.S. and Long, S.P. (1986) Canopy structure and light interception, in *Techniques in Bioproductivity and Photosynthesis*, 2nd edn (eds J. Coombs, D.O. Hall, S.P. Long and J.M.O. Scurlock), Pergamon Press, Oxford, pp. 41–9.

Oechel, W.C. and Strain, B.R. (1985) Native species responses to increased atmospheric carbon dioxide concentration, in *Direct Effects of Increasing Carbon Dioxide on Vegetation* (eds B.R. Strain and J.D. Cure), Department of Energy, Washington DC, pp. 117–54.

Phillips, J. (1974) Effects of fire in forest and savanna ecosystems of sub-saharan Africa, in *Fire and Ecosystems* (eds T.T. Kozlowski and C.E. Ahlgren), Academic Press, New York, pp. 435–81.

Roberts M.J., Long S.P., Tieszen L.L. and Beadle C.L. (1985) Measurement of plant biomass and net primary production, in *Techniques in Bioproductivity and Photosynthesis*, 2nd edn (eds J. Coombs, D.O. Hall, S.P. Long and J.M.O. Scurlock), Pergamon Press, Oxford, pp. 1–19.

Saldarriaga, J.G. (1986) Recovery following shifting cultivation, in *Amazonian Rain Forests – Ecosystem Disturbance and Recovery* (ed. C.F. Jordan), Springer-Verlag, Berlin, pp. 24–33.

Scott, G.A.J. (1986) Shifting cultivation where land is limited, in *Amazonian Rain Forests – Ecosystem Disturbance and Recovery* (ed. C.F. Jordan), Springer-Verlag, Berlin, pp. 34–45.

Singh, J.S. and Joshi, M.C. (1979) Tropical grasslands primary production, in *Grassland Ecosystems of the World*, IBP Vol.18 (ed. R.T. Coupland), Cambridge University Press, Cambridge, pp. 197–218.

Singh, J.S., Lauenroth, W.K. and Steinhurst, R.K. (1975) Review and assessment of various techniques for estimating net aerial primary production in grasslands from harvest data. *Botanical Reviews*, **41**, 181–232.

Singh, J.S., Singh, K.P. and Yadava, P.S. (1979) Tropical grasslands ecosystems synthesis, in *Grassland Ecosystems of the World*, IBP Vol.18 (ed. R.T. Coupland), Cambridge University Press, Cambridge, pp. 231–40.

Singh, J.S. and Yadava, P.S. (1974) Seasonal variation in composition, plant biomass, and net primary productivity of a tropical grassland at Kurukshetra, India. *Ecological Monographs*, **44**, 351–76.

Steven, M.D., Biscoe, P.V. and Jaggard, K.W. (1983) Estimation of sugar beet productivity from reflection in the red and infra-red spectral bands. *International Journal of Remote Sensing*, **6**, 1335–72.

UNEP (1987) *The Greenhouse Gases*, UNEP/GEMS Environment Library No. 1, United Nations Environment Programme, Nairobi.

Vogt, K.A., Grier, C.C. and Vogt, D.J. (1986) Production, turnover, and nutrient dynamics of above- and below-ground detritus of world forests. *Advances in Ecological Research*, **15**, 303–77.

Walter, H., Harnickell, E. and Mueller-Dombois, D. (1975) *Climate-Diagram Maps of the Individual Continents and the Ecological Climatic Regions of the Earth*, Springer-Verlag, Berlin.

Warrick R.A. (1986) Photosynthesis seen from above. *Nature*, **319**, 181.

Wiegert, R.G. (1962) The selection of an optimum quadrat size for sampling the standing crop of grasses and forbs. *Ecology*, **43**, 125–9.

Willis, J.C. (1973) *A Dictionary of Flowering Plants and Ferns*, 8th edn, Cambridge University Press, Cambridge.

World Resources Institute (1990) *World Resources 1990–1991: A Guide to the Global Environment*, Oxford University Press, New York.

2

Savanna at Nairobi National Park, Nairobi

J.I. KINYAMARIO and S.K. IMBAMBA

2.1 INTRODUCTION

Pollen records show that grasslands, similar to those of the present day, have covered much of Africa since the Upper Tertiary period (Van der Hammen, 1983). The African tropical grasslands lie between the rain forests and the deserts on both sides of the equator from about 29°S to about 16°N. These grasslands are characterized by a continuous graminoid stratum, alternating wet and dry seasons, large animal biomass and frequent occurrence of fires that prevent the intrusion of woody species (Menault *et al.*, 1984).

Tropical grasslands have been centres of evolution for many large animal herbivores and are amongst the most productive ecosystems for wildlife in the world (Child *et al.*, 1984). These grasslands have also been centres of evolution of many forage grasses that support the associated large animal biomass. Grasslands, wooded grasslands and bushed grasslands form a major component of African tropical vegetation (Figure 2.1a). They are very important ecosystems in the East African environment where they form major grazing areas (Pratt *et al.*, 1966). Kenya alone has approximately 490 000 km^2 of rangeland which cover 80% of the total land area of the country. These grazing lands carry 60% of the country's estimated 9.0 million head of cattle, 70% of the estimated 8.5 million sheep and goats, nearly 100% of the estimated 1.0 million camels (Ayuko, 1978) and almost all the wildlife population (Talbot and Stewart, 1964; Sinclair, 1975; McNaughton, 1979). Primary production and the factors that influence it have been studied in various grassland ecosystems of the world. Surprisingly, considering their importance, there have been few studies of primary production in tropical African grasslands. These include those of Cassady (1973), Strugnell and Pigott (1978), Ohiagu and Wood, (1979), Owaga (1980), Macharia (1981), Deshmukh and Baig (1983) and

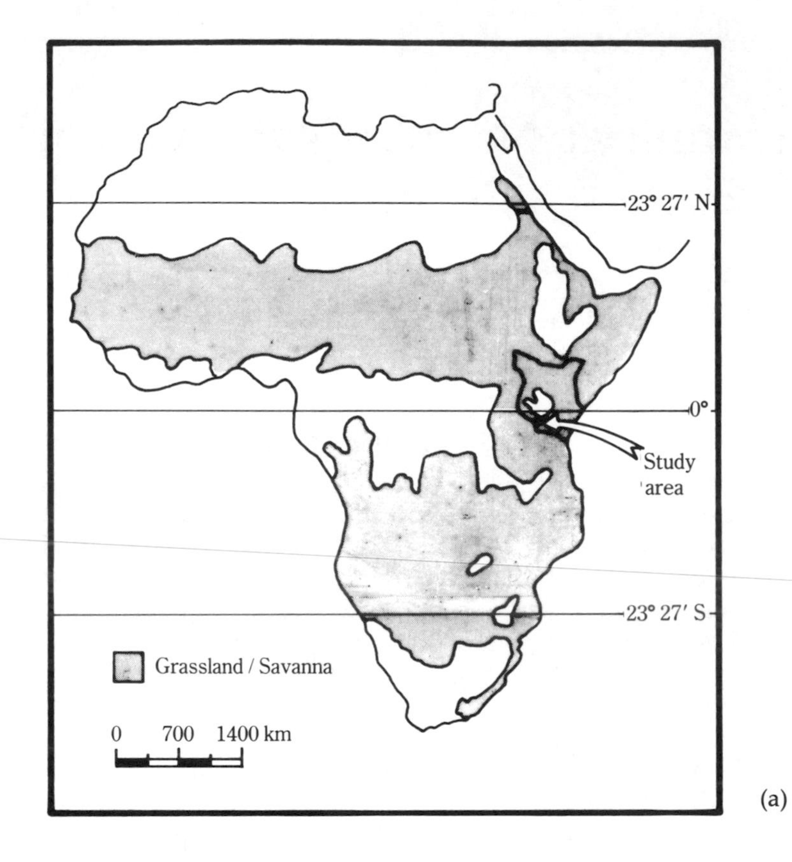
23° 27′ N
0°
Study area
23° 27′ S
Grassland / Savanna
0 700 1400 km

(a)

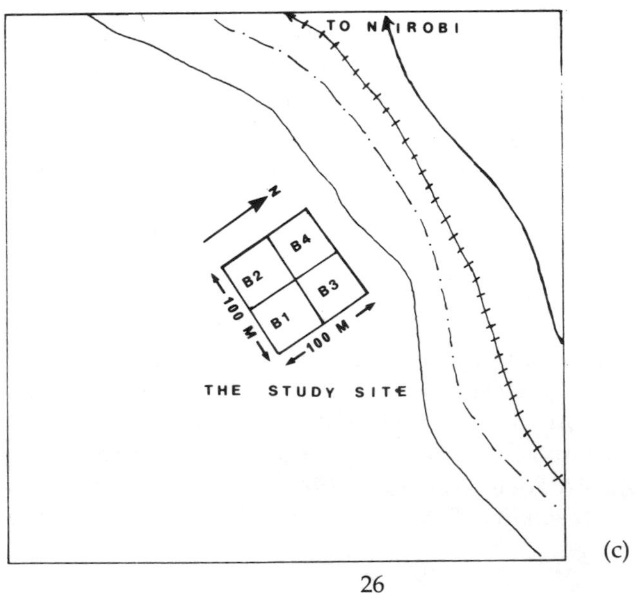
TO NAIROBI
N
B4
B2
B3
B1
100 M
100 M
THE STUDY SITE

(c)

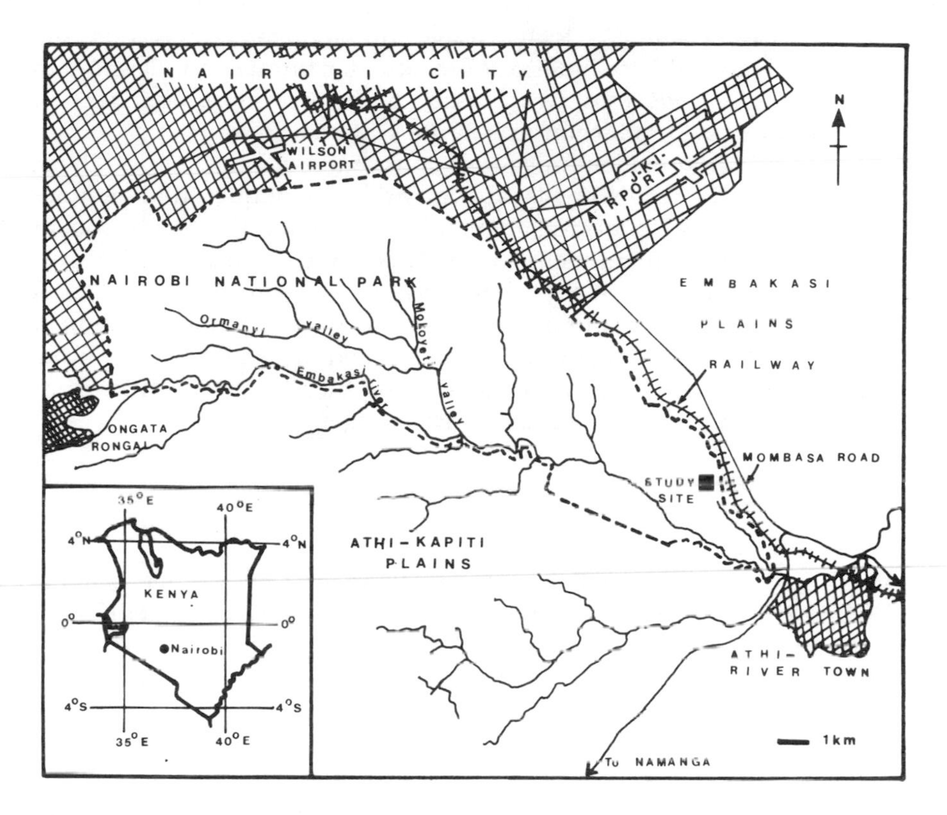

(b)

Figure 2.1. (a) Extent of grassland/savanna vegetation in Africa; (b) Nairobi National Park indicating the location of the study site; and (c) division of the study site into four sampling blocks. The railway line runs to the north of the site.

Deshmukh (1986). Even fewer of these studies have included measurements of such important aspects of primary production as below-ground biomass and decomposition rates of dead materials (Strugnell and Pigott, 1978; Ohiagu and Wood, 1979). Hence, some previous calculations of net primary productivity of tropical grasslands are probably underestimates of the true values. Information on below-ground biomass, canopy architecture and photosynthesis would contribute greatly to the understanding of the structures and production capacities of these ecosystems.

In addition, there is a lack of quantitative information on the contribution of individual species or genera to the primary production of East African grasslands. Macharia (1981), working in various grassland ecosystems of Kenya, found *Sporobolus rangei* Pilg. and

Sporobolus marginatus A. Rich. to be the dominant species in the Amboseli National Park in terms of dry matter production. Deshmukh (1986) found *Themeda triandra* Forssk. to be the most dominant in terms of biomass production in Nairobi National Park.

To provide information on species contributions and photosynthetic rates, noted as requirements above, and to conduct the frequent harvests and litter bag measurements needed for the estimation of primary production as defined in Chapter 1, a nearby site was required. Other requirements were for a site close to a meteorological recording station, one afforded some security against damage to field equipment and one at which grazing could be controlled. Nairobi National Park was chosen as a site meeting these requirements and as a good example of East African savanna vegetation.

2.2 STUDY SITE

The study site was situated in the southern end of Nairobi National Park near Athi River town, about 30 km south-east of Nairobi (Figure 2.1b). Nairobi National Park covers an area of about 112 km^2 south of the city of Nairobi. The park lies close to the equator (1°20′S, 36°50′E) at an altitude of about 1600 m.

This habitat is a typical dry savanna. The mean annual temperature for the Nairobi area is around 19.6°C with monthly maxima and minima in the ranges 23–28°C and 14–11°C, respectively. The highest maximum temperatures occurred during the month of February in each study year (1984–86), while the lowest maximum temperatures occurred during the months of July and August (Figure 2.2a). Minimum temperatures were highest during the long rains and lowest during July and August. The amount of solar radiation closely followed the trend of maximum temperatures.

The daily average total solar radiation (350–2500 nm) of 19.7 MJ m^{-2} received by the plant canopy for the period under study (Figure 2.2b) was within the range for most tropical areas, where daily average total solar radiation ranges from about 14 to 21 MJ m^{-2} d^{-1} depending on cloudiness and elevation (Landsberg, 1965; Cooper, 1975). In East Africa, Strugnell and Pigott (1978) reported daily mean values of 20.2 MJ m^{-2} d^{-1} in Ruwenzori National Park. Muthuri (1985) reported a value of 18.2 MJ m^{-2} d^{-1} in the papyrus swamp of Lake Naivasha.

Rainfall generally occurs in two fairly well defined seasons: from March until May (the long rains) and during October–December (the short rains) (Figure 2.2c). Rainfall is greatly influenced by the inter-tropical convergence zone (ITCZ) which brings in rainfall from the

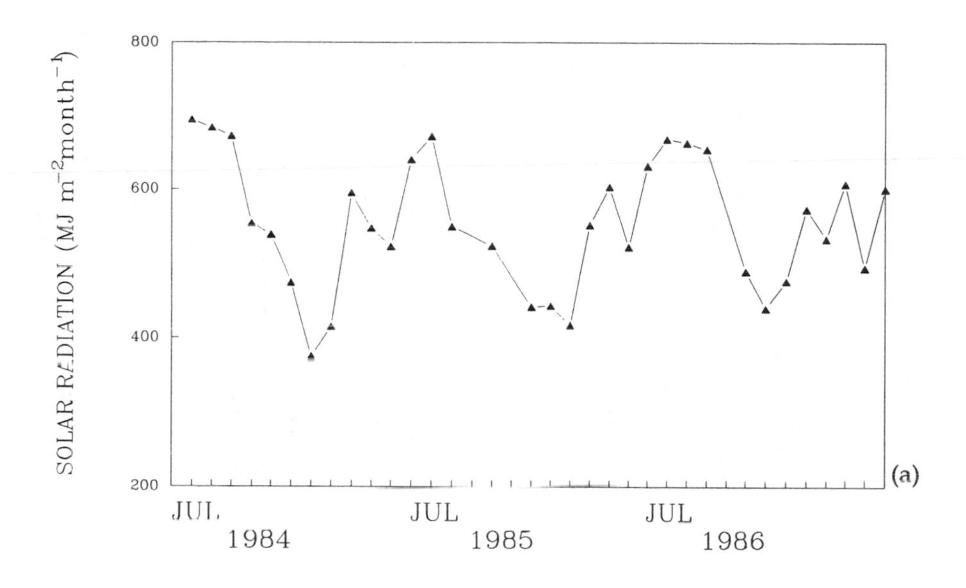

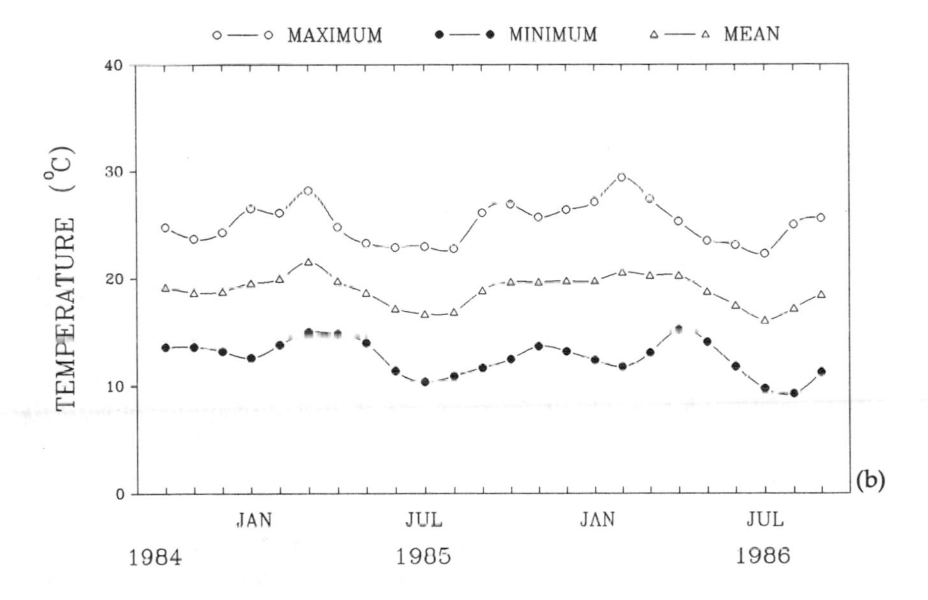

Figure 2.2. Mean weather conditions at the study site for the period October 1984 to October 1986. (a) Average monthly maximum, mean and minimum air temperatures; (b) incident solar radiation

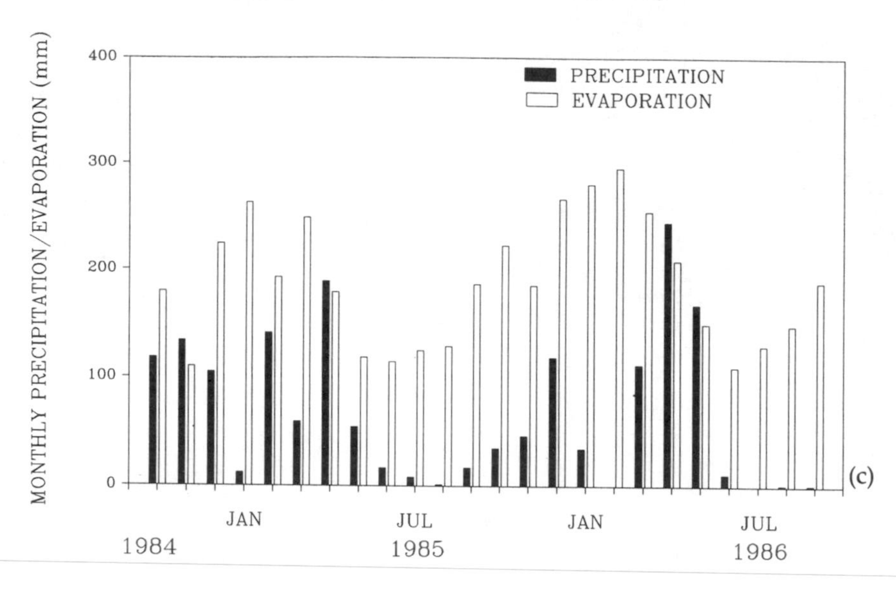

Figure 2.2. *continued* (c) Monthly precipitation and free evaporation

Indian Ocean through the south-easterly monsoon winds. The mean yearly rainfall for the area is about 800 mm. However, large annual variations occur. During the first year of the study, total rainfall was 856 mm. This declined to 775 mm over the following 12 months. Individual seasons may be similarly unpredictable. Thus the long rains in 1984 were very reduced, whilst the dry season of January–March 1985 experienced above average rainfall (Figure 2.2c).

The area was used as a grazing or holding ground for cattle for many years prior to 1946 when Nairobi National Park was established (Lusigi, 1978); such practices have long been discontinued. The park is a concentration area for large herbivorous wild mammals during the dry season, most of whom disperse into the Athi-Kapiti plains to the south and east for the remainder of the year (Deshmukh, 1986).

The study site is a part of the flat plains which cover most of the Park and extend to the Kapiti plains. The underlying rocks are volcanic lavas, tuffs and basement complex (Scott, 1963). They are flat and terminate in bluffs overlooking the Athi in the east.

The soils of the flat plains are: black to dark grey clays, grumsolic soils; shallow yellow to yellow-red friable clays overlying a laterite horizon or rock; shallow soils on steep slopes and alluvial soils. A typical soil profile is shown in Figure 2.3. The black to dark grey clays,

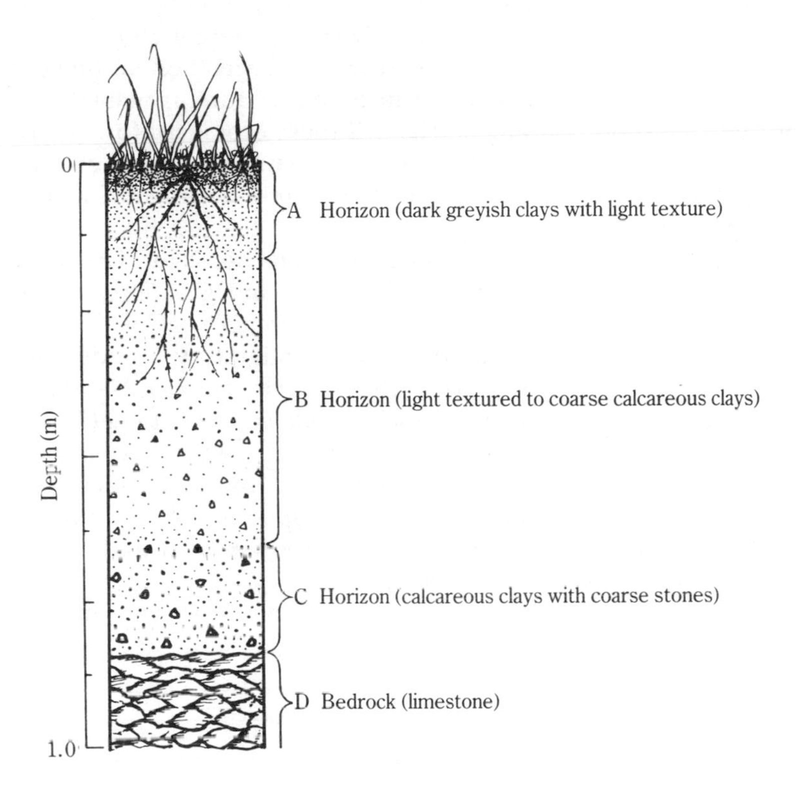

Figure 2.3. Soil profile at the study site illustrating the major horizons.

the grumsolic soils, are the most dominant. These soils crack when dry and are sticky when wet. Because of this, the soils are best utilized in agriculture for dry season grazing.

The vegetation cover of the study site consists of almost pure stands of grasses with a few scattered woody plants. The major grass species are *T. triandra* and *Pennisetum mezianum* Leeke, which together comprise about 60% of the total plant cover. Other grasses, including *Pennisetum stramineum* Peter, *Digitaria macroblephara* (Hack.) Stapf., *Eustachys paspaloides* (Vahl) Lanza and Mattei, *Cymbopogon caesius* (Hook and Arn.) Stapf, *Rhynchelytrum repens* (Willd.) C.E. Hubbard, and *Cenchrus ciliaris* L., make up about 20% of total plant cover. Dicots such as *Indigofera volkensii* Taub. and *Dolichos formosus* A. Rich. contribute 17% of total plant cover. Among the woody species which occur on the plots is *Acacia mellifera* (Vahl) Benth. However, grassland is not considered a climax vegetation for this area (Lusigi, 1978). There is a strong successional trend towards woody vegetation and woodland could be considered to be the potential climax vegetation. The open grassland with scattered woody vegetation is, therefore, the product of several forces which include moisture availability, grazing, burning and poorly drained soils. All these ecological forces have been in functional balance in this area over many hundreds of years. The vegetation, therefore, could be considered a functional sub-climax (Lusigi, 1978).

Such a plant community supports large herds of Thomson's and Grant's gazelle (*Gazella thomsoni* Gunther and *Gazella granti* Brooke), wildebeest (*Connochaetes taurinus* Burchell) Coke's hartbeest (*Alcelaphus buselaphus* Gunther) and migratory herds of eland (*Taurotragus oryx* Lyddeker), zebra (*Equus burchelli* Gray), buffalo (*Syncerus caffer* Sparrman), giraffe (*Giraffa camelopardalis* Matschie) and some waterbuck (*Kobus defassa* Rupell) which occur near river valleys. This ecosystem is also inhabited by birds such as ostrich (*Struthio camelus* Linnaeus), several species of vultures (*Torgos tracheliotus* Forster), eagles (*Polemaetus bellicossus* Daudin) and bustards (*Eupodotis senegalensis* Reichenow).

In 1984 a representative 1 ha plot was chosen as the sampling area, and surrounded with an electric fence to keep out large grazing herbivores (Figure 2.1c).

2.3 METHODS

Measurements, as outlined in Chapter 1, were taken from October 1984 to September 1986, with some modifications as described below. Stratified canopy sampling was also carried out for the assessment of above-ground biomass and leaf area indices. Briefly, the method was as follows: two wooden boards were marked into segments of 20 cm each (0, 20, 40, >40 cm). The 0 cm mark corresponded to the ground or clipping level. The vegetation was trapped between the boards and clipped at its base. No attempt was made to hold the vegetation upright, as this would give a false measurement of biomass in each stratum of the plant canopy. The material on the board was then cut at each marked level with a sharp knife.

For each canopy layer plant material was separated into plant species and categories, i.e. dead material was separated from live, and the live above-ground material separated by organ type (Chapter 1). Projected areas of leaf lamina, live stems and sheaths were determined with the leaf area meter as described in Chapter 1. From these areas, the leaf area index (L) was calculated for each canopy layer.

The depth to which soil cores were taken was determined from a trial investigation which showed that 95–97% of root biomass and dead material was present in the first 15 cm depth. Thus all routine monthly measurements of below-ground biomass used soil cores taken to this depth.

Production was estimated as described in the equations of Chapter 1. Additional values relating to biomass dynamics were calculated as follows.

Turnover rates and turnover times were calculated for the period between July 1984 and July 1986 using the method of Dahlman and Kucera (1965) as follows:

Turnover rate:

$$K_r = W_y / W_{max}$$

where: W_y = annual increment of plant material (g m^{-2})
W_{max} = maximum biomass (g m^{-2})
K_r = turnover rate (y^{-1})

$$K_t = 1 / K_r$$

where: K_t = turnover time in months
K_r = turnover rate (y^{-1})

Annual increments of plant materials were calculated by summing all positive increments over the year.

Stomatal conductance and leaf rates of photosynthetic CO_2 assimilation were measured *in situ* with a portable open gas exchange system. Air was supplied at a constant and monitored flow rate via a mast at 4 m above the canopy (Model ASU, Analytical Development Co., Hoddesdon, UK) to a 10 cm long leaf chamber (Model PLC, Analytical Development Co.). The humidity and CO_2 differentials across the chamber were determined with a capacitance humidity sensor and infra-red gas analyser (Type LCA2, Analytical Development Co.). To determine the light response curve of CO_2 assimilation, the chamber was held in a horizontal attitude at a time of high irradiance and then the photon flux on the leaf varied by placing neutral density filters over the chamber.

For laboratory measurements of CO_2 assimilation, four dominant East African grass species at the study site (*T. triandra, P. mezianum, C. caesius* and *R. repens*) were collected from the field. These were used to examine the photosynthetic capacities of different plant parts. These were potted in their original soil type and placed in a shade house at the Chiromo campus, University of Nairobi. They were watered as necessary. CO_2 assimilation and water loss of attached leaf blade, sheath, stem, flag leaf and inflorescence were monitored in an open gas exchange system (Long and Hällgren, 1985) with an infrared gas analyser (Model 225/2, Analytical Development Co., Hoddesdon, UK) connected to a chart recorder (Model EPR–1FA, TOA Instruments, Tokyo) and a dewpoint hygrometer (Model 880, Cambridge Instruments, Boston, Mass.). Assimilation chamber temperature was maintained at 30°C by circulating thermostatically warmed water through a water jacket surrounding the leaf chamber. Temperature was monitored by thermocouple wires appressed on one side of the plant organ and read with a millivolt meter (Model 60B, Keithley Digital Multimeter). Gas flow rates were monitored by an ADC Sample Selector (Model WA 161, Analytical Development Co.). The assimilation chamber was 8 cm long, 3.5 cm wide and 1 cm deep, with a metal base and acrylic plastic (plexiglass) cover. It was also equipped with a small air stirrer and was surrounded by a water jacket. Light of 1800 μmol m^{-2} s^{-1} was provided by a tungsten halide lamp, and was filtered through 15 cm deep water. The photon flux was measured with a quantum sensor (Model LI 188B, Li-Cor, Lincoln, Nebraska).

Air containing approximately 350 ppm CO_2 was pumped from outside the laboratory, bubbled through cold water to lower the dew point and passed through two pathways: one through the assimilation chamber (as sample air) and the other through the Dewpoint Hygrometer to the analyser (as reference air). The incoming and outgoing dew points were set by using the reference and sample air, respectively. The CO_2 assimilation rates were calculated as described by Long and Hällgren (1985).

Following the determination of CO_2 assimilation rates of different plant organs, the chlorophyll content and the distribution of stomata in these tissues were determined. The chlorophyll extraction method was that of Arnon (1949). Stomatal counts were made from replica slides prepared by painting nail varnish onto the leaf surface (Sampson, 1961).

2.4 RESULTS

2.4.1 Biomass dynamics

Above-ground biomass

Above-ground biomass exhibited a seasonal, bi-modal pattern with peaks during the wet seasons (Figure 2.4a, cf. Figure 2.2). Highest values of biomass were encountered during the long rains (March–June). Biomass ranged from 73 g m^{-2} in October 1984 to 338 g m^{-2} in April 1985. In the first 12 months of study, levels of dead material tended to

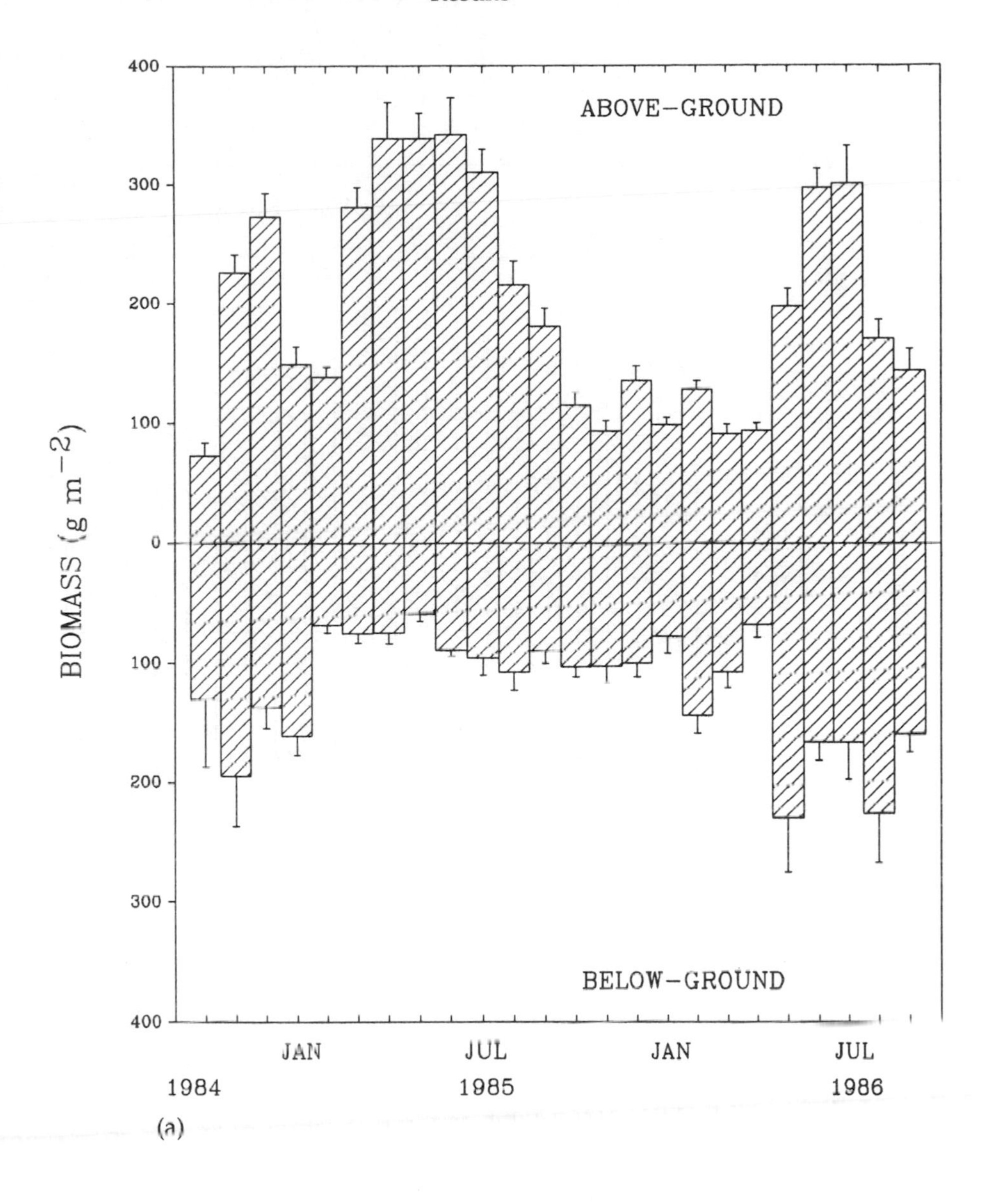

Figure 2.4. (a) Monthly amounts of biomass (dry weight of live vegetation) measured for a unit area of ground, both above and below ground, vertical bars indicate 1 se.

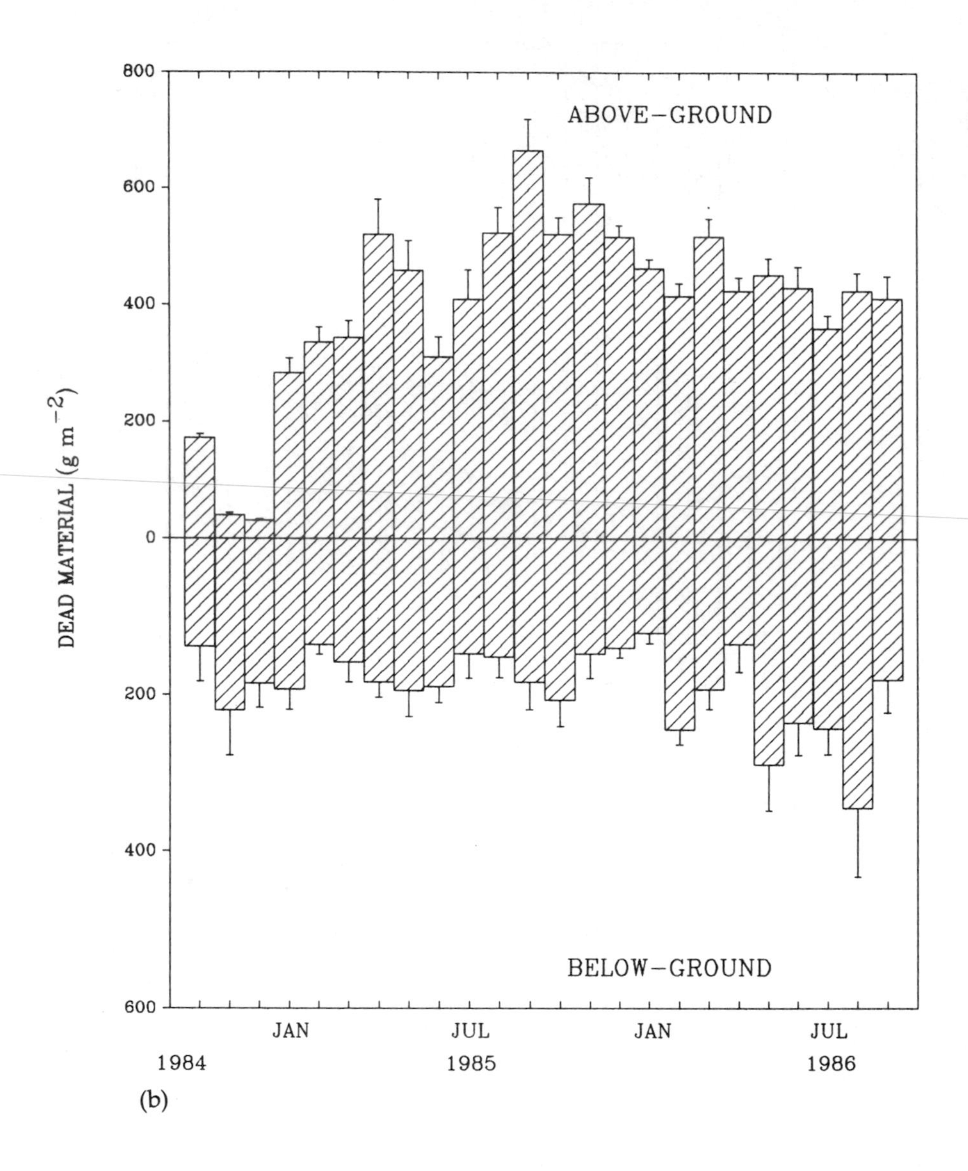

(b)

Figure 2.4. *continued* (b) As for (a) but illustrating amount of dead vegetation.

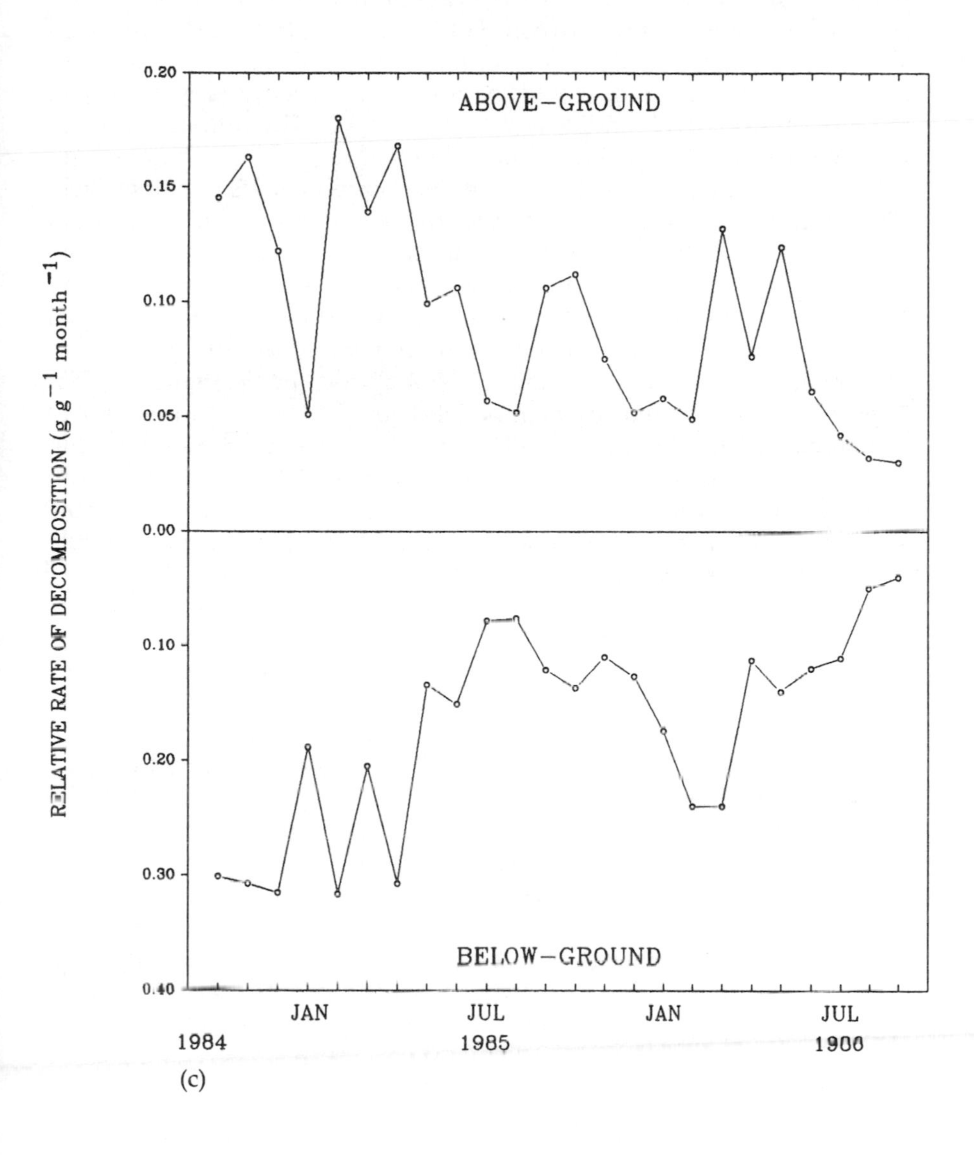

(c)

Figure 2.4. *continued* (c) Monthly relative rates of decomposition of dead vegetation both above and below ground.

be higher in the drier parts of the year, particularly in the longer dry season. A comparison between February and June 1985 revealed significantly higher levels in the former month ($P < 0.001$). However, no significant difference could be found between the wet season months of April 1985 and May 1986 ($P > 0.02$). The amount of dead vegetation steadily increased from a low value of 66 g m^{-2} to a high value of 651 g m^{-2} in December 1984 and September 1985, respectively (Figure 2.4b). From September 1985 the amount of dead vegetation varied less, fluctuating between about 400 and 600 g m^{-2} with some minor troughs and peaks during wet and dry seasons, respectively. The same trend was shown for below-ground dead materials. The level of standing dead material increased from a minimum of 19 g m^{-2} in December 1984 to a maximum of 457 g m^{-2} in September 1985. Thereafter, it fluctuated between *ca* 250 and 450 g m^{-2}. At the start of the study, litter (detached dead material) outweighed the standing dead material, although this was reversed after February 1985. From then on, litter varied between 100 and 190 g m^{-2}. Over the whole study, the mean level of standing dead material was 285 g m^{-2} while that of litter was 116 g m^{-2} (Figure 2.4c).

Below-ground biomass

Below-ground biomass was highest during the dry seasons. However, peak values of biomass occurred just before the end of the growing season or at the start of the dry season. This trend was evident during November 1984 (195 g m^{-2}) and May 1986 (230 g m^{-2}) (Figure 2.4a). Below-ground biomass values differed significantly (*t*-test, $P < 0.001$) between the respective wet and dry season months of June 1986 and March 1986. Dead materials increased during the dry seasons and decreased during the rainy seasons (Figure 2.4b).

Dead below-ground materials fluctuated between 120 and 340 g m^{-2}, with the lowest value (121 g m^{-2}) recorded in January 1986 and the highest value (345 g m^{-2}) recorded during August 1986 (Figure 2.4). On average, dead vegetation below ground exceeded live quantities.

2.4.2 Decomposition rate of dead plant materials

Relative decomposition rate (r) for above-ground dead herbage remained lower on average than for below-ground dead plant material (Figure 2.4c). In both types of plant materials, relative decomposition rates were highest during the rainy seasons and lowest during the dry seasons. Peak values occurred during November and December 1984, February 1985 (when the rains were early), and March 1986 in both types of dead plant materials. Relative decomposition rates ranged from 0.023 to 0.180 g g^{-1} $month^{-1}$ and 0.076 to 0.335 g g^{-1} $month^{-1}$ for above-ground and below-ground dead materials, respectively.

2.4.3 Species composition

Over the whole period of study, *T. triandra* and *P. mezianum* comprised the majority of the biomass (Figure 2.5). Dicotyledons, which included several annuals, are important constituents of this plant community. This was most notable during the period March–September in 1985. Overall sedges contributed the least to biomass (Figure 2.6).

2.4.4 Net primary production

Between October 1984 and September 1986, above-ground net primary production (AP_n) was 1004 g m^{-2} (equivalent to 1.4 g m^{-2} d^{-1}). Below-ground net primary production (BP_n) was 875 g m^{-2} (= 1.2 g m^{-2} d^{-1}). Therefore, the total net primary production (P_n) of the plant community was about 1880 g m^{-2} or 2.6 g m^{-2} d^{-1}. Net primary productivity differed between years. From October 1984 to September 1985 AP_n was 805 g m^{-2} y^{-1} and BP_n 488 g m^{-2} y^{-1}. From October 1985 to September 1986, AP_n was 199 and BP_n 388 g m^{-2} y^{-1}. For the period October 1984 and September 1985, when rainfall was higher, P_n reached 1292 g m^{-2} y^{-1}. However, P_n declined to 587 g m^{-2} y^{-1} between October 1985 and September 1986, the drier year. Monthly values of P_n ranged between 13–308 g m^{-2} above ground and 8–217 g m^{-2} below ground. The trend in monthly P_n closely followed the pattern of rainfall. At the start of the rains, the below-ground compartment accumulated more material than the above-ground compartment, hence there was a positive below-ground P_n during the first or second month of each rainy season. For example, this occurred in October 1984 (96 g m^{-2}), March 1985 (59 g m^{-2}), October 1985 (53 g m^{-2}) and May 1986 (315 g m^{-2}). It is after this increase in below-ground net primary production that above-ground net primary production peaked. This happened during or after the month when the rainfall occurred (Figure 2.2c).

2.4.5 Turnover rate and time

Results of turnover rate and time of different plant materials are presented in Table 2.1. Below-ground biomass had the highest turnover rate (0.95–1.31 y^{-1}) followed by above-ground biomass

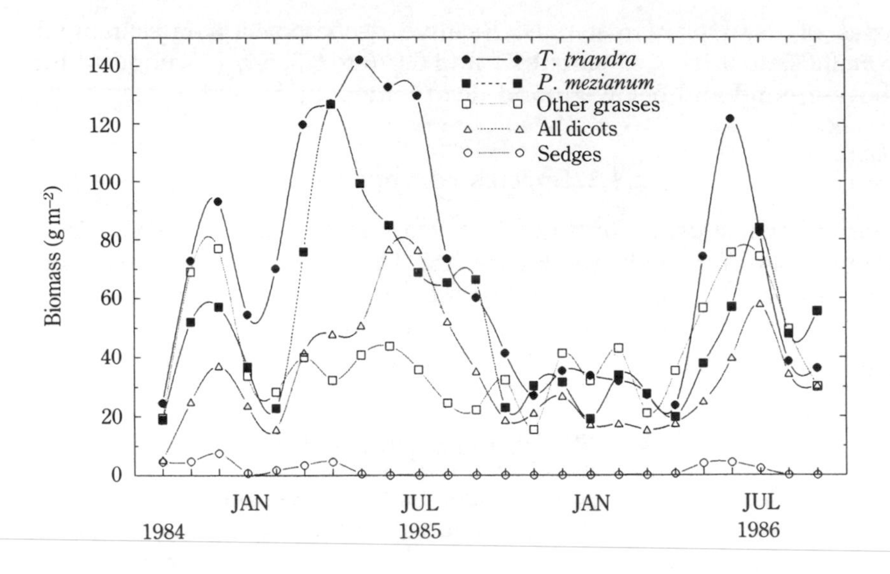

Figure 2.5. The contribution of individual species and species groupings to the total amount of above-ground biomass in each month.

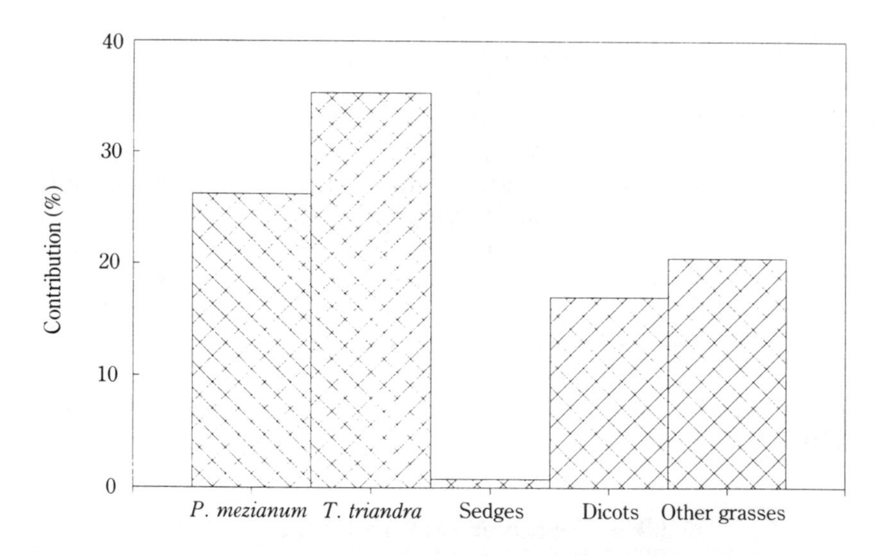

Figure 2.6. The contribution of individual species and species groupings to total above-ground biomass averaged over the study period.

Table 2.1. Turnover rate and time for different plant materials: October 1984 to September 1985 and October 1985 to September 1986

Plant material	*Turnover rate*		*Turnover time (months)*	
	1984/85	*1985/86*	*1984/85*	*1985/86*
Above-ground biomass	1.31	0.92	9.2	13.1
Standing dead	1.18	0.32	10.1	37.3
Litter	0.98	0.59	12.2	20.5
Below-ground biomass	0.95	1.31	12.7	9.1
Below-ground dead	0.92	1.19	13.0	10.1

(0.92–1.31 y^{-1}) and below-ground dead vegetation (0.92–1.19 y^{-1}). Litter had the lowest turnover rate (0.59–0.98 y^{-1}).

From Table 2.1 it can be noted that standing dead material had the longest turnover time of 37 months, followed by litter (21 months).

2.4.6 Structure

Stratified clipping revealed that most of the above-ground biomass occurred in the bottom 0–20 cm canopy layer followed by the middle 20–40 cm layer (Figure 2.7). Least was found in the upper (>40 cm) canopy layer. At the harvest in June 1986, 68% (or 203 g m^{-2}) of total above-ground biomass occurred in the 0–20 cm canopy layer, 23% (or 70 g m^{-2}) in the 20–40 cm canopy layer, and only 9% (or 27 g m^{-2}) occurred in the >40 cm canopy layer. The vertical distribution of leaf area index (L) closely followed that of biomass. In June 1986, about 56% (or 1.73) of the total leaf area index occurred in the 0–20 cm canopy layer, 30% (or 0.9) in the 20–40 cm layer and only 14% (or 0.43) in the top layer (> 40 cm).

Leaf area indices were highest during the wet seasons (Figure 2.9). Values of L ranged from zero (in most dry months) to 3.1 (in May 1986). The trend in L corresponded or lagged closely behind that of rainfall (Figure 2.8). Peak values occurred in November 1984 (1.9), March 1985 (1.5), May 1985 (1.7) and May 1986 (3.1).

Stem and sheath area indices were monitored from March 1985. Stem area index was always lower than leaf or sheath area index (Figure 2.9). Only areas of photosynthetic stems were determined for area indices. Stem area index ranged from zero during the dry season months to 0.95 in May 1986. Sheath area index ranged from zero to 1.53 over the same period. Total area index, i.e. the total of leaf blade, sheath and photosynthetic stems area index, was highest during the

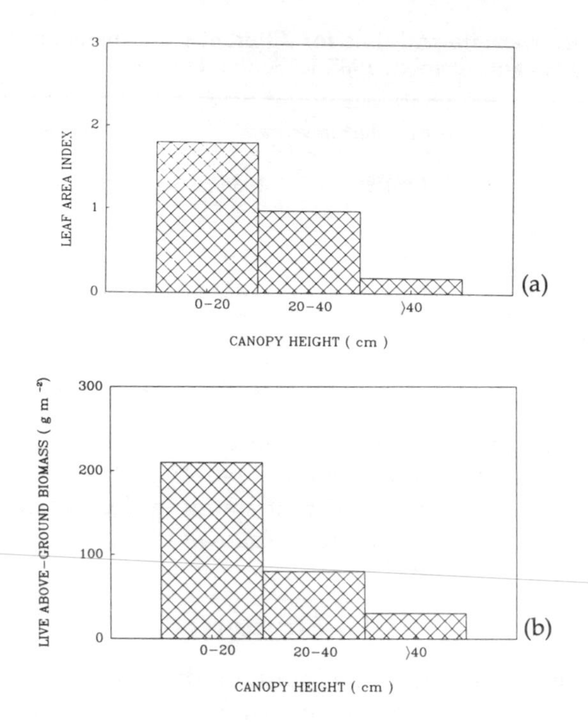

Figure 2.7. The distribution of (a) leaf area index in relation to height above the ground surface and (b) shoot biomass.

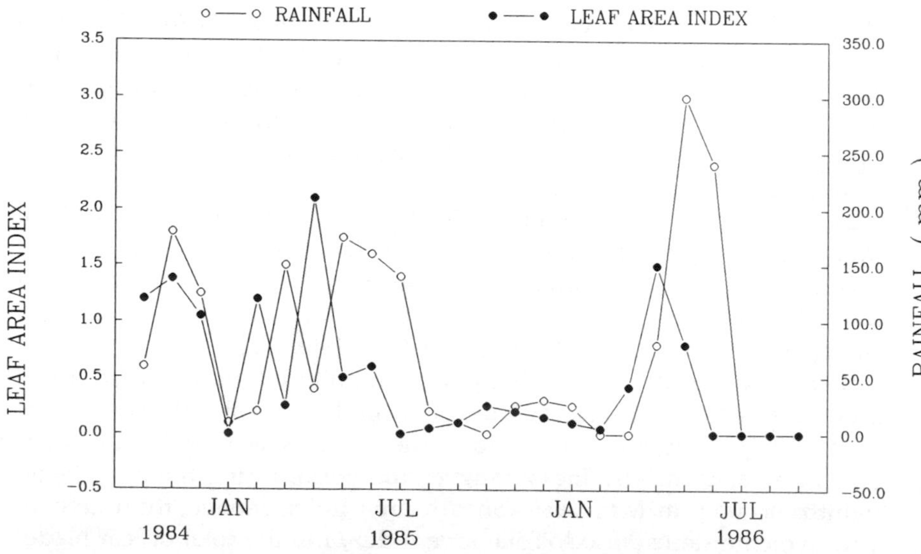

Figure 2.8. Monthly variation in mean leaf (blade) area index in relation to rainfall over the study period.

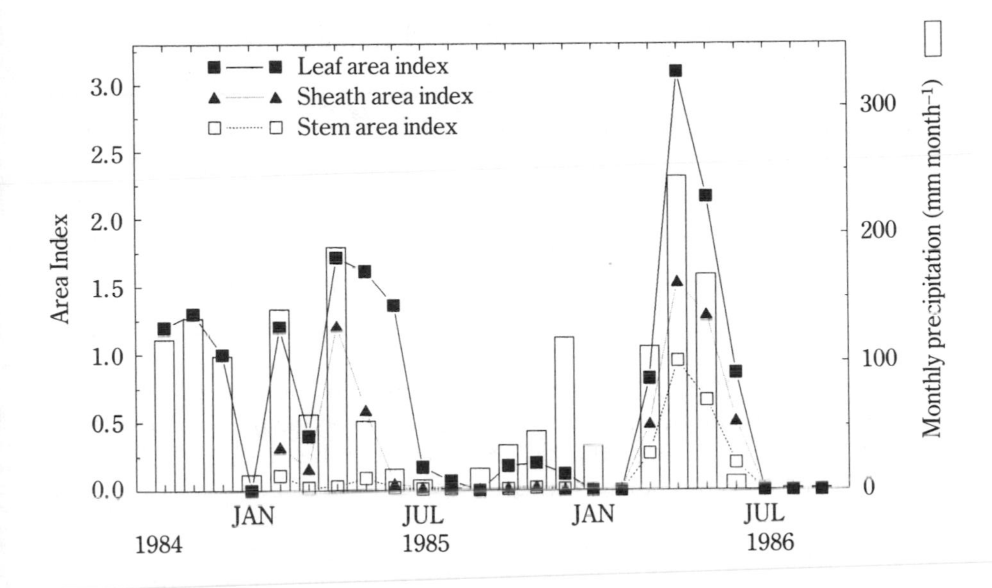

Figure 2.9. Monthly variation in leaf, sheath and stem area index, in relation to rainfall over the study period.

long rainy season months. The highest value was 5.57 in May 1986. The lows of zero were encountered during the dry season months.

2.4.7 Light energy interception by the plant canopy

Total solar radiation

The amount of total solar radiation incident over the period from October 1984 to November 1986 averaged about 19.7 MJ m^{-2} d^{-1} at the top of the plant canopy. Solar radiation showed the expected exponential decline with depth into the canopy. During the March–June 1986 growing season, about 7% of solar radiation was intercepted by the plant canopy at the height of 40 cm from the ground surface (Table 2.2). This amount increased to about 21% at the 20 cm canopy level and about 64% (about 12.6 MJ m^{-2} d^{-1}) at the bottom of canopy. Whilst effort was made to raise the lowest solarimeter above the litter, decrease in solar radiation through the canopy represents both absorption by live and by standing dead shoot material.

Light energy conversion efficiency

Light energy conversion efficiency (ε_c) is a measure of the efficiency with which intercepted solar energy is used to produce new dry matter by a plant community. On a monthly basis, ε_c ranged between

Table 2.2. Mean solar radiation (MJ m^{-2} d^{-1}) received at various canopy levels in various months, 1986

Canopy level	*March*	*April*	*May*	*June*	*July*
0 cm	11.07	7.51	7.23	7.40	7.30
20 cm	20.82	17.30	17.47	15.70	14.61
40 cm	22.35	20.50	17.30	19.20	19.10
Above canopy	23.53	21.93	19.36	19.68	19.32

0 and 0.56 g MJ^{-1}. Highest values occurred in the wet months. For example, ε_c for above-ground biomass was 0.56 g MJ^{-1} in November 1984. A value of 0.48 g MJ^{-1} was calculated for below-ground biomass in May 1986. ε_c was also estimated on an annual basis. The average energy equivalent or calorific value of most plant materials is 17.8 kJ (0.0178 MJ) g^{-1} dry matter (Black, 1956) while the net primary production rate for Nairobi National Park grassland was estimated to be 1292 g m^{-2} y^{-1} during October 1984 to September 1985. Therefore, the amount of energy trapped into plant biomass per annum was 23 MJ m^{-2}. Since the mean solar radiation receipt above the plant canopy averaged 7190 MJ m^{-2} y^{-1}, the coefficient of solar energy conversion was 0.32%. When corrected for energy receipt in the wavelength of photosynthetically active radiation (PAR, 400-700 nm), this equals 0.70%, assuming that PAR was 45% of total radiation (Dykyjova, 1978; Sims and Singh, 1978).

2.4.8 Photosynthesis and related processes

Field measurement of physiological processes – seasonal rates of photosynthesis

Mean rates of net CO_2 assimilation between November 1984 and February 1985 ranged from zero to 27 μmol m^{-2} s^{-1} for upper canopy leaves of both *T. triandra* and *P. mezianum* at 12.00 h (Figure 2.10a). The lowest rates, for these two key species, were recorded during January 1985 when the vegetation was drying out. The highest rates were recorded during December 1984 for both key species (Figure 2.10a).

Seasonal variation in stomatal conductance

Estimates of stomatal conductance around 12.00 h showed that mean rates ranged from 0.038 to 0.213 mmol m^{-2} s^{-1} in *T. triandra* and from 0.059 to 0.44 mmol m^{-2} s^{-1} in *P. mezianum* (Figure 2.10b). In both species, rates increased from November 1984, reached a peak in

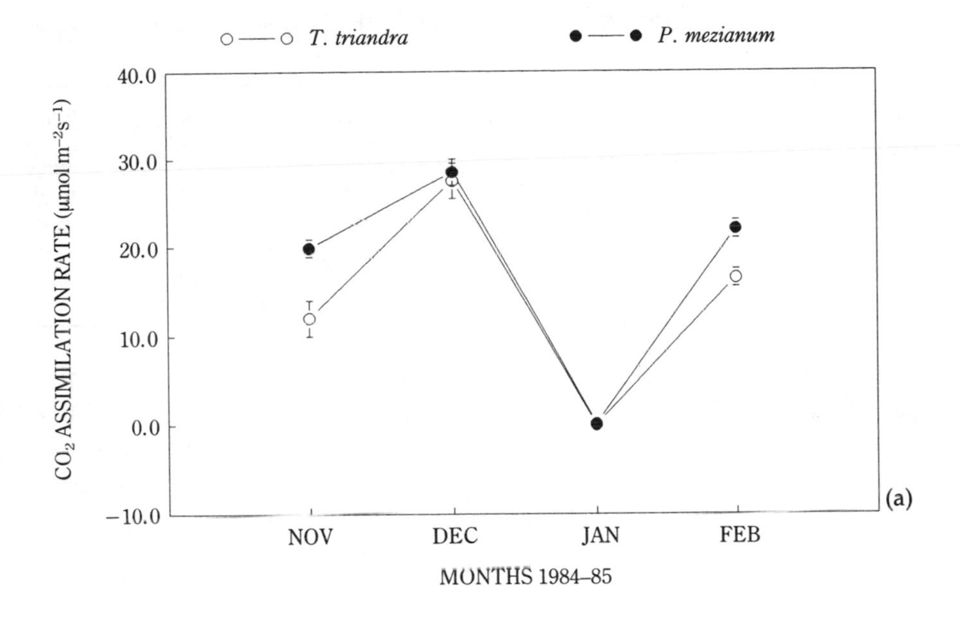

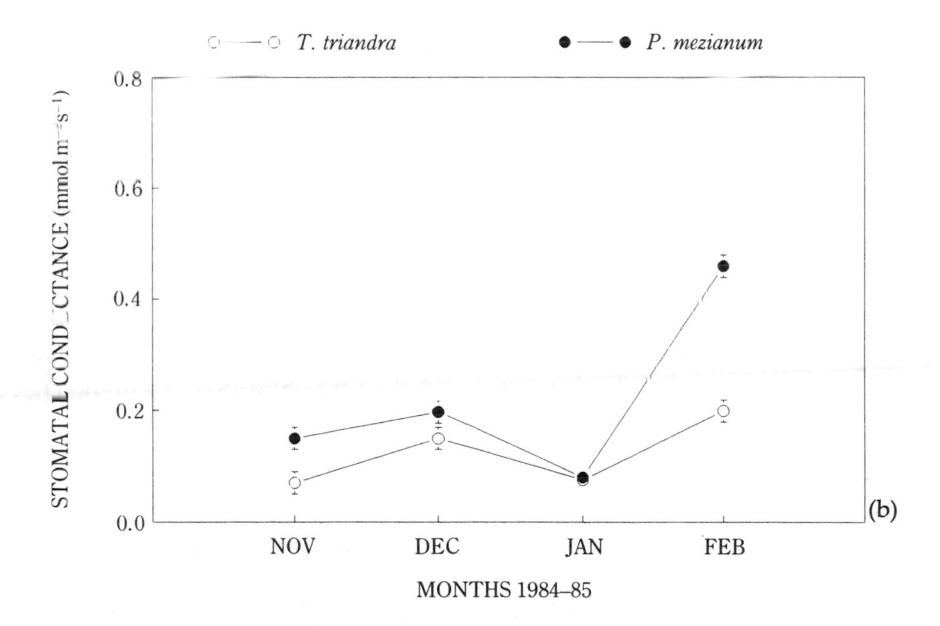

Figure 2.10 (caption overleaf)

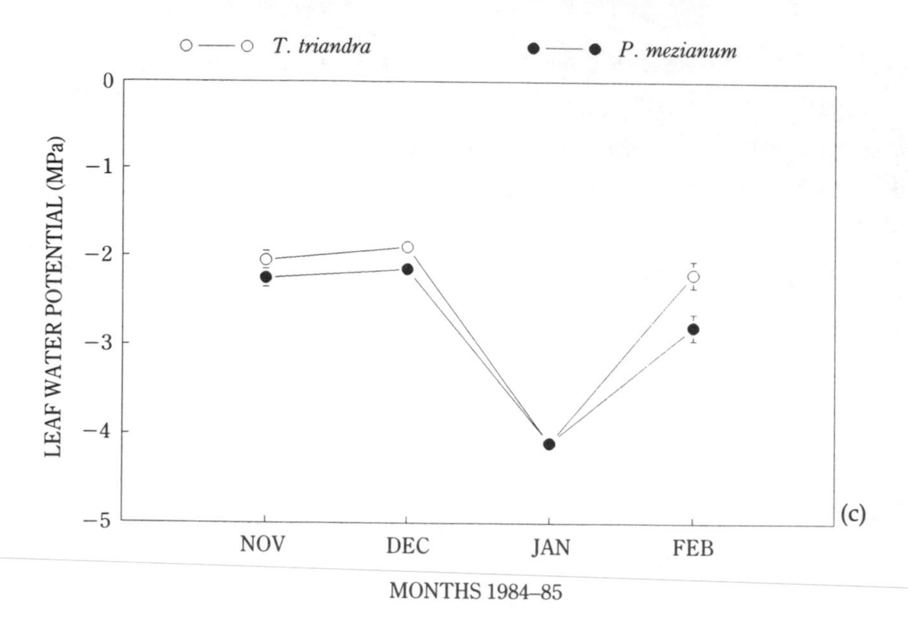

Figure 2.10. Mean (±1 se) (a) leaf CO_2 assimilation rate; (b) stomatal conductance and (c) leaf water potential measured between 11.00 and 13.00 h for upper canopy leaves of *Themeda triandra* and *Pennisetum mezianum*. Measured on four dates in 1984/85, where November and December correspond to the end of one wet season, January a brief dry season and February the beginning of the second wet season.

December 1984, declined again in January 1985 before attaining their maxima in February 1985.

Leaf water potential

Estimates of leaf water potential (ψ_l) for both key species were similar during November and December 1984, at between −2 and −2.5 MPa (Figure 2.10c). Values decreased drastically in January 1985 to below −4.0 MPa at noon in both species. This trend, however, changed in February 1985 when values increased to −2.2 and −2.8 MPa in *T. triandra* and *P. mezianum*, respectively, coinciding with a period of heavy rainfall (Figure 2.2c).

Photosynthetic light response curves of leaves at different canopy levels

Response curves of individual leaves of the key grass species (*T. triandra* and *P. mezianum*) at different canopy levels exhibited typical hyperbolic responses. Leaves of these species from the upper canopy layer exhibited higher rates of CO_2 assimilation at high photon fluxes (Figure 2.11).

Leaves of *T. triandra* from the upper canopy reached a maximum rate of 26.83 μmol m^{-2} s^{-1}, but they did not light saturate even at a photon flux approaching 2000 μmol m^{-2} s^{-1}. Leaves of the middle canopy layers possessed lower maximum photosynthetic rates at high photon fluxes and attained a maximum saturated photosynthetic rate (A_{sat}) of about 12 μmol m^{-2} s^{-1} at a photon flux of *ca* 1000 μmol m^{-2} s^{-1}, nearly half the A_{sat} of the upper canopy layer leaves. The leaves of *T. triandra* from the base of the canopy had the lowest A_{sat} of *ca* 9.6 μmol m^{-2} s^{-1} at a photon flux of *ca* 1000 μmol m^{-2} s^{-1} (Figure 2.11a).

Photosynthetic rates in leaves of *P. mezianum* at different canopy levels were similar to those of *T. triandra* (Figure 2.11b). Leaves from the upper canopy layer showed no evidence of light saturation while leaves from the middle and basal layers were light saturated at a photon flux of *ca* 1000 μmol m^{-2} s^{-1}. Maximum photosynthetic rates (A_{sat}) of about 24, 11 and 8 μmol m^{-2} s^{-1} were recorded for leaves of the upper, middle and lower canopy layers, respectively.

Diurnal courses of photosynthesis, stomatal conductance and leaf water potential

During the growing season in February 1985, CO_2 assimilation rates for both *T. triandra* and *P. mezianum* increased steadily during the morning to reach a maximum of 21.6 μmol m^{-2} s^{-1} for *T. triandra* and 31.9 μmol m^{-2} s^{-1} for *P. mezianum* around solar noon. Rates then declined towards the evening hours (Figure 2.12a). Mean values for *P. mezianum* were highest at around 12.00 h (27 μmol m^{-2} s^{-1}) dropping to a minimum of 7 μmol m^{-2} s^{-1} around 15.00 h (Figure 2.12a). *T. triandra* had lower maximum mean value of about 20 μmol m^{-2} s^{-1} which was attained earlier in the day (11.00 h) than that of *P. mezianum*.

During the start of the dry season in January 1985 both key species exhibited a two-peaked diurnal course in CO_2 assimilation rates (Figure 2.12b). For both species, the first maximum occurred during the morning, around 10.00 h. When this maximum occurred the photosynthetic rate of *T. triandra* was 4.52 μmol m^{-2} s^{-1}, while that of *P. mezianum* was 7.83 μmol m^{-2} s^{-1}. There was a total cessation of photosynthetic activity at midday in both species, due to stomatal closure. During the evening (around 16.00 h) there was some recovery of photosynthetic activity.

During the growing season, mean values of stomatal conductance in

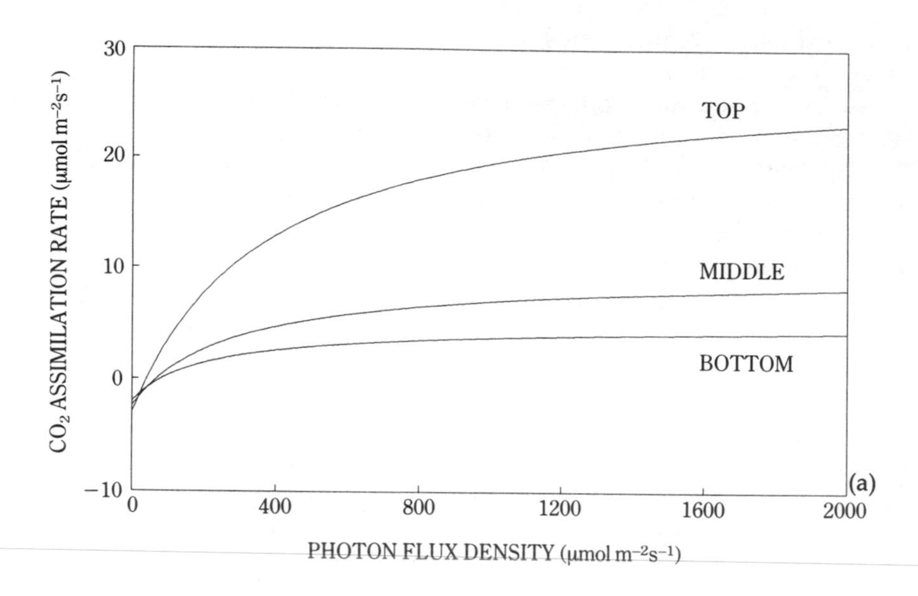

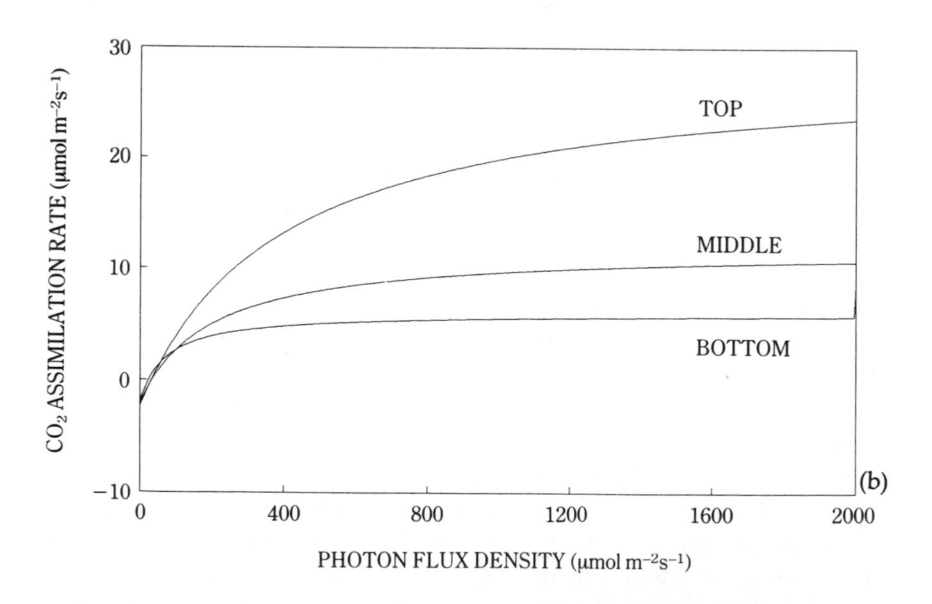

Figure 2.11. Rectangular hyperbolas fitted to the responses of CO_2 assimilation to photon flux for leaves at three canopy levels on 15 February 1985: (a) *Pennisetum mezianum* and (b) *Themeda triandra*. Each curve is the mean of 4–10 individual response curves.

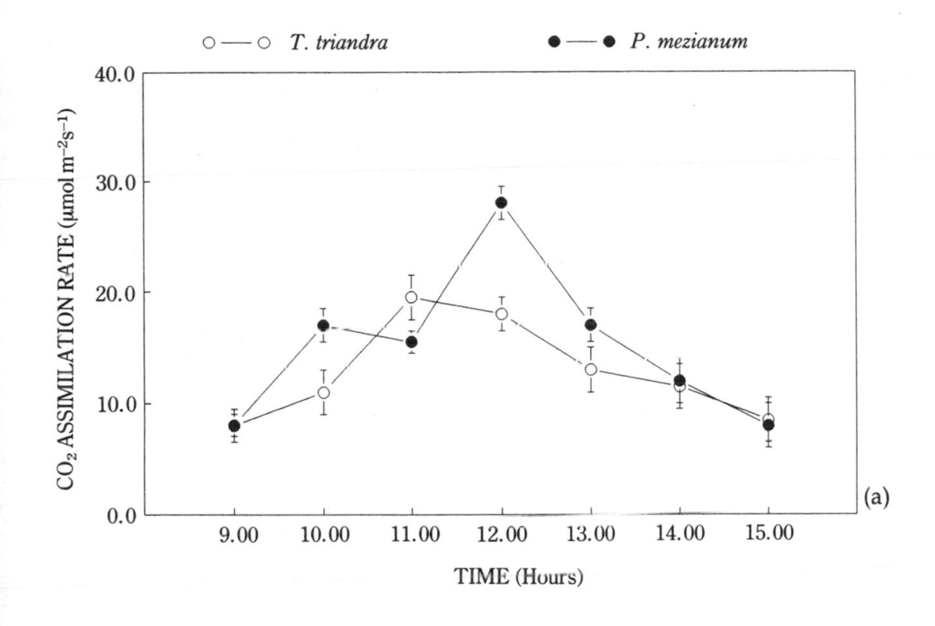

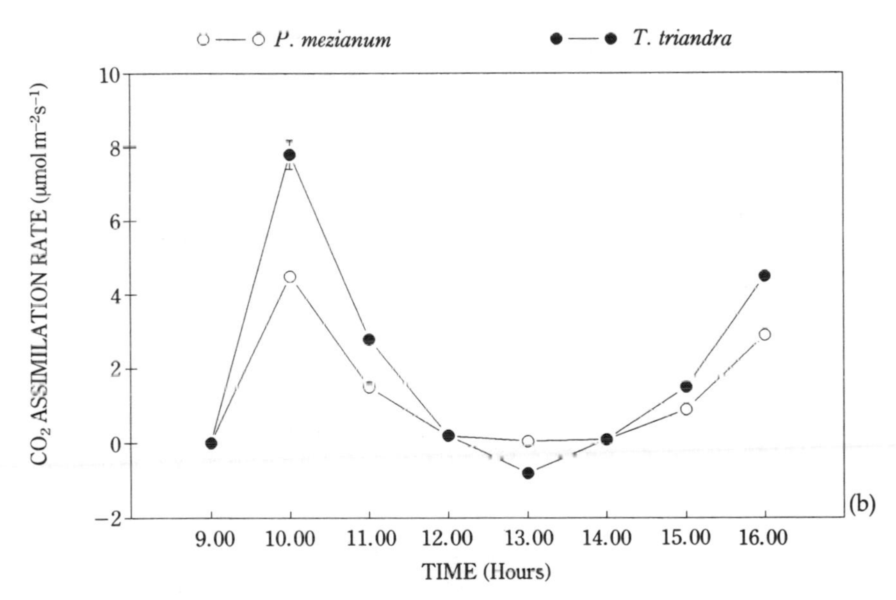

Figure 2.12. Example diurnal responses of CO_2 uptake rates per unit leaf area for upper canopy leaves of *Pennisetum mezianum* and *Themeda triandra* on 2 days with clear skies in (a) February 1985 after the start of the wet season and (b) January 1985 before the end of the dry season. Each point is the mean of *ca* 10 leaves (±1 se).

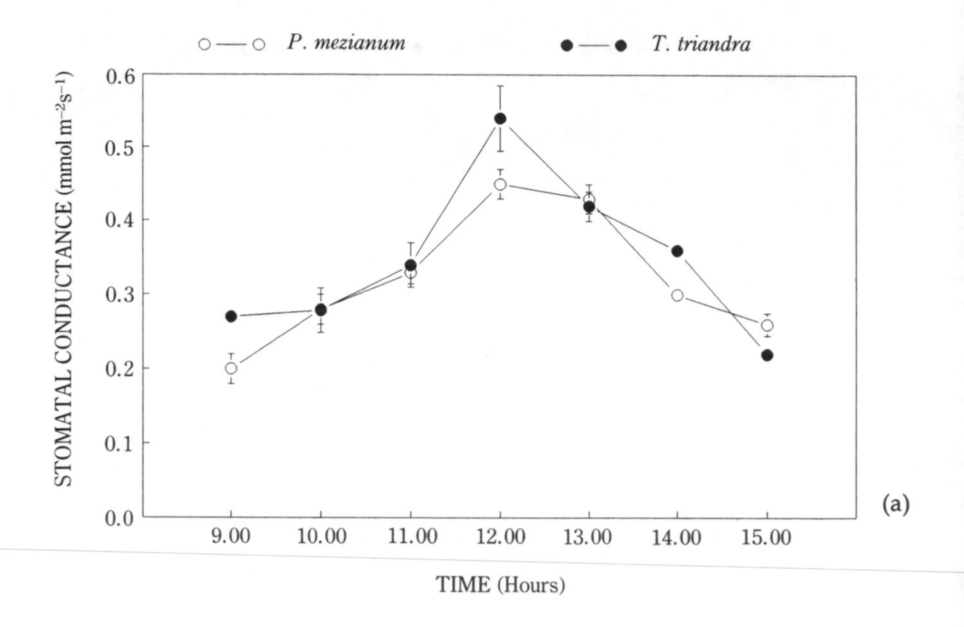

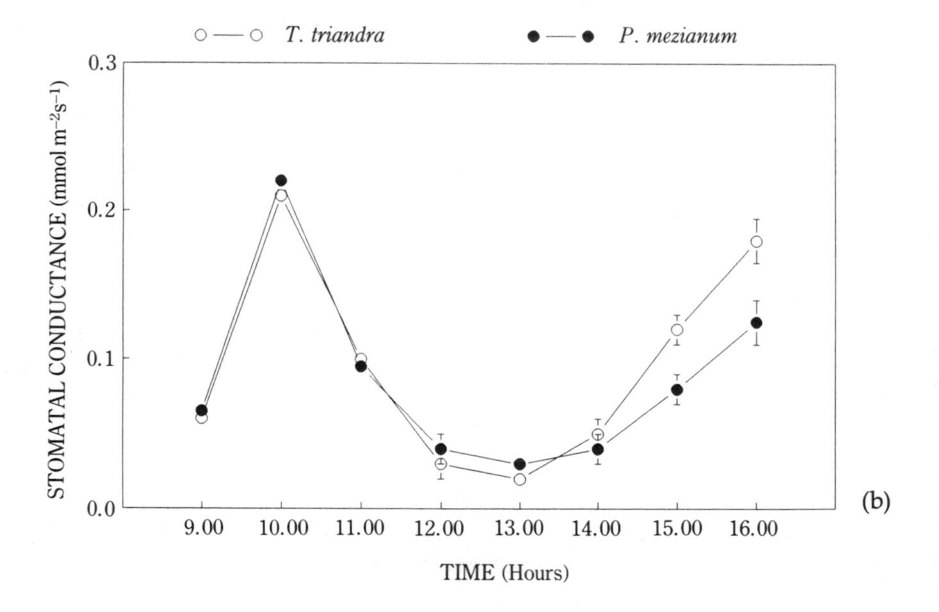

Figure 2.13. Example diurnal responses of stomatal conductance (see Figure 2.12 for details).

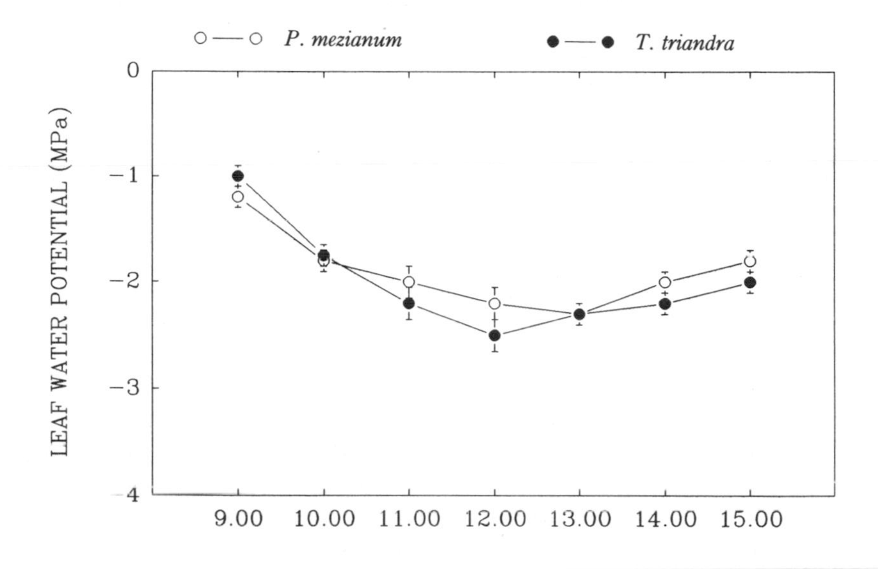

Figure 2.14. Example diurnal response of leaf water potential for upper canopy leaves of *Pennisetum mezianum* and *Themeda triandra* on a day with clear skies at the beginning of the wet season in February 1985. Each point is the mean of *ca* 5 leaves (± se).

both species (Figure 2.13a) increased with the diurnal time course up until 12.00 h, and then decreased towards the evening. In *T. triandra,* stomatal conductance increased from 0.266 mmol m^{-2} s^{-1} at 9.00 h to a maximum of 0.542 mmol m^{-2} s^{-1} at 12.00 h. In *P. mezianum* it increased from 0.201 mmol m^{-2} s^{-1} at 9.00 h to 0.446 mmol m^{-2} s^{-1} at 12.00 h. On average, *T. triandra* showed slightly higher values than *P. mezianum*. During the dry season in January 1985, the two grass species exhibited a two-peaked diurnal pattern of stomatal opening (Figure 2.13b). Stomata of both grass species exhibited maximum conductance during the morning, around 10.00 h. Complete stomatal closure occurred at midday in both species. However, some stomatal opening was recorded in the evening.

During days when grasses were actively growing mean leaf water potential dropped slowly from a value of −1.0 MPa at 9.00 h to a low value of −2.5 MPa at 13.00 h in *T. triandra*. In *P. mezianum* it fell from −1.2 at 9.00 h to −2.2 MPa at 12.00 h (Figure 2.14). Both species recorded their lowest water potentials around midday. Leaf water potential began to increase again towards morning levels in the evening (Figure 2.14).

Table 2.3. Number of stomata per cm^2 in different plant organs of some grass species at the study site (±SE; $n = 8$)

	Leaf blade				
Species	*Abaxial*	*Adaxial*	*Sheath*	*Stem*	*Inflorescence*
T. triandra	29 866 ± 433	7 466 ± 552	6 929 ± 216	836 ± 216	10 513 ± 443
P. mezianum	9 677 ± 584	9 796 ± 564	8 492 ± 335	17 800 ± 786	*
C. caesius	21 742 ± 761	2 150 ± 239	14 574 ± 951	2 031 ± 335	11 923 ± 1544
R. repens	12 185 ± 782	16 844 ± 476	12 066 ± 569	5 801 ± 823	*

* Structures too minute for analysis.

Stomatal counts, total chlorophyll content and photosynthetic rates of different plant organs

The distribution of stomata in different plant parts varied predictably (Table 2.3). Leaf blades of all species possessed stomata, but all species possessed appreciable numbers of stomata on their stems and leaf sheath surfaces.

In all cases the amount of chlorophyll *a* was higher than that of chlorophyll *b* in different plant organs (Table 2.4). The ratio of chlorophyll *a*:*b* exceeded 3:1 in most cases. Leaf blades possessed the highest levels of chlorophyll. There were no appreciable amounts of chlorophyll in the stems of either *T. triandra* or *C. caesius*.

Of all the plant organs examined, leaf blades exhibited the highest levels of CO_2 assimilation rates (Table 2.5). *T. triandra* had the highest rate followed by *P. mezianum*. Sheaths of *P. mezianum* exhibited the highest rate for this part of the leaf, and about two-thirds of the rate recorded for the blade (Table 2.5). Only stems of *P. mezianum* and *R. repens* exhibited some net photosynthesis. The flag leaf blade of *P. mezianum* exhibited the highest rate among the four grass species. Inflorescences of *T. triandra* and *C. caesius* exhibited some net photosynthesis (Table 2.5).

2.5 DISCUSSION

2.5.1 Above-ground biomass and dead plant material

Shoot growth in semi-arid grasses is basically dependent on the availability of soil moisture (Cassady, 1973; Strugnell and Pigott, 1978; McNaughton, 1979). Peak biomass at Nairobi National Park occurred during the long rainy seasons, ranging from about 300 to 338 g m^{-2}.

Table 2.4. Amount of chlorophyll and ratio of chlorophyll *a*:*b* in different species and plant organs (mg g^{-1} fresh weight; ±SE; n = 3)

Species	*Plant organ*	*Chlorophyll a*	*Chlorophll b*	*Total Chl a+b*	*Ratio Chl a:b*
T. triandra	Leaf blade	2.00	0.52	2.52	3.85
	Sheath	0.67 ± 0.01	0.13 ± 0.01	0.80 ± 0.02	5.15 ± 0.18
	Inflorescence	0.52 ± 0.01	0.16 ± 0.01	0.68 ± 0.02	3.25 ± 0.17
P. mezianum	Leaf blade	1.79	0.31 ± 0.01	2.10 ± 0.01	5.77 ± 0.11
	Sheath	0.57 ± 0.01	0.17 ± 0.01	0.74 ± 0.01	3.35 ± 0.07
	Stem	0.36 ± 0.01	0.10	0.46 ± 0.01	3.59 ± 0.07
C. caesius	Leaf blade	1.55	0.42	1.97 ± 0.01	3.69
	Sheath	0.37 ± 0.02	0.15 ± 0.02	0.52 ± 0.02	2.47 ± 0.16
	Inflorescence	0.49 ± 0.01	0.13 ± 0.01	0.62 ± 0.01	3.77 ± 0.09
R. repens	Leaf blade	1.63 ± 0.01	0.32 ± 0.01	1.95 ± 0.03	5.09 ± 0.17
	Sheath	0.53	0.16 ± 0.01	0.69 ± 0.01	3.31 ± 0.20
	Stem	0.14	0.05 ± 0.01	0.19 ± 0.01	2.80 ± 0.17

Table 2.5. Mean photosynthetic rate* (μmol m^{-2} s^{-1}) for different plant parts of some grass species at the study site (±SE)

Species	*Leaf blade*	*Sheath*	*Stem*	*Flag leaf*	*Inflorescence*
T. triandra	28.68 ± 1.14	10.41 ± 1.47	−3.20 ± 0.63	8.85 ± 0.37	8.23 ± 1.54
P. mezianum	19.77 ± 0.37	13.54	4.99 ± 0.98	10.93 ± 2.57	−4.68 ± 0.55
C. caesius	18.57 ± 0.12	6.25 ± 1.48	−2.60 ± 0.37	5.21 ± 2.21	0.38 ± 0.10
R. repens	14.84 ± 2.76	7.29 ± 0.73	9.23 ± 1.33	2.08	−7.14 ± 0.84

* PFD = 1000 μmol m^{-2} s^{-1}; T = 30°C; n = 2.

These ranges were within those reported by several workers in the Nairobi National Park ecosystem. Owaga (1980) reported peak values of about 309 g m^{-2} and Deshmukh (1986) a peak value of 332 g m^{-2} inside exclosures. In other East African grass ecosystems, Sinclair (1975) reported a peak value of 115 g m^{-2} in the Serengeti while in the Ruwenzori National Park, Uganda, Strugnell and Pigott (1978) reported a peak value of 405 g m^{-2}.

Distinct differences in biomass levels attained during the 2 years of measurements were apparent. Despite heavy rainfall during the long rainy season of 1986, biomass levels never matched those of the previous year, even though the long rainy seasons were of similar duration. This phenomenon, termed 'vegetation stagnation' by Tueller and Tower (1979), is sometimes caused by the smothering of shoots as standing dead material accumulates due to the exclusion of grazers from the vegetation. This was evident in this study. Grazing is known to prevent stagnation through stimulation of growth by removing dead vegetation which would otherwise shade the young growth (McNaughton, 1979). On average, over the entire study period, only 32% of the total above-ground production was live. This reflects a large degree of accumulation of dead material after fencing of the study site.

Peak values of dead materials were encountered during the dry seasons. This is in agreement with the findings of Strugnell and Pigott (1978), Macharia (1981), Deshmukh and Baig (1983) and Deshmukh (1986) for East African grasslands. In Nairobi National Park, Deshmukh (1986) estimated a peak in dead material of 374 g m^{-2}, compared to a peak of about 651 g m^{-2} in the present study. Owaga (1980), working in the same ecosystem, reported values of 93 g m^{-2}.

Levels of standing dead material were lowest in December 1984, following the rains. Amounts of dead vegetation steadily increased thereafter, despite a small drop during the May and June 1985 rainy season. In the absence of grazers, much of the standing dead material is broken down by wind and rain to form litter. Rain also leaches soluble nutrients, which form the bulk of the structural material of the standing dead. This process is known to be faster in dead than living plant material (Tukey and Mecklenburg, 1964). After July 1985, standing dead material fluctuated between about 250 and 450 g m^{-2}. This indicated that inside the exclosure its levels had stabilized and changed only according to the time of the rainy periods.

The amount of litter was higher than that of the standing dead in October 1984. This reflected the pre-fencing conditions. Then, most of the standing dead material would have been broken down to litter through trampling by grazing herbivores, enhanced by the fact that this is a dry season concentration area for grazing herbivores (Lusigi, 1978; Owaga, 1980; Deshmukh, 1986). Since after fencing any accumulated standing dead material could only be broken down more slowly, it would be expected to remain at a higher level than litter. This was the case after January 1985.

Stratified clipping revealed that most biomass occurred in the first 20 cm above the ground canopy layer, reflecting the growth pattern of these grasses. Most tropical grasses have co-evolved with grazing herbivores. Because of the persistent effect of grazing down to ground level, much of the foliage (mainly leaves) is found in this first 20 cm above the ground surface, protecting the meristem. It is only during the reproductive phase that current season culms and nodal tillers are formed and elevated above this level. However, since most of these grasses are perennial, those culms formed in the last growing season which are not grazed tend to persist. Morphological evolution leading to meristem protection by physical isolation has been a major feature of grass evolution according to McNaughton (1979).

In any ecosystem, one or several species may dominate depending on the biotic and/or abiotic factors that favour them. Such factors may be fire, soil, rainfall and grazing. Plants that are resistant to grazing and fire pressures dominate in most semi-arid East African grasslands (McNaughton, 1979). Accordingly, it has been reported that *T. triandra* is a dominant species in this and similar plant communities (Strugnell and Pigott, 1978). Lusigi (1978) and Deshmukh (1986) have reported that *T. triandra* contributed the highest amount of biomass in this type of ecosystem. Although dicotyledons made a substantial overall contribution to biomass in this study, this was not the case at the start. Furthermore, most of these dicotyledons, which included *Indigofera volkensii* Taub., *Dolichos formosus* A. Rich., *Sida schimperiana* A. Rich.,*Vigna frutescens* A. Rich., *Clitoria ternata* L. and *Neonotonia wightii* (Wight Arm.) Leckey, were creepers or of short habit. This small stature is probably an adaptation to grazing pressure. Many dicotyledons grow from terminal meristems on extended stems, which are highly susceptible to herbivore destruction through grazing while the grasses grow from basal intercalary meristems that are less accessible to large herbivores (Branson, 1953; Rechenthin, 1956). At the start of the study dicotyledons contributed only a small proportion to biomass. However, after October 1984 their contribution increased. This suggests that without the exclosure dicotyledons were readily grazed, accounting for their low contribution to the overall biomass prior to site fencing.

2.5.2 Below-ground biomass

The highest values of below-ground biomass occurred during the dry seasons. It appears that a larger amount of material is accumulated in the roots during the dry season than during the wet season. This phenomenon has been documented in other grassland ecosystems (Whalley and Davidson, 1968; Singh and Yadava, 1974; Pandey and Sant, 1980). Towards the end of the growing season, there is an accumulation of reserve carbohydrate material in the stem bases and roots. New growth at the beginning of the next growing season draws on these reserves and accordingly below-ground biomass then declines. This process has been termed root 'draw down' by McNaughton (1979) and is a critical feature determining success in these grasslands. Species such as *P. mezianum* which have low abundance under heavy grazing, but become fairly abundant when grazing pressure is removed, probably suffer fatal root reserve depletion under intense grazing pressure (McNaughton, 1979). At the start of the study, *P. mezianum* ranked third in terms of biomass. At the end of the study it ranked second overall after *T. triandra*. This may further explain the dominance of *T. triandra* in these ecosystems, as the ability to maintain a high enough shoot biomass to produce root reserves is of prime importance.

One essential feature of root 'draw down' is that the mobilized food reserves are used in the initial phase of shoot growth, until a fully functional photosynthetic apparatus has been developed. After this phase of development, the shoot is able to manufacture enough material for its subsequent growth. Any extra assimilation after this stage which is not used for growth is again stored in the stem bases and roots. However, one notable phenomenon which occurs in this grassland community is a depression of root biomass at the end of the growing season such as during December 1984 and June 1986. This feature can also be explained by the growth pattern of these semi-arid grasses. At the end of the growing season, plants are in their last phases of growth, i.e. flowering and seed ripening. Therefore, some food reserves may be drawn upon for the formation of seeds (Strugnell and Pigott, 1978). After seed ripening any further assimilation, plus that mobilized from the senescing leaves, is translocated to the roots and stem bases.

The level of below-ground dead material was consistently higher than below-ground biomass, but showed similar trends of accumulation. This indicates that material was dying faster than it was decomposing. This was not reflected in their estimated respective turnover rates, however, which were nearly equal.

Although the magnitude of below-ground biomass in East African grasslands is not well documented, the range of values in our study (197–517 g m^{-2}) are at the lower end of the range of those reported by Strugnell and Pigott (1978). They reported values between 512–2007 g m^{-2} for total root biomass in enclosed sites of Ruwenzori National Park, Uganda. Their higher results may be due to the deep soils on which they worked, permitting root samples to be taken to a depth of 60 cm. Nevertheless, both the present study and that of Strugnell and

Pigott (1978) clearly demonstrate that root dynamics are a critical aspect of East African grasslands, especially where grazing is intense.

2.5.3 Decomposition rate of dead plant materials

Decomposition of plant materials is very much a function of the type of tissue, climatic factors (rainfall, temperature, etc.) and the type of decomposers. In tropical East African grasslands, temperature is never limiting for effective decomposition to take place. However, rainfall usually is limiting, especially during the dry seasons (Abouguendia and Whitman, 1979; Ohiagu and Wood, 1979). In this study, the trend in relative decomposition rates followed that of rainfall, with highest rates during the wet season.

The monthly average relative decomposition rates of between 0.09 and 0.18 g g^{-1} $month^{-1}$ for above-ground materials in this study were much higher than those reported by Macharia (1981) (0.009 – 0.02 g g^{-1} $month^{-1}$). However, our results compare well with those reported in other grassland ecosystems of the world. George and Smeins (1982) reported an average value of 0.07 g g^{-1} $month^{-1}$ for above-ground herbage in Texas. Abouguendia and Whitman (1979) reported values ranging from 0.018 to 0.128 g g^{-1} $month^{-1}$ in ungrazed mixed prairie in western North Dakota, USA. Ohiagu and Wood (1979) in Nigeria reported an average loss of litter weight due to decomposition for an ungrazed plot of 0.132 g g^{-1} $month^{-1}$. None of these studies, however, reported any values for below-ground plant materials. Comparison of decomposition rates between different studies in tropical Africa is difficult because most of these studies fail to specify the mesh size of their litter bags. Ohiagu and Wood (1979) used an unspecified type of wire mesh cage which allowed access to termites.

2.5.4 Net primary production

The above-ground net primary production of 804.74 g m^{-2} y^{-1} (about 2.21 g m^{-2} d^{-1}), during the period from October 1984 to September 1985, was higher than most of those reported by different authors in this ecosystem. Lusigi (1978) reported values of 394.7 g m^{-2} y^{-1}, Owaga (1980) 447.9 g m^{-2} y^{-1}, Macharia (1981) 364 g m^{-2} y^{-1} although Deshmukh (1986) obtained a value of 1071 g m^{-2} y^{-1}. Some of these authors may have grossly underestimated the net primary productivity of this ecosystem (Tables 2.6 and 2.7) because they used either the

Table 2.6. Net primary productivity (g m^{-2} y^{-1}): a comparison of the estimates obtained by taking account of mortality, decomposition and below-ground production, with estimates from above-ground biomass change alone

	Net primary productivity	
Methodology	*1984/85*	*1985/86*
1. Present study method Accounting for mortality and decomposition (including roots and rhizomes)	1292	587
2. IBP standard method (shoots only)	626 (52%)	576 (2%)
3. Maximum–minimum method (shoots only)	401 (69%)	372 (37%)

Values in parentheses are levels of underestimation as a percentage of 1.

'maximum–minimum biomass' method or the 'standard IBP method' of measurement (Table 2.6; Chapter 1).

Since biomass changes as well as losses through death were measured in our study at regular intervals we can use these data to determine what would be obtained with the 'standard IBP method' and the 'maximum–minimum biomass' method. This comparison (Table 2.6) shows that, for our study, the 'standard IBP method' and the 'maximum–minimum' methods would have underestimated productivity by 52 and 69%, respectively during the period October 1984 to September 1985. During the period October 1985 to September 1986, the error of underestimation would have been less (2% by the IBP method, 37% by the 'maximum–minimum method'). This shows that during the normal rainfall years, when productivity is usually higher, these methods will greatly underestimate P_n. It can therefore be concluded that most estimates of P_n in tropical African grasslands are likely to underestimate the true values.

Studies in other tropical African grasslands again show much variability in above-ground net primary production (Table 2.7). This is most probably due both to the variation in methodology used and the inherent climatic variation, particularly rainfall. The daily average AP_n of 2.6 g m^{-2} found in this study included both the dry and wet seasons. However, higher rates were obtained during growing season months. For example, 3.8 g m^{-2} d^{-1} in November 1984, 10.3 g m^{-2} d^{-1} in April 1985 and 6.1 g m^{-2} d^{-1} in May 1986. Total (i.e. AP_n

Table 2.7. Net primary productivity ($g\ m^{-2}\ y^{-1}$) in different African tropical grasslands

Place	*Vegetation type*	*Annual rainfall (mm)*	*Net primary production*	*Author*
Ghana	Savanna woodland	153	763	Nye and Greenland (1960)*
Senegal	Tree savanna	300	42	Moral and Bourliere (1962)*
Cote d'Ivoire	Grassland	1300	550	Cesar (1971)*
Senegal	Tree savanna	435	150	Bille (1973)*
Kenya	*Themeda* grassland	560	500	Cassady (1973)
Kenya	Wooded grassland	680	450	Cassady (1973)
Zaire	Miombo woodland	950	222	Freson (1973)*
Tanzania	Short grassland	613	470	Sinclair (1975)
Tanzania	Long grassland	905	598	Sinclair (1975)
Kenya	*Themeda* grassland	900	395	Lusigi (1978)
Uganda	*Hyparrhenia*/ *Themeda* grassland	600	553	Strugnell and Pigott (1978)
Uganda	*Sporobolus*/ *Chloris* grassland	600	536	Strugnell and Pigott (1978)
Nigeria	Savannah grassland	1115	231	Ohiagu and Wood (1979)
Kenya	*Themeda* grassland	600	402	Owaga (1980)
Kenya	*Themeda* grassland	1034	810	Macharia (1981)
Kenya	*Themeda* grassland	729	364	Macharia (1981)
Kenya	*Themeda* grassland	850	1071	Deshmukh (1986)
Kenya	*Themeda* grassland	800	1880	Present study†

* Cited by Ohiagu and Wood (1979).
† Total net primary production for shoot and root production from October 1984 to September 1986.

+ BP_n) net primary production is very high for these months: 10.5 g $m^{-2}\ d^{-1}$ in November 1984, 12.9 g $m^{-2}\ d^{-1}$ in April 1985 and 17.6 g $m^{-2}\ d^{-1}$ in May 1986. In the Serengeti, McNaughton (1979) estimated an extremely high AP_n over a short period during the wet season, with a peak of 40 g $m^{-2}\ d^{-1}$. During the growing season values were consistently above 20 g $m^{-2}\ d^{-1}$, suggesting that East African grasslands may be among the most productive in the world.

Besides the present study, only that of Strugnell and Pigott (1978) in Table 2.7 measured below-ground production. Strugnell and Pigott (1978) estimated an annual below-ground net primary production (BP_n) of about 1500 g m^{-2}. They emphasized that this value did not include short-term loss and replacement of roots. Their study was based on the 'maximum–minimum biomass' method. The value of 966 g m^{-2} y^{-1} from the present study demonstrates clearly that the below-ground compartment of these grasslands is highly productive and, therefore, an essential component for the functioning of these ecosystems.

The duration of positive values of both AP_n and BP_n varied with time. At the start of the rainy season BP_n was positive. This may be caused by the plants initiating growth in the roots before that of shoots. Thus they may form a large network of roots which would enable them to absorb larger quantities of soil water at the start of the rainy season. Since most of the roots are found in the first 15 cm layer of the soil column, this could be an important survival mechanism in a semi-arid habitat. Taerum (1970) reported that *Eragrostis superba* Peyr. and *T. triandra* send some roots to depths of at least 2 m but retain the bulk of their roots near the soil surface, where they can make maximum use of light rainfall.

2.5.5 Turnover rate and time

The turnover rate of above-ground biomass was high (0.92–1.31 y^{-1}) due to the fact that a lot of live plant material is usually transferred to the standing dead compartment during the dry season (Deshmukh and Baig, 1983). Turnover for the standing dead was lower (0.32–1.18 y^{-1}) as its breakdown to litter by grazing herbivores was removed after fencing (Rath and Misra, 1980). The low turnover of litter (0.59–0.98 y^{-1}) was due to slow microbial breakdown, a process which was only rapid during the wet seasons. Other authors have found a more or less similar pattern and magnitude in turnover rate and time for other tropical grasslands (Misra and Misra, 1979; Rath and Misra, 1980) (Table 2.8).

2.5.6 Canopy structure

Maximum growth of plants occurred during the rainy seasons. It is during these periods that shoot growth is maximal due to the greater availability of soil moisture in most East African grasslands

Table 2.8. Turnover rate of different grassland communities

Place	*Plant community*	*Turnover rate (%)*					*Authority*
		*LAB**	*SDB*	*Litter*	*LBB*	*DBB*	
USA	Prairie	97	—	45	—	—	Ovington *et al.* (1963)†
USA	Savannah	98	—	45	—	—	
USA	*Andropogon* grassland	90–92	—	—	—	—	Golley (1965)†
USA		—	—	—	—	—	Dahlman and Kucera (1965)
India	*Dichanthium* grassland	—	—	—	—	—	Singh (1967)†
India	*Heteropogon* grassland	—	—	—	—	—	Jain and Misra (1972)†
India	*Dichanthium* grassland	—	—	82	—	—	Misra (1973)†
Uganda	Grassland	174–188	—	—	—	—	Strugnell and Pigott (1978)
India	Grassland	77	—	119	—		Misra and Misra (1979)
Kenya	Savannah	92–131	32–118	59–98	95–131	92–119	Present study

* LAB, above-ground biomass; SDB, standing dead; LBB, below-ground biomass; DBB, dead below-ground material; TBB, total below-ground material.
† Cited by Misra and Misra (1979).

(McNaughton, 1979). Therefore, during these times shoots bear the largest amounts of foliage, and hence highest leaf area indices are realized (Strugnell and Pigott, 1978).

The range in leaf area indices values (0–3.1) is similar to that reported by Strugnell and Pigott (1978), which was 0–3.0 for a *T. triandra* dominated grassland in Ruwenzori National Park, Uganda, in an enclosed site. In our study, the highest values were recorded during the long wet season each year. The stratified pattern of leaf area reflected the growth habit of these grasses as discussed previously. The area indices of stems and sheaths were in the range 0–1.53, and they contributed greatly to the total photosynthetic area index. In May 1986 stems and sheaths provided 40% of this total, thus providing important additional photosynthetic surfaces.

2.5.7 Photosynthesis and related processes

Maximum photosynthetic rates for both *T. triandra* and *P. mezianum* compared well with laboratory results for African and South American C_4 grasses quoted by Medina (1986) (2.2–43.9 μmol m^{-2} s^{-1}).

T. triandra and *P. mezianum* are both C_4 grasses. However, our results are much higher than those obtained in the field by Medina (1986), which ranged from 2.2 to 7.9 μmol m^{-2} s^{-1}.

Seasonal rates of CO_2 assimilation for both key species in the present study were highest during the rainy season and lowest during the dry season. This related to the decrease in stomatal conductance. Stomatal conductances were highest during the rainy season and lowest during the dry season, for both species. This reflected a decrease in leaf water potential from December (growing season) to January (dry season), following a decrease in the internal water status of the plants. However, after the late season rains of February 1985, the water status of plants improved. Photosynthetic rates increased for both species during the same period. Moreover, most of the photosynthetic tissue available during January 1985 was either mature or senescent, and leaves are known to lose a substantial amount of their photosynthetic capacity with age (Ticha *et al.*, 1985). High seasonal leaf water potential values coincided with high values of photosynthetic rates and stomatal conductances.

In the North American tallgrass prairie, Knapp (1985) found that during severe seasonal water stress (leaf water potentials of < -6.00 MPa), photosynthesis decreased to zero, yet following a substantial late season rainfall photosynthesis increased from about 28 to 48% of early season rates.

Although the leaves used in our photosynthetic measurements were more or less of the same age, they had developed under different environmental conditions due to variations in the microclimate within the plant canopy. Leaves of the upper canopy can be regarded as sun leaves and bottom canopy layer leaves as shade leaves. Both will behave differently in their photosynthetic activity at higher photon flux values (Bolhar-Nordenkampf, 1985). Woledge and Parsons (1986) noted that it was the difference in the light received during leaf development and expansion that caused the difference in the photosynthetic capacity between sun and shade leaves. Prioul *et al.* (1980) found that chloroplast structure and Rubisco activity, both important determinants of photosynthetic capacity, appeared to be controlled by the amount of light directly reaching the leaf during leaf development. Upper canopy leaves in our study did not reach saturation at high photon fluxes (>1800 μmol m^{-2} s^{-1}) and were therefore adapted to a high light intensity environment. The lower canopy leaves could not fully utilize high light fluxes. Light interception is a function of canopy architecture. This may determine photosynthetic production of whole grass stands by either limiting light interception of young shoots or by allowing light to penetrate more or less easily into the dense foliage of older plant stands. Studies have shown that both stomatal and mesophyll conductances are usually greater in leaves grown in bright light than those grown in dim light (Woledge, 1977). Plants grown in high light levels also possessed more stomata per unit leaf area, more mesophyll cells, more and larger chloroplasts and greater Rubisco activity (Prioul *et al.*, 1980). This is correlated with higher photosynthetic activity in leaves in the upper canopy layer.

Results of the present study show that, due to shading, there were reductions of about 50% in CO_2 assimilation rates in the middle canopy leaves and about 70% in bottom canopy leaves of the three grass species. Slightly lower reductions (up to 30%) in canopy photosynthesis due to leaf shading have also been reported in temperate grasslands (Woledge and Leafe, 1976). The shading of young shoots by standing dead material has previously been reported to reduce production in grasslands (Old, 1969; Woledge, 1973; Woledge and Leafe, 1976; Knapp, 1984). Thus as the grass stands age and accumulate standing dead material, photosynthetic production will decline as a greater proportion of light is intercepted by dead or senescent tissue (Woledge and Parsons, 1986).

The diurnal course of CO_2 assimilation during the rainy season for both *T. triandra* and *P. mezianum* exhibited a steady increase during the morning, a peak around midday and a decline towards evening. An increase in leaf stomatal conductance up to midday favoured increased rates of CO_2 assimilation. However, plants did suffer a decreased water status around noon, but this was apparently insufficient to limit CO_2 assimilation, during the growing season. Monomodal daily courses of photosynthesis of both C_3 and C_4 plants are frequently observed with well-watered plants. Pearcy *et al.* (1974) reported such diurnal courses in leaf conductance in *Phragmites communis* Trin. growing in a moist habitat but under extremely high evapo transpirational demands. In irrigated *Tidestromia oblongifolia*, Björkman *et al.* (1972) also noted a monomodal daily course of CO_2 uptake and transpiration rates. Jones (1987) noted a similar trend in *Cyperus papyrus* L. in Lake Naivasha, Kenya.

During the dry season, in January 1985, plants in our study exhibited a morning and a late afternoon peak in stomatal conductance and CO_2 assimilation, accompanied by a severe midday depression. Similar patterns have been noted before in C_3 and C_4 plants, especially in arid environments (Pearcy *et al.*, 1974). CO_2 uptake increases to high rates early in the morning, but stomatal closure at midday occurs to such a degree that CO_2 assimilation is restricted. Stomata open again in the late afternoon, resulting in a second peak of CO_2 uptake.

Photosynthetic capacity of different plant organs

All plant organs studied possessed appreciable numbers of stomata and chlorophyll, except the stems of *T. triandra* and *C. caesius*. Thus, in addition to the leaves, other organs of these grasses are likely to be significant sources of the products of photosynthesis. It is, therefore, not enough to base the production of these species on leaf blade photosynthesis alone. There is no similar work with which to compare these findings. However, data from Tieszen and Imbamba (1978) indicated that the inflorescences of finger millet carried out about 6% of the total photosynthesis of the leaf blades. Other studies have dealt with agronomic crops such as barley (Thorne, 1959, 1965), wheat and oats (Nalborczyk *et al.*, 1981). The latter authors found that the flag leaf in wheat contributed the most in photosynthetic activity while in oats, both the flag leaf and the second leaves exhibited similar photosynthetic activity at certain stages of plant development. The combined contribution of the stem and sheath ranged from about 20% in barley to 60% in rye.

2.6 CONCLUSIONS

This study has shown that East African grasslands are more productive than may have previously been appreciated. Production is mainly limited by lack of rainfall during the dry seasons. The study also showed that other green plant structures (stems, sheaths and inflorescences) contributed greatly to plant photosynthesis. Below-ground biomass production has also been shown to add very significantly to the total net primary production of these grasslands. Hence, it is clear that most of the available primary productivity data which have omitted both turnover and below-ground production are gross underestimates of the true values in most tropical grasslands.

This study has also shown the importance of decomposition in estimating net primary production. Decomposition was shown to be high during the growing season in these grasslands, with the result that any estimates of net primary production based on biomass dynamics alone would inevitably be an underestimate. This degree of underestimation varied markedly between the 2 years.

2.7 SUMMARY

1. Changes in amounts of live and dead vegetation, decomposition, leaf canopy development and leaf photosynthesis were measured and

primary production estimated at monthly intervals over 2 years for a *Themeda triandra/Pennisetum mezianum* dominated area of grassland in Nairobi National Park, Kenya.
2. Above-ground biomass varied from 73 g m^{-2} to 338 g m^{-2}, on a dry weight basis. Changes in biomass correlated closely with the bimodal distribution of rainfall in each year. *P. mezianum* and *T. triandra* constituted over 60% of the total above-ground biomass. Dead vegetation above ground increased from 66 to 651 g m^{-2}, possibly reflecting the exclusion of fire and large herbivores during this period. Decomposition rates varied from 0.02 g g^{-1} $month^{-1}$ in the dry season to 0.18 g g^{-1} $month^{-1}$ in the wet season.
3. Below-ground biomass showed peaks at the end of the major wet seasons of *ca* 200 g m^{-2}. Significant increases in quantities of both live and dead material occurred over the 2 years. Decomposition rates ranged from 0.05 to 0.32 g g^{-1} $month^{-1}$, again the maxima correlated closely with variation in rainfall.
4. Leaf area indices varied from zero in the driest periods to a peak of 3 during the wettest period. Leaf photosynthetic rates showed a peak of *ca* 30 μmol m^{-2} s^{-1}, despite a decline in leaf water potential to <-2 MPa on these days. Rates were severely depressed at midday during dry periods, when midday water potentials were *ca* −4.0 MPa and stomatal closure occurred.
5. Net primary production values, determined from monthly increments in live and dead vegetation, corrected for losses through decomposition were 1290 and 587 g m^{-2} y^{-1} in 1984/85 and 1985/86, respectively. Values for 1984/85 were more than double those estimated by the earlier International Biological Programme (IBP) methods. The degree of underestimation by the IBP methods differed markedly between the 2 years.

REFERENCES

Abouguendia, Z.M. and Whitman, W.C. (1979) Disappearance of dead plant material in a mixed grass prairie. *Oecologia*, **42**, 23–9.

Arnon, D.I. (1949) Copper enzymes in isolated chloroplasts. Polyphenoloxidase in *Beta vulgaris*. *Plant Physiology*, **24**, 1–15.

Ayuko, L.J. (1978) Management of rangelands in Kenya to increase beef production, Socio-economic constrains and policies, in *Proceedings of 1st International Rangeland Congress*, Denver, Colorado, USA, pp. 82–6.

Björkman, O., Pearcy, R.W., Harrison, A.T. *et al.* (1972) Photosynthetic adaptation to high temperature: a field study in the Death Valley, California. *Science*, **175**, 786–9.

Black, J.W. (1956) The distribution of solar radiation over the earth's surface. *Archiv für Meteorologie, Geophysk und Bioklimatologie, Serie B*, **7**, 165–89.

Bolhar-Nordenkampf, H.R. (1985) Shoot morphology and leaf anatomy in relation to photosynthesis, in *Techniques in Bioproductivity and Photosynthesis*, 2nd edn (eds J. Coombs, D.O. Hall, S.P. Long and J.M.O. Scurlock), Pergamon Press, Oxford, pp. 62–94.

Branson, F.A. (1953) Two new factors affecting resistance of grasses to grazing. *Journal of Range Management*, **6**, 165–71.

Cassady, J.T. (1973) The effect of rainfall, soil moisture and harvesting intensity on grass production on two rangeland sites in Kenya. *East African Agricultural and Forestry Journal*, **39**, 26–36.

Child, R.D., Heady, H.F., Hickey, W.C. *et al*. (1984) *Arid and Semi-arid Lands: Sustainable Use and Management in Developing Countries*. Winrock International, Morrilton, Arkansas.

Cooper, J.P. (1975) Control of photosynthetic production in terrestrial systems, in *Photosynthesis and Productivity in Different Environments*, IBP Vol.3 (ed. J.P. Cooper), Cambridge University Press, Cambridge, pp. 593–621.

Dahlman, R.C. and Kucera, C.L. (1965) Root production and turnover in native prairie. *Ecology*, **46**, 84–9.

Deshmukh, I. (1986) Primary production of a grassland in Nairobi National Park. *Journal of Applied Ecology*, **23**, 115–23.

Deshmukh, I.K. and Baig, M.N. (1983) The significance of grass mortality in the estimation of primary production in African grasslands. *African Journal of Ecology*, **21**, 19–23.

Dykyjova, D. (1978) Determination of energy content and net efficiency of solar energy conversion by fishpond halophytes. *Ecological Studies*, **28**, 216–33.

George, J.F. and Smeins, F.E. (1982) Decomposition of common curly-mesquite herbage on Edwards Plateau, Texas. *Journal of Range Management*, **35**, 104–6.

Hall, D.O. and Rao, K.K. (1981) *Photosynthesis*. Edward Arnold, London.

Jones, M.B. (1986) Wetlands, in *Photosynthesis in Contrasting Environments* (eds N.R. Baker and S.P. Long) Elsevier, Amsterdam, pp. 103–38.

Jones, M.B. (1987) The photosynthetic characteristics of papyrus in a tropical swamp. *Oecologia*, **71**, 355–9.

Knapp, A.K. (1984) Post-burn differences in solar radiation, leaf temperature and water stress in influencing production in a lowland prairie. *American Journal of Botany*, **71**, 220–7.

Knapp, A.K. (1985) Effect of fire and drought on the ecophysiology of

Andropogon gerardii and *Panicum virgatum* in a tallgrass prairie. *Ecology*, **66**, 1309–20.

Landsberg, H.E. (1965) Global distribution of solar and sky radiation, in *World Maps of Climatology*, 2nd edn, Springer-Verlag, Berlin, pp. 1–6.

Long, S.P. and Hällgren, J.-E. (1985) Measurement of CO_2 assimilation by plants in the field and the laboratory. in *Techniques in Bioproductivity and Photosynthesis* 2nd edn (eds J. Coombs, D.O. Hall, S.P. Long and J.M.O. Scurlock), Pergamon Press, Oxford, pp. 62–94.

Lusigi, W.J. (1978) *Planning Human Activities on Protected Natural Ecosystems*, Strauss and Cramer, Germany.

Macharia, J.N.M. (1981) *Bioproductivity in Relation to Photosynthesis in Four Grassland Ecosystems in Kenya*. MSc Thesis, University of Nairobi, Nairobi.

McNaughton, S.J. (1979) Grassland-herbivore dynamics, in *Serengeti: Dynamics of an Ecosystem* (eds E. Rodenwalt and H.J. Jusatz), University of Chicago Press, Chicago, pp. 46–81.

Medina, E. (1986) Forests, savannas and montane tropical environments. in *Photosynthesis in Contrasting Environments* (eds N.R. Baker and S.P. Long) Elsevier, Amsterdam, pp. 139–71.

Menault, J.C., Barbault, R., Lavelle, P. and Lepage, M. (1984) African savannas: biological systems of humidification and mineralization, in *Ecology and Management of the World's Savannas*. (eds J.J. Tothill and J. Mott) Commonwealth Agricultural Bureaux, Melbourne, Australia, pp. 114–33.

Misra, M.K. and Misra, B.N. (1979) Turnover rates of a grassland community. *Comparative Physiology and Ecology*, **4**, 150–3.

Muthuri, F.M. (1985) *The Primary Productivity of Papyrus* (Cyperus papyrus L.) *in Relation to Environmental Variables*. PhD Thesis, University of Nairobi, Nairobi.

Nalborczyk, E., Nalborczyk, T. and Wawrzonowska, B. (1981) Models of photosynthetic activity in cereals, in *Photosynthesis*. VI. *Photosynthesis and Productivity, Photosynthesis and Environment* (ed. G. Akoyunoglou), Balaban International Science Services, Philadelphia, USA, pp. 97–106.

Ohiagu, C.E. and Wood, T.G. (1979) Grass production and decomposition in southern Guinea savanna, Nigeria. *Oecologia*, **40**, 155–65.

Old, S.M. (1969) Microclimate, fire and plant production in an Illinois prairie. *Ecological Monographs*, **39**, 355–84.

Owaga, M.L.A. (1980) Primary productivity and herbage utilisation by herbivores in Kaputei plains, Kenya. *African Journal of Ecology*, **18**, 1–5.

Pandey, D.D. and Sant, H.R. (1980) Plant biomass and net primary production of the protected and grazed grasslands of Varanasi, India. *Journal of Ecology*, **7**, 77–83.

Pearcy, R.W., Berry, J.A. and Bartholomew, B. (1974) Field photosynthetic performance and leaf temperature of *Phragmites communis* under summer conditions in Death Valley, California. *Photosynthetica*, **8**, 104–8.

Pratt, D.J., Greenway, P.J. and Gwynne, M.D. (1966) A classification of East African rangeland, with an appendix to terminology. *Journal of Applied Ecology*, **3**, 369–82.

Prioul, J.L., Brangeon, J. and Reyss, A. (1980) Interaction between external and internal conditions in the development of photosynthetic features in a grass leaf. I. Regional responses along a leaf during and after low-light or high- light acclimation. *Plant Physiology*, **66**, 762–9.

Rath, S.P. and Misra, B.N. (1980) Effect of grazing on the turnover rates of a tropical grassland. *Geobios*, **7**, 109–11.

Rechenthin, C.A. (1956) Elementary morphology of grass growth and how it affects utilisation. *Journal of Range Management*, **9**, 167–70.

Sampson, J. (1961) A method of replicating dry or moist surfaces for examination by light microscopy. *Nature*, **191**, 932–3.

Scott, R.M. (1963) *Soils of the Nairobi-Thika-Yatta-Machakos Area*. Department of Agriculture, Kenya, pp. 1–60.

Sims, P.L. and Singh, J.S. (1978) The structure and function of ten western North American grasslands. IV. Compartmental transfer and energy flow within the ecosystems. *Journal of Ecology*, **66**, 983–1009.

Sinclair, A.R.E. (1975) The resource limitation of trophic levels in tropical grassland ecosystems. *Journal of Animal Ecology*, **44**, 497–520.

Singh, J.S. and Yadava, P.S. (1974) Seasonal variation in composition, plant biomass, and net primary productivity of a tropical grassland at Kurushetra, India. *Ecological Monographs*, **44**,351–76.

Strugnell, R.G. and Pigott, C.D. (1978) Biomass, shoot production and grazing of two grasslands in the Ruwenzori National Park, Uganda. *Journal of Ecology*, **66**, 73–97.

Taerum, R. (1970) Comparative shoot and root growth studies on six grasses in Kenya. *East African Agricultural and Forestry Journal*, **36**, 94–113.

Talbot, L.M. and Stewart, D.R. (1964) First wildlife census of the entire Serengeti-Mara region, East Africa. *Journal of Wildlife Management*, **28**, 815–27.

Thorne, G.N. (1959) Photosynthesis of lamina and sheath of barley leaves. *Annals of Botany*, **23**, 365–70.

Thorne, G.N. (1965) Photosynthesis of ears and flag leaves of wheat and barley. *Annals of Botany*, **29**, 317–29.

Ticha, I., Catsky, J., Hodanova, D., Pospililova, J., Kase, M. and Sestak, Z. (1985) Gas exchange and dry matter accumulation during leaf development, in *Photosynthesis During Leaf Development*. (eds Z. Sestak, S. Catsky and P.G. Jarvis), Dr W. Junk Publishers, The Hague, pp. 157–215.

Tieszen, L.L. and Imbamba, S.K. (1978) Gas exchange of finger millet inflorescences. *Crop Science*, **17**, 495–8.

Tueller, P.T. and Tower, J.D. (1979) Vegetation stagnation in three-phase big game exclosures. *Journal of Range Management*, **32**, 258–63.

Tukey, H.B. and Mecklenburg, R.A. (1964) Leaching of metabolites from foliage and subsequent reabsorption and distribution of the leachate in plants. *American Journal of Botany*, **51**,737–42.

Van der Hammen, T. (1983) The palaeoecology and palaeogeography of savannas, in *Tropical Savannas* (ed. F. Bourlière), Elsevier, Amsterdam, pp. 19–35.

Whalley, R.D.B. and Davidson, A.A. (1968) Physiological aspects of drought dormancy in grasses. *Proceedings of the Ecological Society of Australia*, **3**, 17–19.

Woledge, J. (1973) The photosynthesis of ryegrass leaves grown in a simulated sward. *Annals of Applied Biology*, **73**, 229–37.

Woledge, J. (1977) The effects of shading and cutting treatments on the photosynthetic rate of ryegrass leaves. *Annals of Botany*, **41**, 1279–86.

Woledge, J. and Leafe, E.L. (1976) Single leaf and canopy photosynthesis in a ryegrass sward. *Annals of Botany*, **40**, 773–83.

Woledge, J. and Parsons, A.J. (1986) Temperate grasslands, in *Photosynthesis in Contrasting Environments* (eds N.R. Baker and S.P. Long) Elsevier, Amsterdam, pp. 173–97.

3

Saline grassland near Mexico City

E. GARCIA-MOYA and P. MONTANEZ CASTRO

3.1 INTRODUCTION

Salt-affected plant communities occupy between 3.7 and 9.5 million square kilometres worldwide (Flowers *et al.*, 1986). In Mexico, 32 000 ha of saline grassland are situated in the old Texcoco lake-bed (Secretaria de Recursos Hidraulicos, 1971). These grasslands are largely semi-natural and dominated by varieties *mexicana* (Beetle) of the halophytic C_4 grass, *Distichlis spicata* (L.) Greene, known locally as 'zacate salado'. It is a dioecious perennial, which may grow up to 30 cm in height. It has long scaly rhizomes and sometimes forms stolons. The stems are erect and rigid with dense panicles and few florets. This species is widely distributed in the American continents. It occurs across a broad range of latitudes and altitudes, from southern Canada to Chile and Argentina and from sea level to elevations of more than 2000 m. Habitats range from coastal beaches and marshes to deserts and regions of inland drainage. However, a common feature of all locations is a high salinity or alkalinity in the soil and ground water. In addition, most sites at which the species is found are subject to occasional water-logging. The species is adapted to high levels of electrolytes in the rooting medium by: the presence of salt glands in the leaves and shoots which secrete salts; a xeromorphic morphology; and a capacity to accumulate organic osmotica. The species is adapted to water-logging by the presence of pronounced root aerenchyma in which air spaces occupy *ca* 70% of the root cross-sectional area (Long and Mason, 1983).

Several authors, including Rzedowski (1957), Evans (1978) and Cabrera (1988), have described the botanical composition of the Texcoco lake-bed. As noted above, the dominant species is *D. spicata*, for which five varieties have been reported in Mexico. The variety found in Texcoco is var. 'mexicana' (Beetle, 1987). *D. spicata* commonly accounts for 98% of the vegetation cover of the lake-bed, and often forms monospecific stands. Otherwise it may occur with small quantities of *Suaeda torreyana* Watson and *Sporobolus pyramidatus* (Lam.) Hitch. Fifty-eight other species of rare occurrence have been recorded in this vegetation (Cabrera, 1988).

There has been marked human influence on the Texcoco Lake, from the foundation of Tenochtitlan (now Mexico City) in 1325 until the drainage of the lake commenced in 1911. Intentional fires are common practice to improve forage quality during the dry season, preventing the establishment of woody species.

Within the tropics and sub-tropics, saline land areas amount to around 15 000 km^2 (Long and Baker, 1986). Saline communities have been the subject of many previous productivity studies. These have, however, largely concerned temperate communities, with the exception of mangroves (Long and Mason, 1983). The closest comparisons, therefore, to the high altitude, saline grasslands studied here are with the saline communities of warm temperate regions. Waisel (1972) reported an above-ground productivity for halophytic ecosystems between 10 and 400 g m^{-2} y^{-1}. However, Glenn *et al.* (1982) reported higher above-ground yields of *Salicornia europaea, Beta maritima* and seven *Atriplex* species. The most productive species were irrigated at 12 h intervals with 40 kg m^{-3} sea water and produced 900–1360 g m^{-2} y^{-1}. Odum (1974), suggested that *Spartina alterniflora* on marshes of the south-east coast of the USA has a productivity of 750–4000 g m^{-2} y^{-1}. However, if account is taken of the turnover of the vegetation above and below ground, then the productivity may be as high as 6650 g m^{-2} y^{-1} (Wiegert, 1979). No figures for the productivity of *D. spicata* in the tropics are available. However, Mudie (1974) reported that *D. spicata* in the Gulf of California will produce up to 660 g m^{-2} y^{-1}. Bellis and Gaither (1985) have reported shoot productivities of 250–1000 g m^{-2} y^{-1}, with below-ground biomass increments of some 10 times these values. Even outside of the tropics, virtually all of these previous studies have concerned single years of study and thus provide no information on the year-to-year variability in production and the correlation of this variability with climatic variables.

The primary objective of this study was to assess the net primary production of a tropical saline grassland using techniques that would take full account of losses through death over 3 years. Secondary objectives were to assess the relationship between production and selected environmental variables, radiation interception and seasonal changes in leaf photosynthetic characteristics.

3.2 STUDY SITE

3.2.1 Location

The Chapingo experimental area was located in Montecillos, Texcoco, Mexico state (19°27′30″N and 98°54′30″W), at an elevation of 2241 m (Figure 3.1).

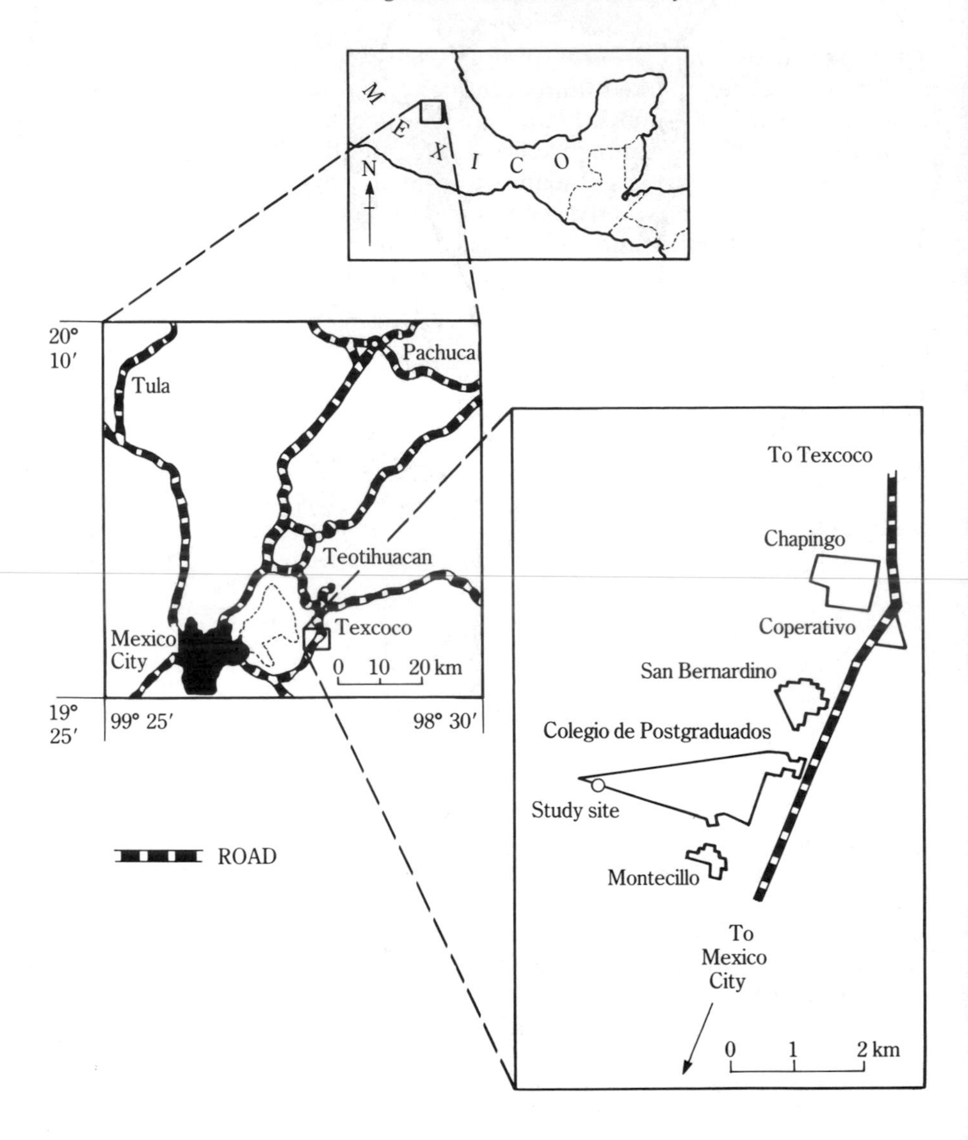

Figure 3.1. Location of the Chapingo study site in Mexico.

3.2.2 Climate

Long-term meteorological data, over 25 years of observation, were supplied by the Chapingo meteorological station (19°30′N, 98°51′W, altitude 2241 m) (Figure 3.2a,c). This station is located 5.2 km north-east of the study area. Interception of solar radiation (400–2000 nm) by

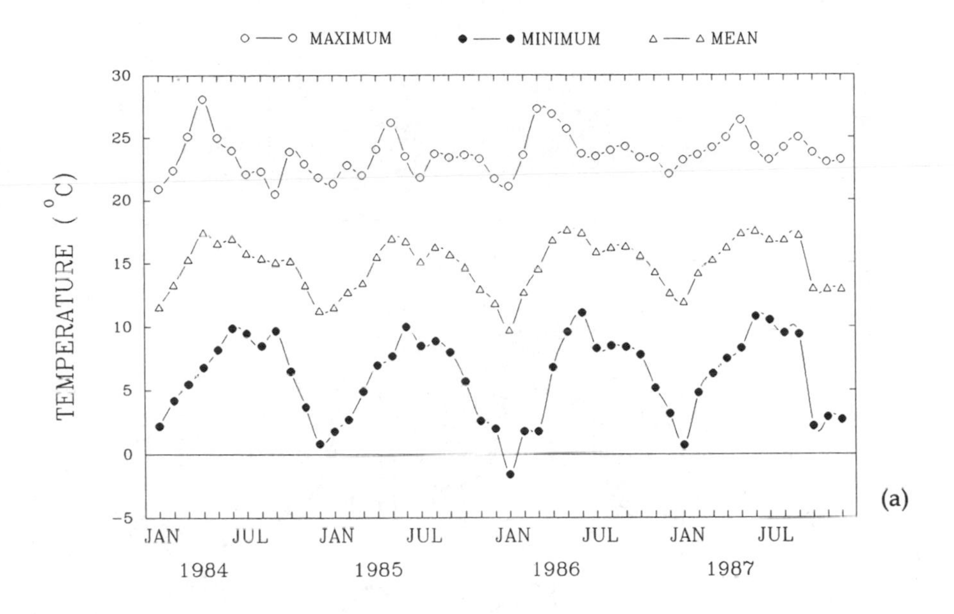

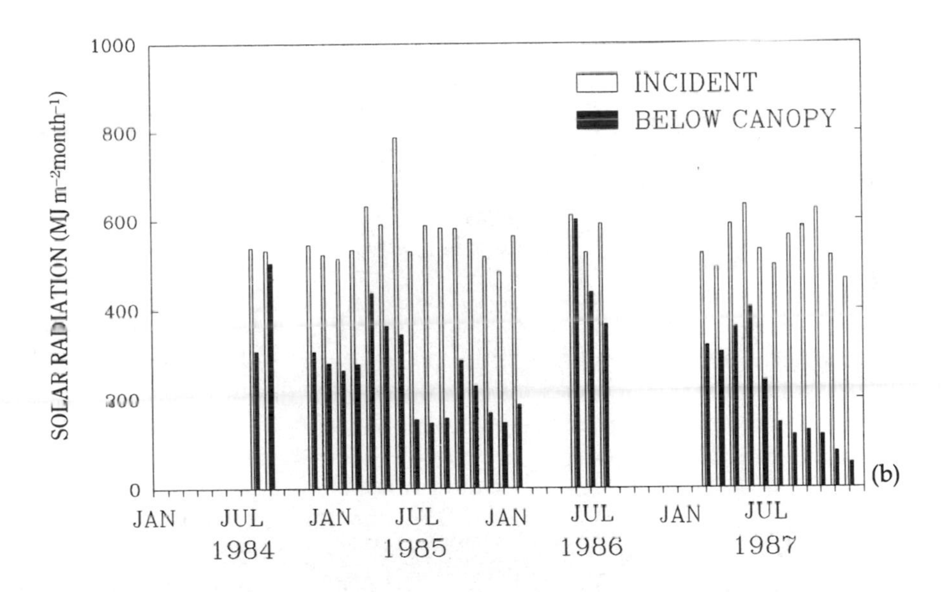

Figure 3.2. (a) Average monthly maximum, mean and minimum temperatures at the Chapingo study site; (b) monthly totals of solar radiation incident and below the canopy.

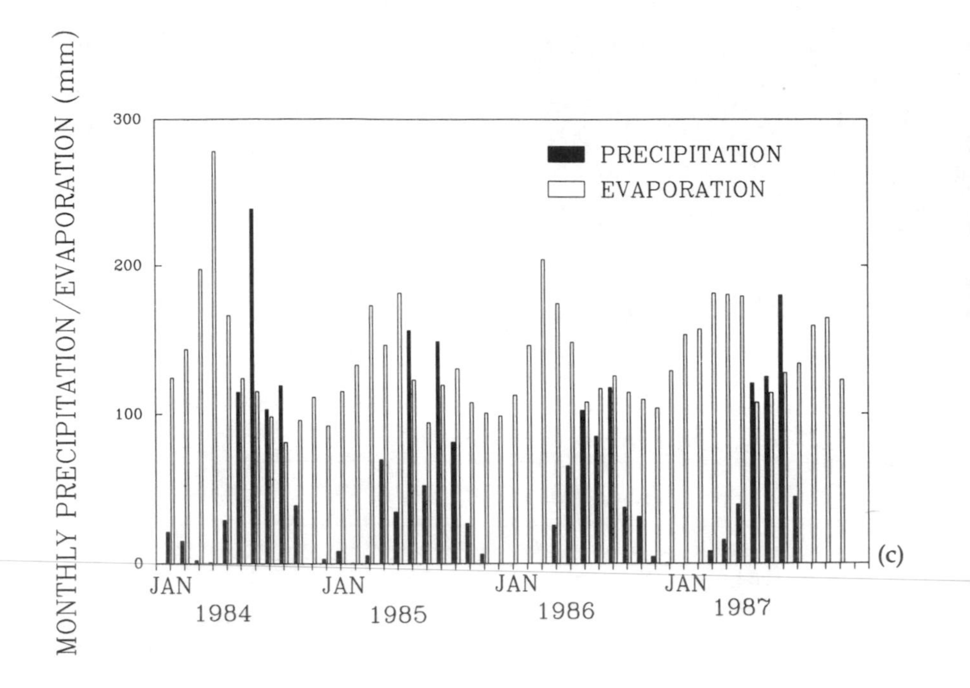

Figure 3.2. *continued* (c) Monthly totals of precipitation and pan evaporation at the Chapingo site.

the canopy was measured using tube solarimeters and integrators (TLS Delta-T Devices, Cambridge, UK). Observations were disrupted in 1986 following a fire. Under the Köppen system as modified by García (1973) for Mexican conditions the climate is classified as CWo(w)b(i′)g; the driest of the temperate subhumid climatic group, with long and mild summers. A 6-month rainy period occurs from May to October, and the mean annual precipitation is 579.4 mm. The mean annual temperature is 15.1°C. The coldest month has a mean temperature of 11.6°C. There is little annual variation of the mean monthly temperatures which vary between years by around 0.1–6.1°C. The mean maxima and minima over the 25 years are 26.8°C (April and May) and 1°C (January). The mean annual evapo-transpiration is 1706.5 mm. Higher rates occur in spring (198.7, 192.4 and 185.4 mm in March, April and May, respectively), and lower rates throughout the rest of the year.

The climate during the actual period of study from September 1984 to December 1987 was as follows. The annual temperature cycle was broadly similar throughout (Figure 3.2a). Maximum temperatures showed little variation, with a range of 20.5–27.3°C. The highest mean maximum monthly temperature of 27.3°C occurred in March 1986. The minimum monthly temperatures ranged from −1.6°C in January 1986 to 11.1°C in June 1986. Annual mean temperature was an almost constant 15°C.

From the available data, highest incident monthly solar radiation receipts (Figure 3.2b) occurred in May 1985 (787 MJ m^{-2}). Annual receipts were 6907 MJ m^{-2} in 1985 and 6574 MJ m^{-2} in 1987. The lowest monthly levels were 483 and 466 MJ m^{-2} in December 1985 and December 1987, respectively. At the base of the canopy, solar radiation varied from 601 MJ m^{-2} in March 1986 to only 55 MJ m^{-2} during December 1987.

Precipitation was concentrated into the period from April to September (Figure 3.2c). The highest levels occurred in June 1985 (156 mm) and in August 1986 and 1987 (117 and 180 mm respectively). The lowest figures occurred from October to March, with no precipitation being recorded in many months: November 1984; December 1985 to March 1986; January to February and October to December 1987. Combining observations from the study period with previous records, 1984 was the wettest year with a total precipitation of 684 mm. The subsequent 3 years were all drier, annual precipitation falling to 589 mm in 1985 and 469 mm in 1986 before rising a little to 533 mm in 1987.

Months with high evaporation rates tended also to have low precipitation, i.e. October to May (Figure 3.2c). The highest evaporation rates were observed in May 1985 (181.5 mm) and in March 1986 and 1987 (204 and 181.6 mm, respectively). The lowest rates were 81.3 mm in September 1984, 94.9 mm in July 1985 and 107.8 mm in June 1987. Annual evaporation was always much higher than annual precipitation, but also declined somewhat over the period of study. Thus evaporation totalled 1628 mm in 1984, 1523 mm in 1985 and 1592 mm in 1986. The highest total evaporation (1783 mm) occurred in 1987.

Overall, then, there was little difference in the annual pattern and range of maximum and minimum temperature during the study. Incident solar radiation receipts were fairly similar for 1985 and 1987. The main differences between years were in amounts of precipitation and evaporation.

3.2.3 Relief and substrate

The site shows very little variation in surface height (<10 cm). Alluvial lacustrine deposits of the Quarternary age are the main substrate, with intrusive and extrusive igneous material eroded from the surrounding volcanoes and their deposits (ashes, basalts, andesites, tuffs and breccias) present on the lake-bed as loess and alluvial deposits.

3.2.4 Soil

The soil is an entisol exhibiting little or no discernible horizons and saturated with water during certain seasons. The soil is also characterized by high concentrations of salts. As a proportion of the cations, Na (78%) dominates in all the strata, with K (18%), Ca (2.5%) and Mg (1.5%) also present. The depth of the water table during the dry period is more than 4.8 m, but during the rainy season it reaches the surface (Del Valle, 1983). Figure 3.3 shows a typical soil profile from the experimental plot. Apart from the upper layers of clay/sandy loam and volcanic sand, the rest of the profile is differentiated only by changes in pH.

3.3 METHODS

The methods followed those outlined in chapter 1. The study area was enclosed with a barbed wire fence to prevent access to grazers. The site remained undisturbed throughout the period of investigation, apart from the fire of February 1986. Monthly harvests were carried out on 20 randomly selected 0.25 × 1.0 m plots to estimate both above-ground biomass and dead material of *D. spicata*, other grasses and herbaceous species over the period September 1984 to December 1987. Live material (biomass) was separated from dead, and then separated into two components: leaves, and stems with inflorescences. Projected areas of both components were measured using an electronic area meter (Delta-T Devices, Cambridge, UK). The below-ground plant organs were sampled by taking 10 cm deep × 7.5 cm diameter soil cores in the centre of the selected plots. Initial sample size was five cores per month. The size of the corer was increased to 15 cm × 7.5 cm at the beginning of 1985, and eight cores were taken per month. The sample size was increased to 40 per month in January 1986, for the remainder of the study period. The roots were washed with running water, stained with tetrazolium chloride, sorted into live and dead, dried and weighed.

Decomposition was estimated by the use of litter bags. Initially, 10 bags of 2 mm mesh, measuring 3 cm × 2 cm, were placed in the field each month, each containing 1 g of dead material. This was increased to 20 bags measuring 8 × 8 cm with 3 g per bag in February 1985 and for the remainder of the investigation. Bags were located at ground

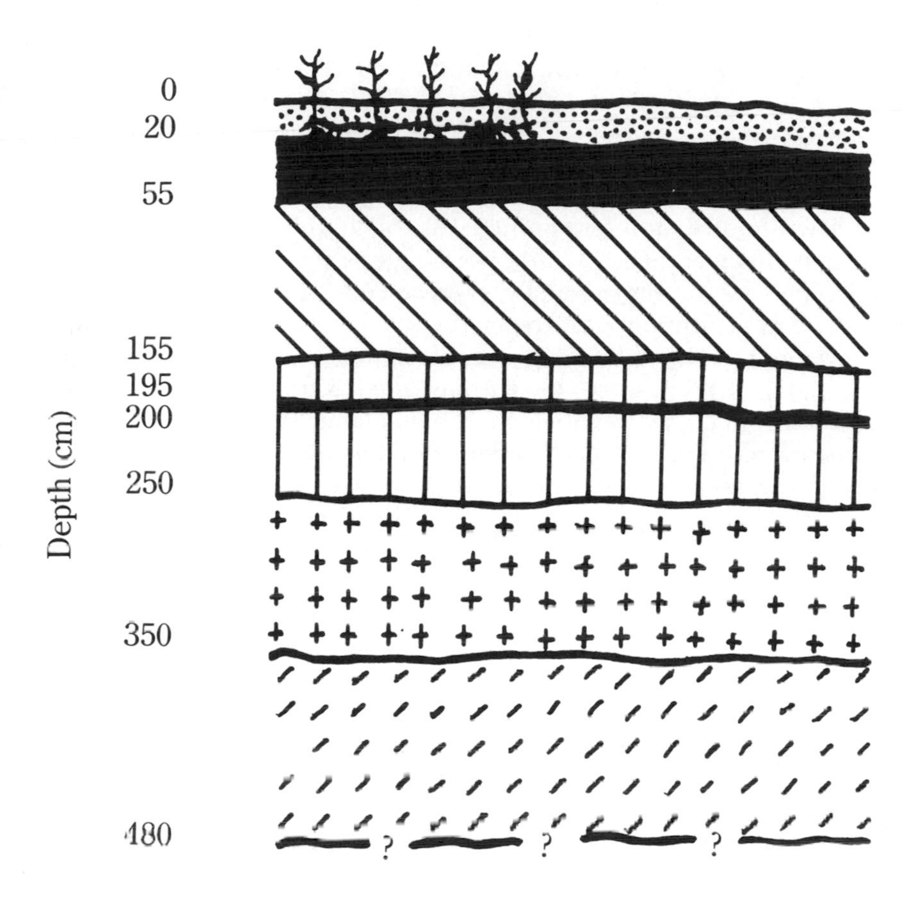

Figure 3.3. Soil profile for the study site. Clay and sandy loam (wet) dark greyish brown (10 yr 4/2) (dry) dark grey (2.5 y 5/2) 95% of the roots. pH = 9; Fine volcanic sand, slightly compacted (wet), black (2.5 YR 2.5/0) (dry dark grey (2.5 y 4/0) 5% of the roots. pH = 9.4; \ \ \ Clay with microfossils and inclusions (wet) very dark greyish brown (2.5 y 3/2) (dry) light grey (5 y 6.5/1). pH = 9.3; | | | Clay with microfossils and inclusions (wet) dark reddish brown (5 YR 3/3) (dry) reddish brown (5 YR 5/3). pH = 9.0; + + + Clay with microfossils and inclusions (wet) dark olive grey (5 Y 3/2) (dry) dark grey (5 Y 4/1). pH = 9.4; / / / Clay with microfossils and inclusions (wet) dark brown (10 YR 3/3) (dry) greyish brown (2.5 Y 5/2). pH = 9.2.

level and at 5 cm depth for above- and below-ground measurements respectively. Productivity was calculated according to the equation of Roberts *et al.* (1985) (Chapter 1).

Interception of incident solar radiation (400–2000 nm) by the canopy was measured using tube solarimeters (TLS) and integrators (MVI, Delta-T Devices, Cambridge, UK). These observations were disrupted in February 1986 following the fire.

The uptake of CO_2 under field conditions was measured diurnally, from dawn to dusk at monthly intervals, on 10 fully expanded leaves. Measurements were made with a portable open gas exchange system, comprising: an infra-red gas analyser (type LCA2), a leaf chamber (type PLC), and an air supply unit (type ASU, Analytical Development Co., Hoddesdon, UK). In addition light response curves were determined on in late May 1988 in the upper (30 cm), middle (15 cm) and bottom (5 cm) of the canopy of *D. spicata* with a photon flux range of 3 – 1791 μmol m^{-2} s^{-1}, generated by placing neutral density filters over the window of the chamber. Simultaneously, the light response curves of *Sporobolus pyramidatus* and *Suaeda torreyana* were also determined.

3.4 RESULTS

3.4.1 Biomass

Above-ground biomass exhibited a unimodal pattern in each year of study. Peak values occurred during, or immediately after, the rainy season (i.e. July to October) and lowest values in the drier, colder part of the year (Figure 3.4a). Levels ranged from 5 g m^{-2} in April 1986 to 430 g m^{-2} in October 1985. Although highest amounts of shoot biomass were encountered in 1984, that year showed the least variation in month-to-month mean values. Following 1984, above-ground biomass levels were much reduced, particularly after the fire in 1986 when they fell to zero. However, recovery was swift, with levels returning to over 200 g m^{-2} by the following August.

The major contribution to total above-ground biomass was provided by *D. spicata* (>79%), whilst the contribution of forbs and other grasses amounted to less than 21%. In general, monthly variation in shoot biomass of each species component followed that of total biomass, with increased levels during the rainy season. As *D. spicata* was the major contributor, it will be discussed in some detail. Its leaf biomass was less than its stem biomass for most of the study period, except during April to June 1986, and July 1987 (Figure 3.4b). Stem biomass ranged from 2.1 g m^{-2} in April 1986 to 296 g m^{-2} in October

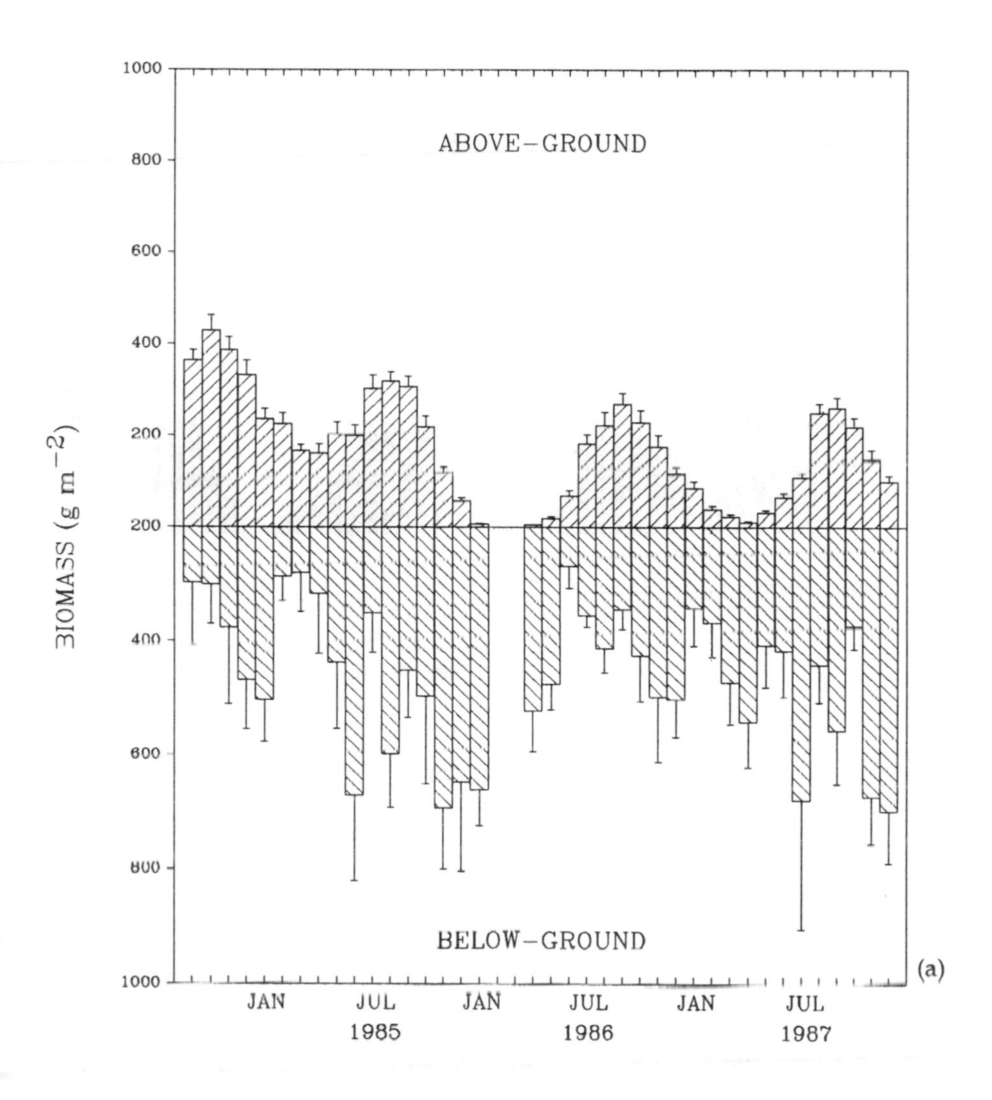

Figure 3.4. Monthly means (±1se) of (a) above- and below-ground biomass (dry weight); (b) distribution of above-ground biomass between stems and leaves of *Distichlis spicata*, other grasses, and forbs; (c) above- and below-ground quantities of dead vegetation; (d) above- and below-ground relative decomposition rate; and (e) above- and below-ground absolute quantities of dead vegetation lost through decomposition.

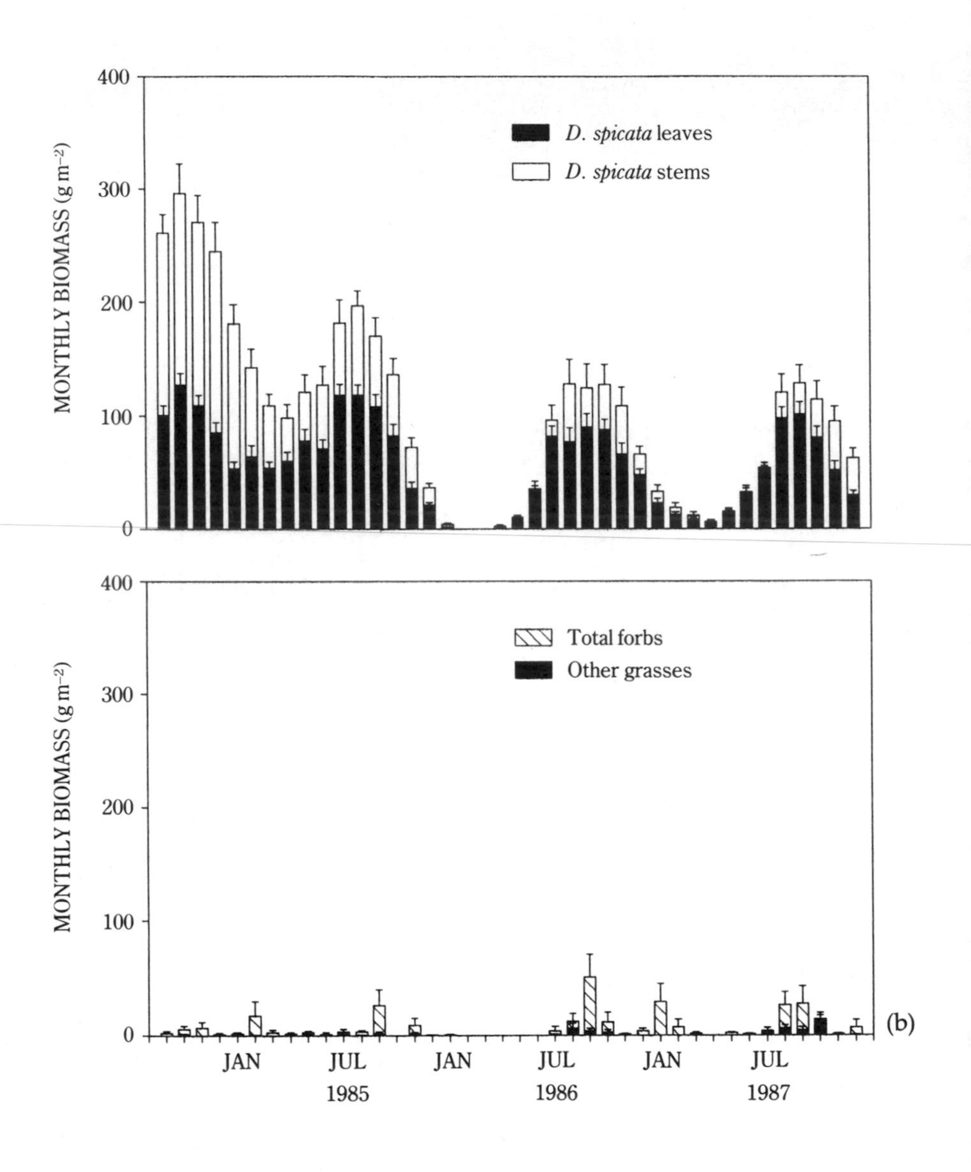

Figure 3.4. *continued*

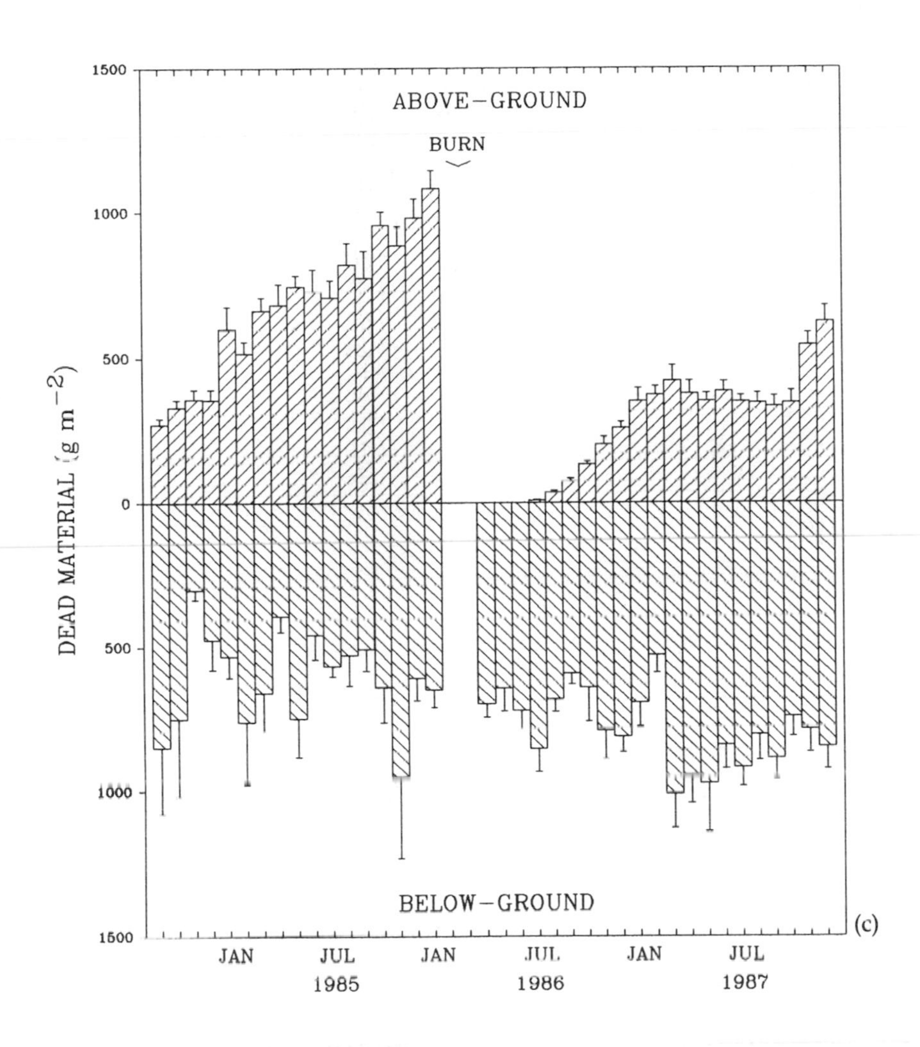

Figure 3.4. *continued*

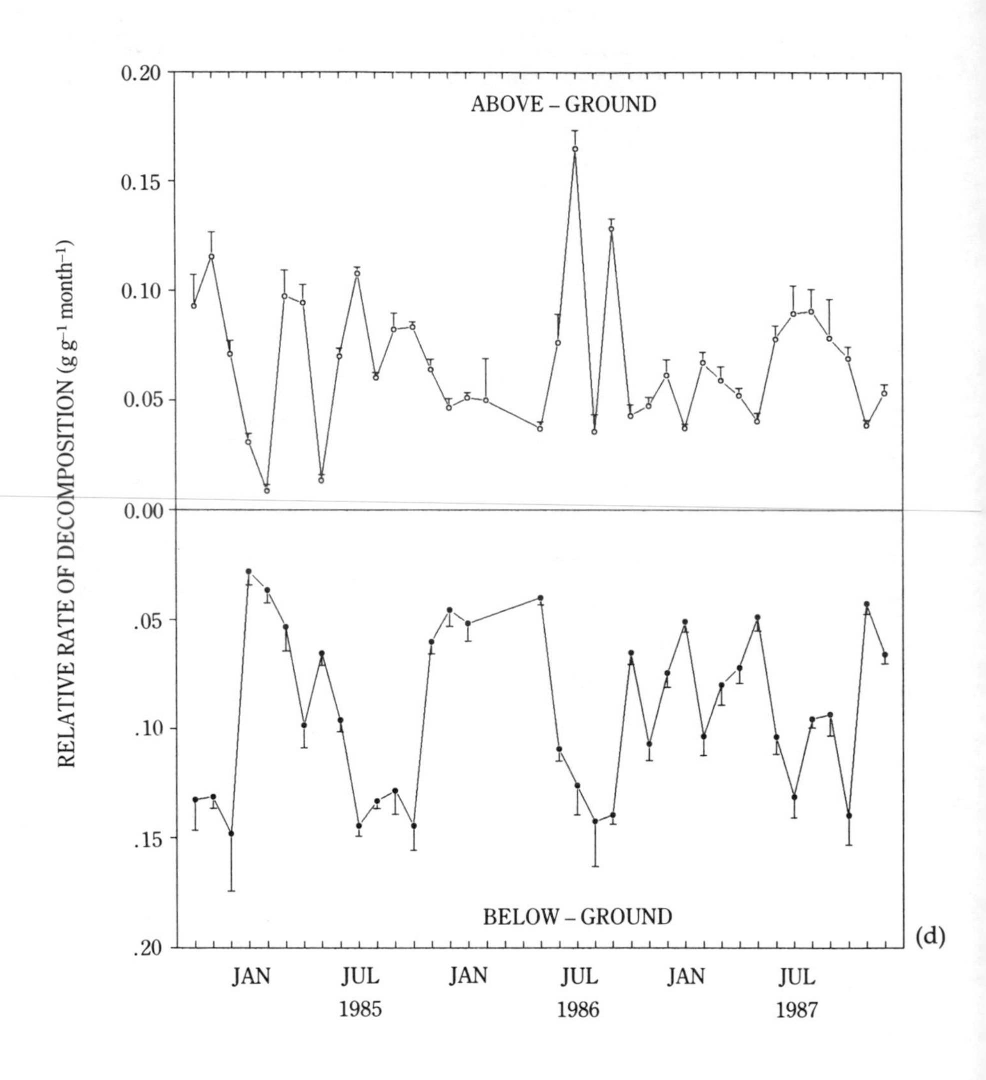

Figure 3.4. *continued*

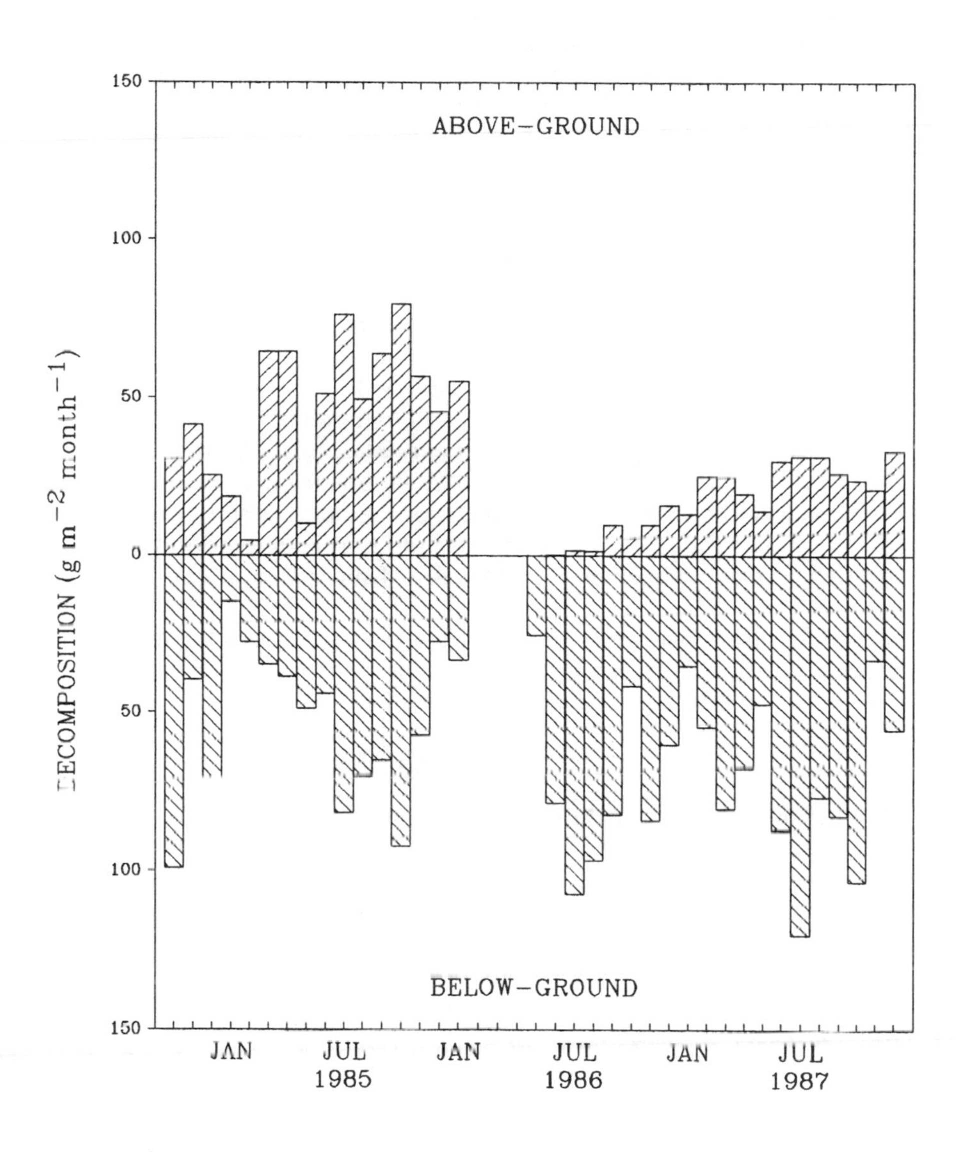

Figure 3.4. *continued*

1984. Levels for 1984 were on average 2 to 5 times higher than in subsequent years. As with total biomass, stem biomass tended to increase towards late summer (August to October), and general levels declined after 1984. Leaf biomass did not show such an extreme range as stem biomass, but still varied between 2 g m^{-2} in January 1986 and 127 g m^{-2} in October 1984. High levels of leaf biomass generally corresponded to high levels of stem biomass, except in August 1986. Again, there was a general decline after 1984.

For the other herbaceous species in the community, total biomass peaked in September in 1985, 1986 and 1987, and varied between zero in several months and 51 g m^{-2} in September 1986. The other grasses constituted a consistently low level proportion of the total biomass, ranging from 0 to 14 g m^{-2}.

The below-ground live biomass did not exhibit any strongly defined trend (Figure 3.4a). However, higher values tended to occur in winter. The minimum was 268 g m^{-2} in June 1986, whilst the highest value (698 g m^{-2}) occurred in December 1987. Below ground, biomass levels were consistently much higher than above ground, nearly three times greater on average. It is interesting to note that below-ground biomass declined immediately after the 1986 fire, from 522 g m^{-2} in April to 268 g m^{-2} in June. This may be indicative of the transfer of metabolites from the roots to the shoots to re-construct the canopy. Apart from this, the below-ground biomass was not much affected by the fire, normal levels being regained by July or August.

3.4.2 Dead vegetation

Above ground, dead vegetation accumulated throughout the period of study (from 338 g m^{-2} in June 1984 to 1083 g m^{-2} in January 1986), until it was eliminated by fire in February 1986 (Figure 3.4c). Accumulation was quickly resumed and continued until March 1987 (422 gm^{-2}), after which levels remained relatively constant until they began to rise again the following winter. At the end of the study, in December 1987, above-ground dead vegetation had reached 624 g m^{-2}. Below-ground, levels of dead material were not depressed by the fire in 1986, and if anything showed a slight increase for a few months. Again, the general trend was towards accumulation, from a low of 294 g m^{-2} in November 1984 to 844 g m^{-2} in December 1987. The highest level attained was 1009 g m^{-2} in March 1987.

3.4.3. Decomposition

The above- and below-ground relative decomposition rates showed similar patterns to that observed for the above-ground biomass, with higher rates during or immediately after the warm, rainy season

(Figure 3.4d, cf. Figure 3.2). The above-ground maximum was 0.16 g g^{-1} $month^{-1}$ in July 1986, whilst the comparable figure below-ground (0.14 g g^{-1} $month^{-1}$) was attained on several occasions. Lower rates tended to occur in the drier months, with minima of 0.03 g g^{-1} $month^{-1}$ (below ground) and 0.01 g g^{-1} $month^{-1}$ (above ground) in January and February 1985, respectively. Overall, mean rates below ground (0.09 g g^{-1} $month^{-1}$) were higher than those above (0.07 g g^{-1} $month^{-1}$).

Regression analysis of relative decomposition rates and corresponding monthly meteorological data revealed several statistically significant correlations, at a probability level of $P<0.05$. Both above- and below-ground relative rates of decomposition were most highly correlated with precipitation ($r = 0.57$, df = 38 for above ground, $r = 0.42$, df = 36 for below ground) and air temperature ($r = 0.48$, df = 34 for above ground, $r = 0.49$, df = 35 for below ground). The most significant correlations, however, were with the difference between precipitation and evaporation, used as a crude indicator of moisture availability ($r = 0.60$, df = 37 for above ground, $r = 0.62$, df = 37 for below ground).

Monthly absolute rates of decomposition peaked in the second half of each year (Figure 3.4c). The highest values recorded were in September 1984 at 104 g m^{-2} $month^{-1}$ above ground and 124 g m^{-2} $month^{-1}$ below. Minimum rates occurred at the start of 1985, with 17 g m^{-2} $month^{-1}$ below ground in January and only 5.1 g m^{-2} $month^{-1}$ above ground in February. Again, higher values were observed below ground, where the overall mean of 70 g m^{-2} $month^{-1}$ was twice that above (36 g m^{-2} $month^{-1}$). Above ground, absolute decomposition was greatly affected by the 1986 fire, with rates falling to zero immediately afterwards. Dead material below ground was not so affected and, consequently, absolute rates of decomposition did not decline so markedly, although the minimum for that year (7 g m^{-2} $month^{-1}$) was registered in May.

3.4.4 Net primary production

In order to eliminate random monthly fluctuations, the mean biomass, dead vegetation and relative decomposition rate were 'smoothed', using a weighted moving average over a 3-month period:

$$x'_i = 0.25x_{i-1} + 0.5x_i + 0.25x_{i+1}$$

Where x'_i is the 'smoothed' mean for month i and x_i is the untransformed mean for month i.

Net primary production (P_n) was calculated for the community as a whole from these 'smoothed' estimates. Above ground, high values of P_n were obtained in mid-summer (Figure 3.5a): 122 g m^{-2} $month^{-1}$ in July 1985, 88 g m^{-2} $month^{-1}$ in July 1986 and 91 g m^{-2} $month^{-1}$ in August 1987. However, the vegetation at the site in January 1986 also showed a high productivity (63 g m^{-2}). It is probable that the fire in the following month depressed above-ground P_n for the rest of that year, but there was a definite increase noted during each mid-winter.

On an annual basis, 1985 was the most productive year (895.35 g m^{-2}) with respect to above-ground P_n, values declining in 1986 to 537.72 g m^{-2} before recovering to 574.41 g m^{-2} in 1987. Statistically significant correlations ($P<0.05$) were found between above-ground P_n and minimum temperature (r = 0.49, df = 39), mean temperature (r = 0.35, df = 35), precipitation (r = 0.4, df = 39) and the difference between precipitation and evaporation (r = 0.43, df = 39).

In order to calculate yearly turnover, annual P_n was divided by the mean biomass. Above-ground turnover rose from 4.3 in 1985 to 5.0 in the fire affected vegetation of 1986 and then again to 5.2 in 1987.

It is difficult to detect any definite trend in below-ground P_n (Figure 3.5b), although negative values tended to occur during the winter months: −20 g m^{-2} $month^{-1}$ in March 1985, −14 in January 1986 and −28 g m^{-2} $month^{-1}$ in January 1987. The highest productivities were 261 g m^{-2} $month^{-1}$ in November 1985, 253 g m^{-2} $month^{-1}$ in May 1986 and 331 g m^{-2} $month^{-1}$ in March 1987. Annual values were much higher than those above ground (1099 g m^{-2} y^{-1} in 1985, 884 g m^{-2} y^{-1} in 1986 and 1037 g m^{-2} y^{-1} in 1987), once again underlining the importance of the below-ground compartment. Below-ground P_n was found to be significantly correlated ($P<0.05$) with minimum temperature (r = 0.36, df = 39) and mean temperature (r = 0.4, df = 39). In contrast to the above-ground situation, below-ground turnover declined from 2.3 in 1985 to 2.0 in 1986, before rising slightly to 2.1 in 1987.

Total annual P_n reflected the yearly variation in above- and below-ground P_n, with 1985 being the most productive year (1995 g $^{-2}$ y^{-1}). Total P_n fell to 1422 g m^{-2} y^{-1} in 1986 and then rose to 1611 g m^{-2} y^{-1} in 1987. Total turnover was 2.9 in 1985, 2.5 in 1986 and finally 2.6 in 1987.

3.4.5 Leaf area index and solar radiation interception

The total leaf area index (L) of the canopy exhibited a similar trend to that shown by the above-ground biomass (Figure 3.6a). Again, higher values were attained from July to October (1.7 in August 1984, 1.6 in July and August 1985, 1.1 in September 1986 and August 1987). Low values occurred from November to June (0.5 in June 1984, 0.1 in

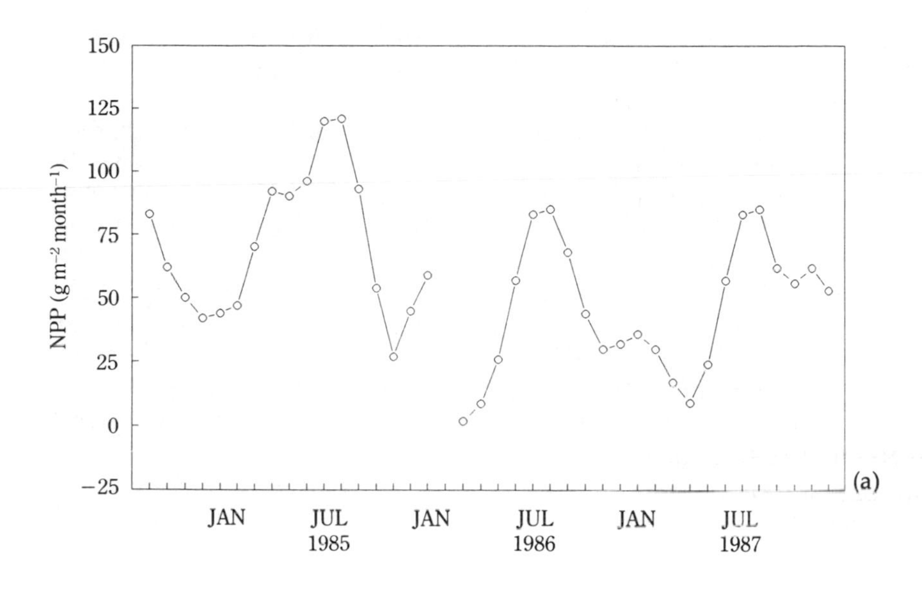

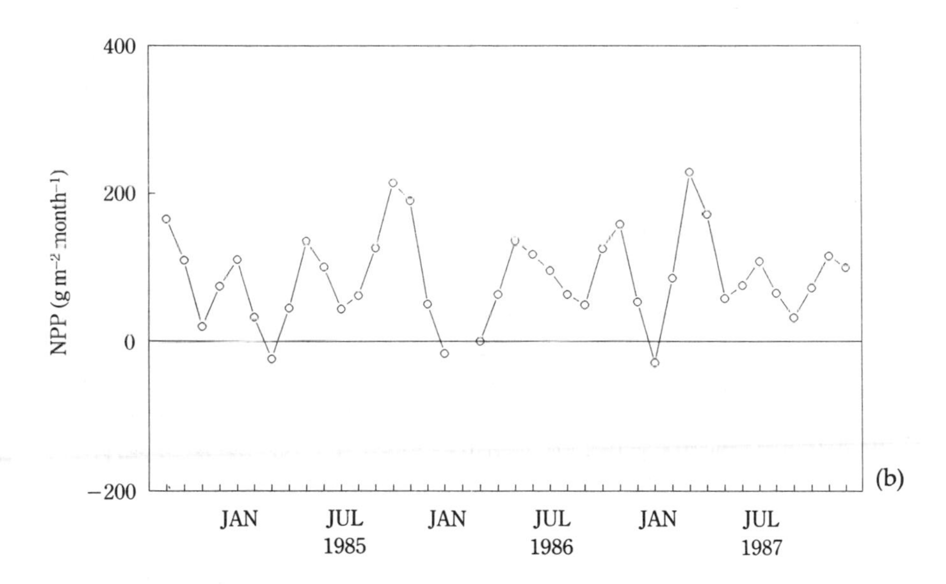

Figure 3.5. Net monthly primary production: (a) above ground (shoots); and (b) below ground (roots).

December 1985 and 0 from January to April 1986 and in April 1987). The year 1984 exhibited higher and more consistent values of L in comparison to the following years.

Stem area index of *D. spicata* presented a similar trend to L (Figure 3.6b), with higher values being recorded in mid- to late summer. In addition, stem area indices in 1986 and 1987 were generally only 50% of those reported for 1985. Thus the highest measurement of 0.73 was encountered in both July and August 1985. Maxima for 1986 and 1987 both occurred in the September, 0.39 and 0.36 for each respective year. *D. spicata* accounted for 63–100% of total leaf area indices throughout the study period (Figure 3.6a).

Leaf area duration (*LAD*), a measure of the persistence of the canopy (Table 3.1), was calculated for 1985 and 1987 from the equation given by Beadle (1985). Intercepted solar radiation for these 2 years was 57 and 62% of that incident on the canopy. The efficiency of conversion of intercepted solar radiation into plant biomass (ε_c) was calculated from:

$$\varepsilon_c = \frac{\text{net primary production (energetic value)}}{\text{sum of intercepted radiation}}$$

where: ε_c = efficiency of conversion of intercepted solar radiation into net primary production.

Roberts *et al.* (1985) state that, for most plant material, the energy content per unit organic biomass will fall within the range 17–20 MJ kg^{-1}. ε_c was therefore calculated for the upper and lower reaches of this range. These values, together with above-, below-ground and total P_n, are combined in Table 3.1.

3.4.6 Photosynthesis

The reported results of leaf photosynthesis are limited to February 1985, September 1987, and May 1988. The assimilation rates of *D. spicata* changed on a seasonal basis (Figure 3.7a). In February 1985, a saturated assimilation rate (A_{sat}) of 4.17 $\mu mol\ m^{-2}\ s^{-1}$ was observed. In May 1988 and September 1987 the corresponding rates were 15.1 and 16.7 $\mu mol\ m^{-2}\ s^{-1}$, respectively.

When leaf photosynthesis was measured in the various canopy strata of *D. spicata* they exhibited different photosynthetic characteristics (Figure 3.7b). Leaves at the top of the canopy never became light saturated, even at a photon flux approaching 2000 $\mu mol\ m^{-2}\ s^{-1}$. The highest assimilation rate of these leaves was 9.6 $\mu mol\ m^{-2}\ s^{-1}$ at a photon flux of 1724 $\mu mol\ m^{-2}\ s^{-1}$. Middle and lower canopy leaves both became light saturated at about 650 $\mu mol\ m^{-2}\ s^{-1}$. However, the

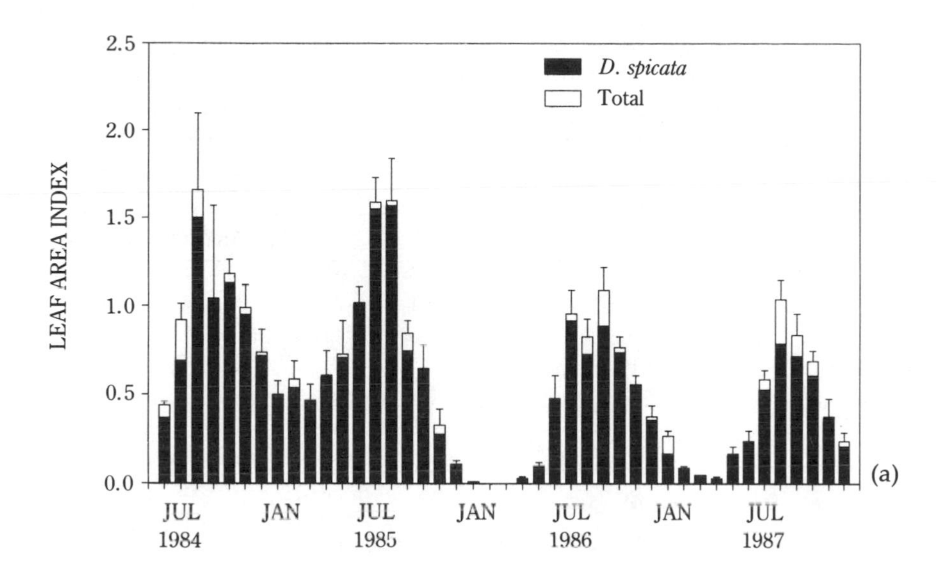

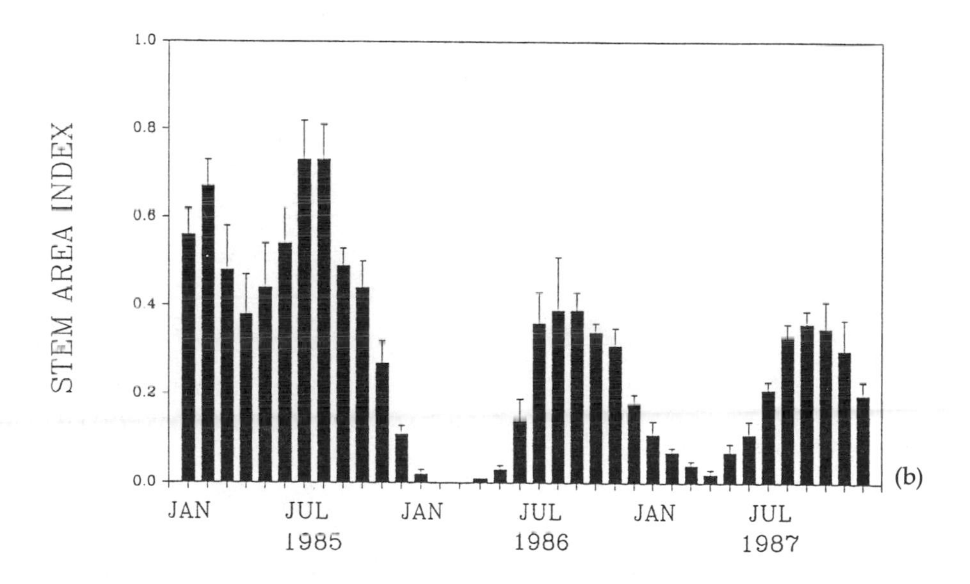

Figure 3.6. (a) Leaf area index of *Distichlis spicata* and other species; and (b) stem area index of *Distichlis spicata*.

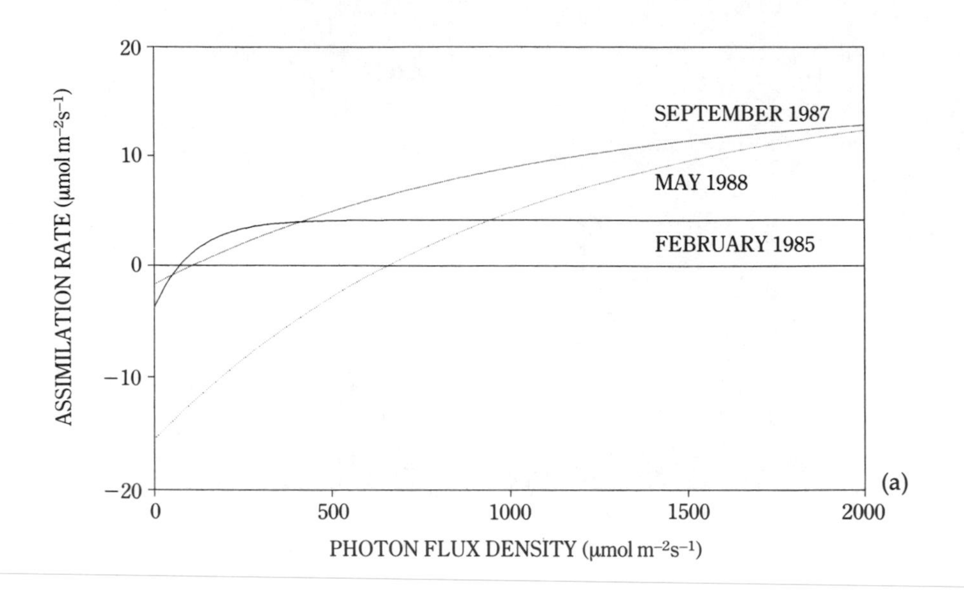
ASSIMILATION RATE (μmol m-2s-1)
20
10
0
−10
−20
SEPTEMBER 1987
MAY 1988
FEBRUARY 1985
(a)
0
500
1000
1500
2000
PHOTON FLUX DENSITY (μmol m-2s-1)

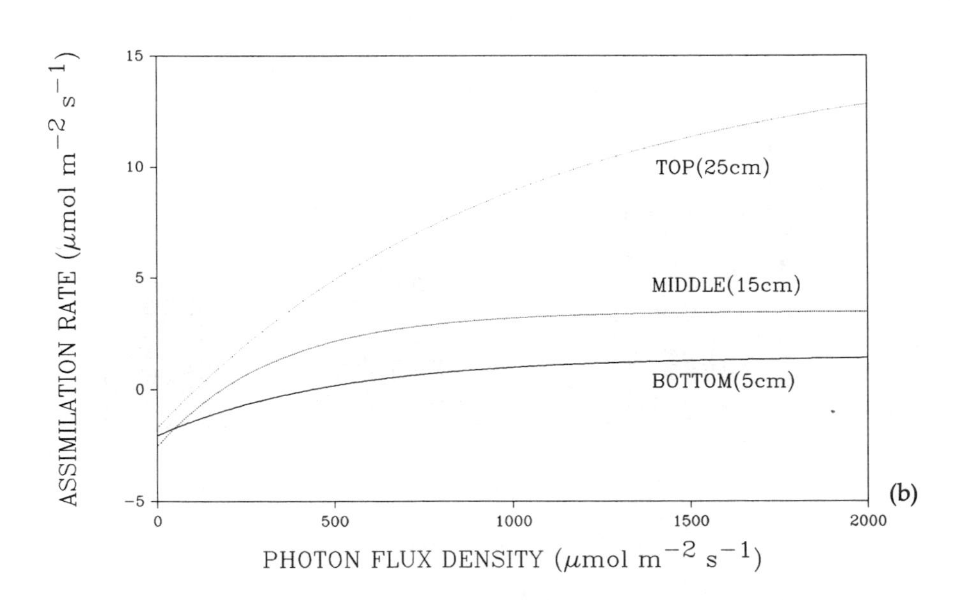
ASSIMILATION RATE (μmol m-2 s-1)
15
10
5
0
-5
TOP(25cm)
MIDDLE(15cm)
BOTTOM(5cm)
(b)
0
500
1000
1500
2000
PHOTON FLUX DENSITY (μmol m-2 s-1)

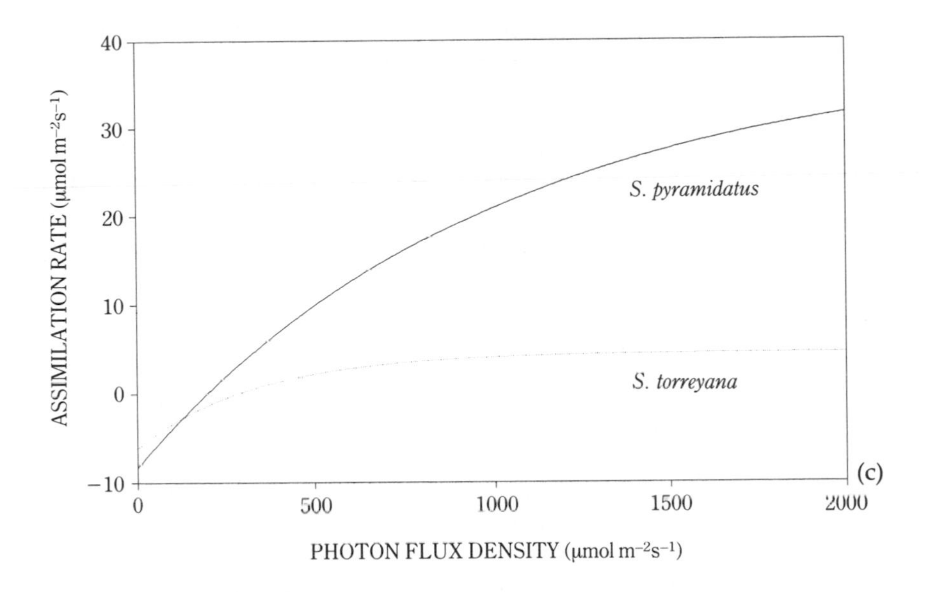

Figure 3.7. (a) Rectangular hyperbolas fitted by the maximum likelihood method to the light response curves of photosynthesis, expressed as net CO_2 uptake per unit leaf area, for leaves of *Distichlis spicata* in different months as indicated. Curves are based on data for *ca* 10 leaves in the upper canopy; (b) rectangular hyperbolas fitted to the light response curves of photosynthesis for *Distichlis spicata* at three canopy levels in late May 1988; and (c) rectangular hyperbolas fitted to the light response curves of photosynthesis for mature leaves of *Sporobolus pyramidatus* and *Suaeda torreyana*, measured in late May 1988.

mid-canopy leaves attained much higher assimilation rates, with a maximum of 3.6 μmol m^{-2} s^{-1} compared to only 0.32 μmol m^{-2} s^{-1} for the basal leaves.

In studies carried out in May 1988 on photosynthesis in other species in the study area, *Sporobolus pyramidatus*, a perennial prostrate grass, did not show light saturation at 1322 μmol m^{-2} s^{-1}, with an assimilation rate of 23 μmol m^{-2} s^{-1} (Figure 3.7c). At the same time, *Suaeda torreyana* reached light saturation at 637 μmol m^{-2} s^{-1} and assimilated CO_2 at a rate of 3.2 μmol m^{-2} s^{-1} (Figure 3.7c).

3.5 DISCUSSION

One of the main points to emerge from this study is the general reduction in above-ground biomass after 1984 (Figures 3.4 a,b), apart

Table 3.1. A comparison of the years 1985 and 1987 at the experimental site

Year	Solar radiation		L	LAD	AP_n	BP_n	P_n	Efficiency (ε_c)
	Total incident	Intercepted						
1985	6906.6	3940.6	1.60	285.6	895	1099	1994	0.86–1.01
1987	6574.5	4083.1	1.05	143.4	574	1037	1611	0.67–0.79

Solar radiation = MJ m^{-2} y^{-1}.
L, leaf area index; *LAD*, leaf area duration (days); AP_n above-ground annual net primary production (g m^{-2} y^{-1}); BP_n, below-ground annual net primary production (g m^{-2} y^{-1}); and P_n, total annual net primary production (g m^{-2} y^{-1}).

from that caused by the fire in 1986. It would seem that this was primarily due to the variation in precipitation between years, which declined from a peak in 1984 to a low in 1986 (Figure 3.2c). Net primary production seemed to be similarly linked to annual rainfall, with both above-, below-ground and total P_n being highest in 1985, falling in 1986 and then rising again in 1987 (Figure 3.5 a,b; Table 3.1).

Also of importance was the rapid regrowth of the shoots after being totally removed by the fire. It is interesting to note that above-ground biomass declined to only 6 g m^{-2} in the month before the fire, and to 10 g m^{-2} in April 1987, when presumably the fire no longer exerted any effect. *D. spicata* therefore seems well adapted to the complete renovation of its photosynthetic apparatus after perturbation by fire or water stress and this is, no doubt, related to its large underground reserve of biomass. Abrams *et al.* (1986) report higher values of above-ground biomass (422 g m^{-2}) for an annually burned tallgrass prairie in Kansas than an unburned one (364 g m^{-2}).

From the beginning of the study until just before the fire, the *D. spicata* community accumulated nearly 620 g m^{-2} of organic matter, in its combined living and dead pools. Such accumulation indicates that this community functions as a net sink for carbon from the atmosphere. This accumulation is most marked in the below-ground biomass (Figure 3.4a), illustrating that with a decreased incidence of fire, a significant accumulation of carbon in dead organic matter could occur.

Estimates of P_n from this study (Table 3.1) are comparable to those referred to in the Introduction, although ours are not as great as some of the highest estimates for tropical grasslands, e.g. 3538 g m^{-2} y^{-1} at Kurukshetra, India (Singh and Yadava, 1974), 8529 g m^{-2} y^{-1} for

Brachiaria mutica pastures (Almeida, 1981), 4308 g m^{-2} y^{-1} for *Digitaria decumbens* at Cardenas Tabasco, Mexico, and 9900 g m^{-2} y^{-1} for *Echinochloa polystachya* on a central Amazon floodplain (Chapter 6). However, production at the site of the present study was greater than that reported for temperate grasslands by Sims and Singh (1978) (54–523 g m^{-2} y^{-1}) and Sala *et al.* (1988) (100–700 g m^{-2} y^{-1}), and for savanna grasses (Chapter 2). The present results fall within the range 126–3396 g m^{-2} y^{-1} reported for grassland communities by Van Dyne *et al.*, (1978) and the range 200–2000 g m^{-2} y^{-1} suggested by Golley and Lieth (1972).

Because of differences in methodologies between this and other studies, below-ground P_n values are more difficult to compare. Reference can be made to data reported on related species. Bellis and Gaither (1985) reported biomass increments of up to 3250 g m^{-2} for *D. spicata* over a 5-month period (October 1982 to February 1983), and a mean annual standing crop of 8040 g m^{-2}. These figures were considered by Bellis and Gaither (1985) to be the highest observed for this species. Hackney and de la Cruz (1986) studied the below-ground productivity of the related halophytic grass *Spartina cynosuroides* (L.) Roth in a marsh on the Mississippi River. Yields of 1900 g m^{-2} y^{-1} were calculated by the maximum–minimum procedure, or 4000 g m^{-2} y^{-1} using Smalley's (1960) technique. Ellison *et al.* (1986) calculated monthly below-ground P_n of *Spartina alterniflora* at Rhode Island, USA, by subtracting the minimum monthly from the maximum monthly below-ground biomass. Values obtained ranged from 1670 to 2995 g m^{-2} y^{-1}. The values of P_n reported in this study are quite similar to those obtained for tropical grasslands in Thailand (Chapter 4) and in Kenya (Chapter 2) using the same methodology. The present values are higher than both the average P_n reported by Murphy (1975) for tropical grasslands (1080 g m^{-2} y^{-1}) and the values reported by Sims and Singh (1978) for temperate grasslands (225–1425 g m^{-2} y^{-1}).

Seasonal below-ground production of the temperate grass *Festuca pratensis* was estimated by Hansson and Andren (1986) to be 360 to 460 g m^{-2} y^{-1}; approximately one-half to one-third of the values reported in the present study. Sims *et al.* (1978) measured the root biomass for 10 major grasslands in North America, obtaining values equivalent to 156 to 2000 g m^{-2} over a year of study. Sims and Singh (1978) estimated below-ground P_n in the same communities at 148–641 g m^{-2} y^{-1}. Van Dyne *et al.* (1978) gave values of between 18.2 and 1465 g m^{-2} y^{-1} for below-ground P_n, in North American grasslands similar to those of the two reports above.

An important point in the present study relates to the contribution of the below-ground compartment of the community to total P_n. This is an aspect which has been very much neglected by many other workers, due to the sampling difficulties involved. This study has demonstrated that the below-ground portion can contribute more than one-half (in fact *ca* 65 %) of total P_n, a comparable figure to those reported in other chapters of this volume. Even higher values, 91–93%, have been reported for the contribution of below-ground organs to P_n in a *Stipa tenuis/Piptochaetium napostaenese* community of central Argentina (Distel and Fernandez, 1986).

When the P_n values calculated in this report are compared to average P_n for agricultural crops, such as described by Whittaker and Likens (1975) for cultivated land (650 g m^{-2} y^{-1}), or by Fischer and Turner (1978) for the semi-arid zones (250–1000 g m^{-2} y^{-1}), our values are much higher. Pettersson *et al.* (1986) reported total P_n (ash free) for *Medicago sativa* of between 960 and 1240 g m^{-2} y^{-1} for the first and second years of study, and between 1840 and 2260 g m^{-2} yl^{-1} for the third year.

In the present study, the greatest quantities of biomass, both above- and below-ground, and the highest levels of P_n seemed to be associated with the warmer and wetter part of the year. It would seem that the community is highly dependent on precipitation. This is in accordance with the work of Sims and Singh (1978), who found that annual precipitation explained between 46 and 58% of the variation in total P_n in temperate grasslands. Moreover, Sala *et al.* (1988) found a close correlation between annual precipitation and AP_n at 9500 sites throughout the central USA, such that AP_n could be calculated from annual precipitation using the following equation with a value for r^2 of 0.9:

$$AP_n = -34 + 0.6\ APPT$$

where AP_n is above-ground net primary productivity (g m^{-2}) and $APPT$ = annual precipitation (mm).

Inserting the annual precipitation observed at Chapingo for 1985 and 1987, this equation would predict the respective amounts of AP_n to be 319 and 286 g m^{-2} y^{-1}. From Table 3.1, it can be seen that both of these values appear gross underestimates. However, it should be borne in mind that the regression of Sala *et al.* (1988) is based on estimates of AP_n derived simply form maximum standing crops. On this basis the estimate of AP_n using this equation at our study site is not very different from the peak levels of above-ground biomass (Figure 3.4a). It might be assumed that different relationships between AP_n and $APPT$ would hold for such widely differing sites as those in North America and Mexico. The higher leaf area index values of *D. spicata* were 1.1–1.7 corresponding to mid-summer (Figure 3.6a). The highest value (1.7), was less than that found at the study site in Kenya

(Chapter 2). The values of the turnover rate of *D. spicata* for the below-ground component are very similar for the 3 years of the study at a little over 2.0 y^{-1}. These values are quite different from those extrapolated from biomass changes alone. These were 0.97 y^{-1} for Indian grasslands (Singh and Yadava, 1977), 0.84–1.07 for a *Trachypogon* savanna of the Venezuelan Llanos (San Jose *et al.*, 1982) and 0.10–0.49 y^{-1} for North American prairie (Sims and Singh, 1978).

The photosynthetic rate, reported during the dry period, was 4.17 μmol m^{-2} s^{-1}, only one-fourth of the values of 15.1 and 16.7 μmol m^{-2} s^{-1} recorded in May and September, respectively. These last values however correspond to the lower half of the range commonly reported for C_4 mesophytic leaves (Long and Hällgren, 1985).

3.6 CONCLUSIONS

Total net primary production, when estimates based on dry matter changes are corrected for the decomposition of dead vegetation, was 1994 g m^{-2} y^{-1} in 1985 and 1611 g m^{-2} y^{-1} in 1987. In both years below-ground production (BP_n) constituted over 55% of total P_n. These figures are significantly greater than those suggested by the standard IBP methods which do not take direct account of vegetation turnover.

Above ground, biomass and decomposition rates exhibited a unimodal pattern in each year of study. The major contribution (>79%) to total above-ground biomass was provided by *D. spicata*. Below-ground biomass was more variable in time and space, but showed a significant rise up to a fire on the site in 1986. Above-ground dead vegetation accumulated throughout the study period, until it was eliminated by fire in February 1986. However, it recovered rapidly.

Below ground, dead vegetation levels were not depressed by the fire in 1986. Above- and below-ground relative decomposition rates were highly correlated ($P<0.05$) with precipitation and air temperature. However, a better correlation was found with the difference between precipitation and evaporation, used as a crude indicator of moisture availability.

The average absolute quantities of dead vegetation decomposed in the soil were twice (70 g m^{-2} $month^{-1}$) those of the dead vegetation on and above the ground surface (36 g m^{-2} $month^{-1}$). Decomposition of below-ground dead vegetation was recorded in every month of the year illustrating the errors which result when P_n is estimated without measurement of decomposition. Although a tropical grassland, production and growth at this high altitude are positively correlated with temperature.

3.7 SUMMARY

1. Changes in the quantities of live and dead vegetation, and decomposition, were monitored at monthly intervals from September 1984 to December 1987 for *Distichlis spicata* dominated saline grassland near Mexico City. Although within the tropics, this high altitude site shows a distinct winter in which plant growth is inhibited by low temperature and lack of moisture.
2. Biomass levels ranged from 336 g m^{-2} in June 1986 to 917 g m^{-2} in August 1985. On average 73% of biomass was below-ground. Quantities of dead vegetation increased from 653 g m^{-2} in November 1984 to 1836 g m^{-2} in November 1985. A fire in February 1986 removed all dead vegetation from the site surface which then increased to 624 g m^{-2} at the end of 1987.
3. Mean relative rates of decomposition varied from 0.01 g g^{-1} $month^{-1}$ in February 1985 to 0.16 g g^{-1} $month^{-1}$ in July 1986 for above-ground material. On average, relative decomposition rates below-ground were roughly 1.4 times the values for above-ground litter, whilst absolute quantities decomposed were twice as great. Decomposition rate showed positive correlations with temperature and with the difference between precipitation and evaporation, used as an indicator of moisture availability.
4. Monthly net primary production varied form −79 g m^{-2} $month^{-1}$ in January 1987 to 350 g m^{-2} $month^{-1}$ in March of the same year. In each year below-ground net primary production accounted for *ca* 65% of the total. Annual net primary production varied from 1422 g m^{-1} in 1986 to 1995 g m^{-1} in 1985, and could be related to variation in precipitation between years and the occurrence of an accidental fire at the beginning of 1986 which corresponded with a 29% decrease in P_n in that year.
5. The values of P_n observed in this study are high relative to many previous studies of semi-arid tropical grasslands and we consider that this is more likely to be a result of methodological differences than site differences. The present study took account of both below-ground production and corrected for losses of dead vegetation through decomposition. The community intercepted 60% of the annual average of 6700 MJ of solar radiation received, and converted this into plant

biomass with an average efficiency of between 0.7 and 1.0%. In the absence of fire there was a net accumulation of organic matter, which amounted to over 600 g m^{-2} from September 1984 to January 1986. This community may therefore be seen as a net sink removing carbon from the labile pools of the global carbon cycle.

REFERENCES

Abrams, M.D., Knapp, A.K and Hulbert, L.C. (1986) A ten year record of aboveground biomass in a Kansas tallgrass prairie: effects of fire and topographic position. *American Journal of Botany*, **73**, 1509–15.

Almeida, M.R. (1981) Productividad primaria de tres praderas de especies tropicales: para (*Brachiaria mutica*), grama amarga (*Paspalum conjugatum*) y pangola (*Digitaria decumbens*). *Boletina de la Sociedade Botanica de Mexico*, **41**, 3–18.

Beadle, C.A. (1985) Plant growth analysis, in *Techniques in Bioproductivity and Photosynthesis*, 2nd edn (eds J. Coombs, D.O. Hall, S.P. Long and J.M.O. Scurlock) Pergamon Press, Oxford, pp. 20–5.

Beetle, A.A. (1987) *Las gramineas de Mexico. II.* Secretaria de Agricultura y Recursos Hidraulicas, Subsecretaria de Desarrollo y Fomento Agropecuario y Ferestal, Direccion General de Normatividad Pecuaria, Cotecoca.

Bellis, V.J. and Gaither, A.C. (1985) Seasonality of above-ground and below-ground biomass for six salt marsh plant species. *Journal of the Elisha Mitchell Scientific Society*, **101**, 95–109.

Cabrera, P.E.M. (1988) *Respuesta de* Distichlis spicata *(L.) Greene var.* mexicana *Beetle, 'zacate salado', al fuego en el ex-lago de Texcoco*. Tesis Profesional, Universidad Michoacana de San Nicolas de Hidalgo, Morelia, Michoacan.

Del Valle, C.H.F. (1983) *Los procesos de accumulacion de sales en intemperismo en cubetas lacustres del ex-Lago de Texcoco*. Tesis de Maestria en Ciencias. Centro de Edafologia, Colegio de Postgraduados. Chapingo, Mexico.

Distel, R.A. and Fernandez, O.A. (1986) Productivity of *Stipa tenuis* Phil. and *Piptochaetium mapostaenese* (Speg.) Hack in semi-arid Argentine. *Journal of Arid Environments*, **11**, 93–6.

Ellison, M.A., Bertness, M.D. and Miller, T. (1986) Seasonal patterns in the belowground biomass of *Spartina alterniflora* (Graminae) across a tidal gradient. *American Journal of Botany*, **73**, 1548–54.

Evans, de C.M. (1978) *Factores limitantes para la colonizacion vegetal del lecho del ex-Lago de Texcoco, Mexico*. Tesis de Maestria en Ciencias, Colegio de Postgraduados, Chapingo, Mexico.

Fischer, R.A. and Turner, N.C. (1978) Plant productivity in the arid and semiarid zones. *Annual Review of Plant Physiology*,**29**, 277–317.
Flowers, T.J., Hakibgheri, M.A. and Clipson, N.J.W. (1986) Halophytes. *The Quarterly Review of Biology*, **62**, 313–37.
García, E. (1973) *Modificaciones al Sistema de Clasificacion de Köppen*, Instituto de Geografia, Universidad Nacional Autonama de Mexico, Mexico.
Glenn, E.P., Fontes, M.R. and Yensen, N.P. (1982) Productivity of halophytes irrigated with hypersaline seawater in the Sonoran desert, in, *Biosaline Research: A Look to the Future* (ed. A. San Pietro), Plenum Press, New York. pp. 491–4.
Golley, F.B. and Lieth, H. (1972) Bases of organic production in the tropics, in *Tropical Ecology with Emphasis on Organic Production* (eds P.M. Golley and F.B. Golley), University of Georgia, Athens, Georgia, pp. 1–125.
Hackney, C.T. and De la Cruz, A.A. (1986) Belowground productivity of roots and rhizomes in a giant cordgrass marsh. *Estuaries*, **9**, 116–22.
Hansson, A-C. and Andren, O. (1986) Below-ground plant production in a perennial grass ley (*Festuca pratensis* Huds.) assessed with different methods. *Journal of Applied Ecology*, **23**, 657–66.
Long, S.P. and Baker, N.R. (1986) Saline terrestrial environments, in *Photosynthesis in Contrasting Environments* (eds N.R. Baker and S.P. Long), Elsevier, Amsterdam, pp. 63–102.
Long, S.P. and Hällgren, J.-E. (1985) Measurements of CO_2 assimilation by plants in the field and laboratory, in *Techniques in Bioproductivity and Photosynthesis*, 2nd edn (eds J. Coombs, D.O. Hall, S.P. Long and J.M.O. Scurlock). Pergamon Press, Oxford, pp. 62–94.
Long, S.P. and Mason, C.F. (1983) *Saltmarsh Ecology*. Blackie, Glasgow.
Mudie, P.J. (1974) The potential economic uses of halophytes. in *Ecology of Halophytes* (eds R.J. Reimold and W.H. Queen). Academic Press, New York. pp. 565–97.
Murphy, P.G. (1975) Net primary productivity in tropical terrestrial ecosystems, in, *Primary Productivity of the Biosphere*, (eds H. Lieth and R.W. Whittaker), Springer-Verlag, Berlin, pp. 217–31.
Odum, E.P. (1974) Halophytes, energetics and ecosystems, in *Ecology of Halophytes* (eds R.J. Reimold and W.H. Queen), Academic Press, New York, pp. 599–602.
Petterson, R., Hansson, A.-C., Andren, O.Y. and Steen, E. (1986) Above- and below-ground production and nitrogen uptake in lucerne (*Medicago sativa*). *Swedish Journal of Agricultural Research*, **16**, 167–77.
Roberts, M.J., Long, S.P., Tieszen, L.L. and Beadle, C.L. (1985)

Measurements of plant biomass and net primary production, in *Techniques in Bioproductivity and Photosynthesis*, 2nd edn (eds J. Coombs, D.O. Hall, S.P. Long, and J.M.O. Scurlock). Pergamon Press, Oxford, pp. 1–19.

Rzedowski, J. (1957) Algunas asociaciones vegetales de los terrenos del Lago de Texcoco. *Boletin de la Sociedad Botanica de Mexico*, **21**, 19–33.

Sala, O.E., Parton, W.J., Joyce, L.A. and Laurenroth, W.K. (1988) Primary production of the central grassland region of the United States. *Ecology*, **69**, 40–5.

San Jose, J.J., Berrade, F. and Ramirez, J. (1982) Seasonal changes of growth, mortality and disappearance of below-ground root biomass in the Trachypogon savanna grass. *Acta Oecologia Plantarum*, **3**, 347–58.

Secretaria de Recursos Hidraulicos (1971) *Estudio Agrologico Especial del ex-Lago de Texcoco*, Publicacion No. 2, Serie Estudios, Edo. de Mexico, Mexico.

Sims, P.L. and Singh, J.S. (1978) The structure and function of ten western North American grasslands. III. Net primary production, turnover and efficiencies of energy capture and water use. *Journal of Ecology*, **66**, 573–97.

Sims, P.L., Singh, J.S. and Laurenroth, W.K. (1978) The structure and function of ten western North American grasslands. I. Abiotic and vegetational characteristics. *Journal of Ecology*, **66**, 251–85.

Singh, J.S. and Yadava, P.S. (1974) Seasonal variation in composition, plant biomass and net primary productivity of a tropical grassland at Kurukshetra, India. *Ecological Monographs*, **44**, 351–76.

Smalley, A.E. (1960) Energy flow of a saltmarsh grasshopper population. *Ecology*, **41**, 672–7.

Van Dyne, G.M., Smith, F.M., Czapleswski, R.L. *et al.* (1978) Analysis and syntheses of grassland ecosystems dynamics, in *Patterns in Primary Productivity in the Biosphere* (ed. H. Lieth), Dowden, Hutchinson & Ross, Stroudsberg, Pennsylvania, pp. 199–204.

Waisel, Y. (1972) *The Biology of Halophytes*. Academic Press, New York.

Whittaker, R.H. and Likens, G.E. (1975) The biosphere and man, in *Primary Productivity of the Biosphere*. (eds H. Lieth and R.H. Whittaker), Springer-Verlag, Berlin, pp. 305–28.

Wiegert, R.G. (1979) Ecological processes characteristic of coastal *Spartina* marshes of the south-eastern USA, in *Ecological Processes in Coastal Environments* (eds R.L. Jefferies and A.J. Davy), Blackwell, Oxford, pp. 467–90.

4

Monsoon grassland in Thailand

APINAN KAMNALRUT and J.P. EVENSON

4.1 INTRODUCTION

In tropical, monsoonal South-East Asia, there are still vast areas of evergreen forests which are regarded as the climatic climax vegetation (Myers, 1980). However, large tracts of grassland are interspersed with the forest, forming an important aspect of the natural vegetation (Dobby, 1973). Peninsular Thailand and Malaysia (Figure 4.1) have prolonged periods of rainfall around the monsoons interrupted by only 2–3 months of dry weather. Such conditions, in association with poor drainage, infertile soils and periodic fires produce a moist savanna type of vegetation. In southern Thailand, this grassland, according to the definition of Bourlière and Hadley (1970), shows the characteristics of tropical savanna. It contains a continuous grass/sedge stratum surrounding isolated trees and clumps of trees rising to 5–12 m. The original primary vegetation is considered to have been evergreen forest, replaced by savanna on poorer soils following ancient burning associated with shifting agriculture (Dobby, 1973; Whitmore, 1984). It is current practice in these areas of southern Thailand peninsula for local villagers to set fire to the vegetation during the dry season to encourage the regrowth of young shoots for grazing at the start of the next rainy season.

The tropical forests of the region have been the subject of previous studies, but so far no studies of the productivity of the grasslands in South-East Asia have been reported (Murphy, 1975; Yabuki *et al.*, 1985; Misra, 1979). However, previous work by Caldwell (1975), Murphy (1975) and Singh and Joshi (1979) suggests that net primary production (P_n) for this biome should range from 40 to 3810 g m^{-2} y^{-1}.

This study aimed to assess the total net primary production of a semi-natural grassland in southern Thailand by monitoring monthly changes in biomass, dead material and decomposition. The amount of incident solar radiation intercepted by the canopy was also monitored and the grassland's 'photosynthetic' conversion efficiency calculated.

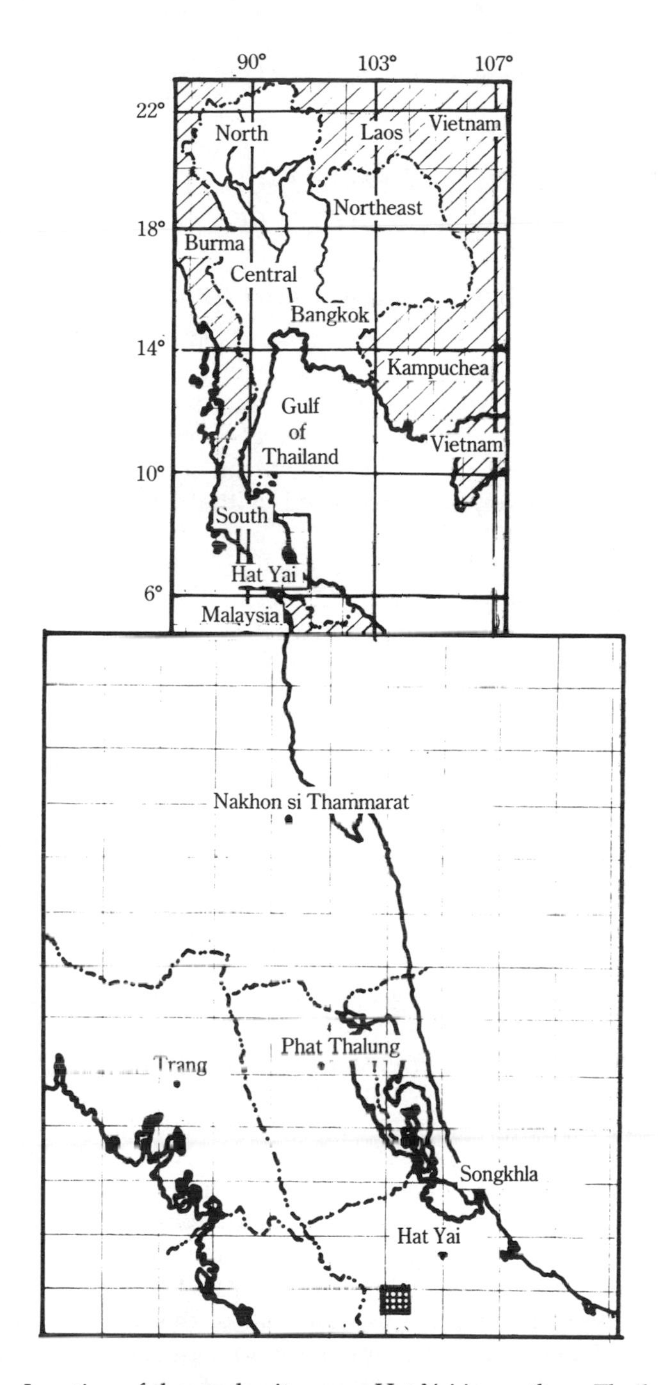

Figure 4.1. Location of the study site, near Hat Yai in southern Thailand. The study site location is indicated by the square.

In addition, accidental fires allowed study of the recovery of production.

4.2 STUDY SITE

The study site was located at Ban Klong Hoi Khong, Hat Yai district, Songkhla Province in southern Thailand, latitude 6°20′N, longitude 100°56′E and 30 m above mean sea level (Figure 4.1). This area of Thailand is climatically classified as wet monsoonal tropics. During the study period, macro-climatic data were collected from the meteorological equipment set up at the site. Mean monthly rainfall and pan evaporation, however, were recorded at the meteorological station at Hat Yai airport, 10 km south of the study site. Maximum and minimum temperatures show only small seasonal fluctuation. Monthly mean temperatures varied from 25.5°C in January 1986 to 29.1°C in April of the same year (Figure 4.2a). Solar radiation was measured with a tube solarimeter (Delta-T Devices, Cambridge, UK) sited above the grass canopy. Monthly incident receipts varied from 207.1 MJ m^{-2} in December 1985 to 618.8 MJ m^{-2} in September 1984 (Figure 4.2b). The highest percentage of incident radiation intercepted by the canopy was 90% in December 1984, with the lowest, 8.6%, in March 1986. From 1984 to 1986, annual rainfall was found to be lower than the 10 year average, especially during the last quarter of each year (Figure 4.2c; Table 4.1). This was particularly noticeable in 1984, when only 40 mm of rain fell in November. In 1984 and 1985, the January to May periods showed above-average rainfall. The rainfall pattern exhibits extreme seasonal variability as has been described previously (Evenson, 1983). The probability of obtaining a month in which evaporation exceeds precipitation between January and March has been estimated as greater than 75%, thus constituting a dry season. Soil water reserves would normally be expected to have fallen almost to zero by the end of February. Intermediate rainfall of between 110 and 150 mm $month^{-1}$ occurs during April to September. This period can be further divided into further 'wet' and 'dry' seasons, as somewhat drier conditions may prevail after June. From October to December the north east monsoon produces an average of 1000 mm of rain in the second 'wet' season (Evenson, 1983). However, less than half of this amount was received in 1984, and even in 1986 rainfall for this period was only 697 mm (Table 4.1). It is only in this last quarter of the year that rainfall equals or exceeds potential evaporation (Figure 4.2c). Water stress is therefore

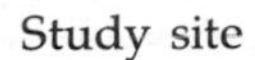

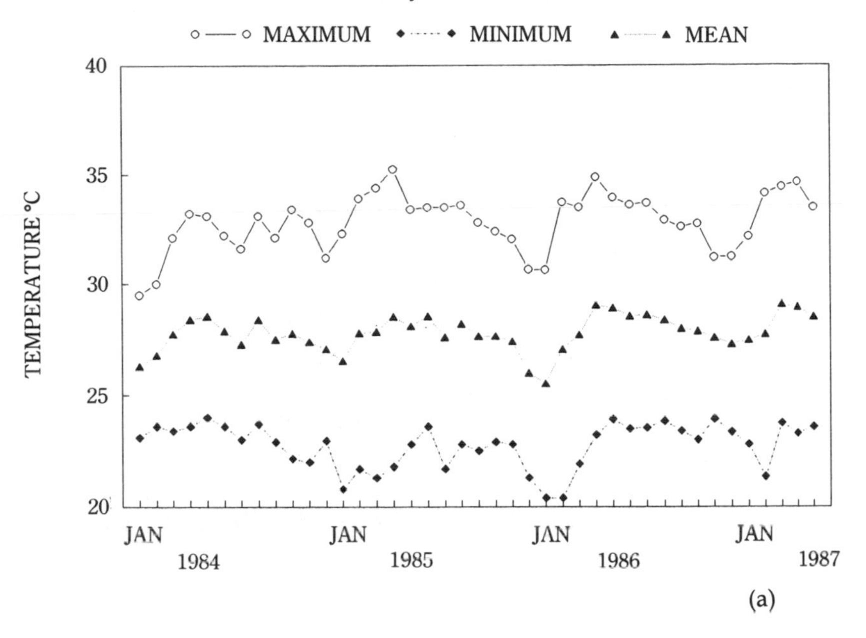

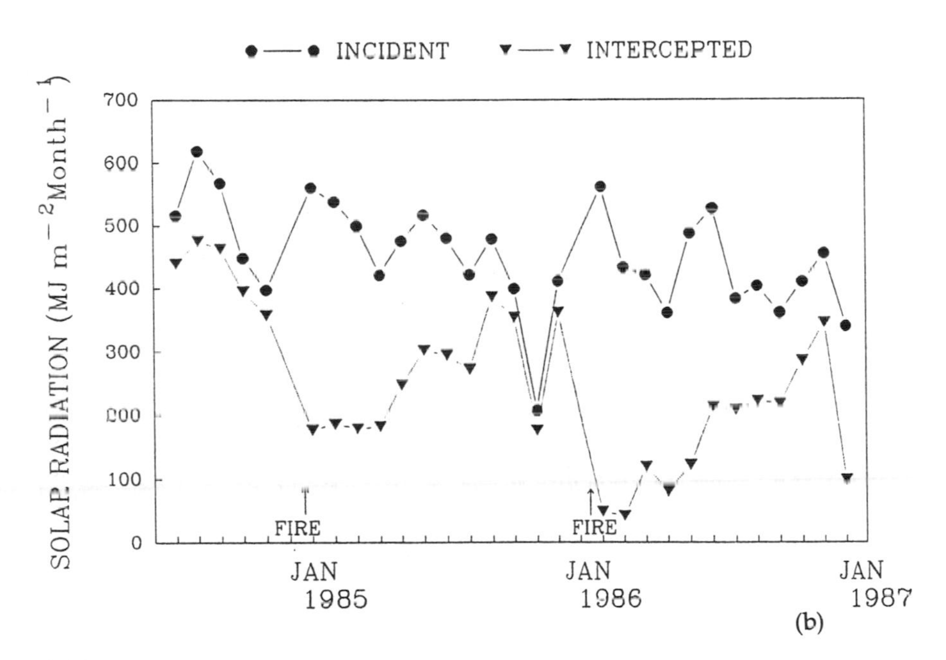

Figure 4.2. (a) Average monthly maximum, mean and minimum temperatures at the Hat Yai study site (January 1984 to May 1987); (b) monthly totals of solar radiation incident above the canopy and intercepted by the canopy (August 1984 to February 1987).

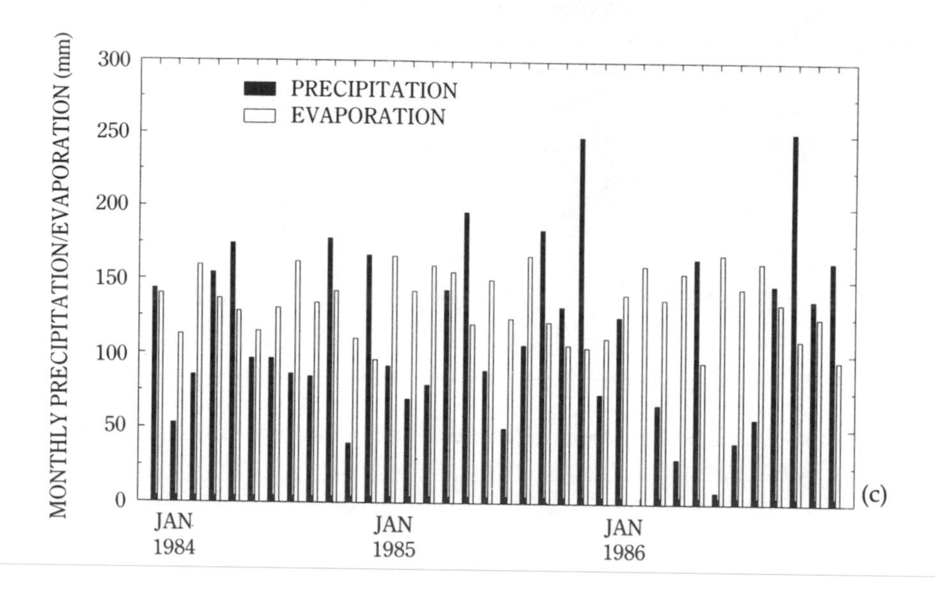

Figure 4.2. *continued* (c) Monthly totals of precipitation and pan evaporation at the Hat Yai study site (January 1984 to December 1986).

a major, if not the key, factor affecting plant production from January to September (Evenson, 1983).

The soil at the study site is a humic gley, with an average *A* horizon depth of 40 cm (Figure 4.3). It is formed from a parent material of old alluvium and is poorly drained with slow runoff. At the end of the 'wet' seasons the soil is often sodden or flooded in places and free standing water may collect. In the dry period the water table can drop 4 m below the surface. In some places there is a gravel layer in the A_3 horizon. The soil has an average pH of 4.0 and is markedly deficient in N, P, K, Ca, S and Cu (Yabuki *et al.*, 1985).

The vegetation of the site is continuous grassland with scattered trees. Following the species nomenclature of Gilliland (1971) and Smitinand and Larsen (1972), the most abundant grass species are *Eulalia trispicata* [Schult.] Henr. and *Lophopogon intermedius* A. Cam. Grasses constitute more than 60% of the community cover. Sedges (Cyperaceae) account for about 10% of cover, with *Fimbristylis tristachya* R. Br. the most abundant. The commonest dicotyledonous species of the savanna floor community is *Dillenia hookerii* Pierre (Dilleniaceae). *Dipterocarpus obtusifolius* Teijsm. ex Mig. var. *obtusifolius* (Dipterocarpaceae) is the major tree species in the area, and

Table 4.1. The distribution of rainfall (mm) on a seasonal basis

Seasonal period	*1984*	*1985*	*1986*	*10-year average*
Dry 1 (J,F,M)	281	240	191	197
Wet 1 (A,M)	328	339	195	277
Dry 2 (J,J,A)	279	245	105	351
Wet 2 (S,O,N,D)	469	636	696	1 207
	1 359	1 462	1 188	2 032

has a thick, fire resistant bark. The trees are evergreen and some of the ground flora persists during the dry season. According to local villagers, the site has been in its present state for at least the past 60 years. It has been very occasionally grazed by domestic cattle, especially on the new growth of grasses following a fire.

4.3 METHODS

Sampling of biomass and dead vegetation was as described previously (Chapter 1) and commenced in April 1984. The live vegetation sampled in each month for each quadrat was separated into four categories: *E. trispicata*; *L. intermedius* and other grasses; *F. tristachya*; and *D. hookerii*. Preliminary analysis had demonstrated that the bulk (>90%) of the underground biomass was within 10 cm of the surface in both 'wet' and 'dry' seasons, so cores were taken to only 15 cm depth. Thirty litter bags both above and below ground were filled and placed at monthly intervals, and collected after 1 month. A mesh size of 1 mm was used during 1984, and replaced with a mesh size of 2 mm from 1985 onwards. After part of the site was burned in February 1986 additional sets of litter bags were filled and analysed on the burned areas.

A fire in January 1985 burnt about 40% of the study area. Sampling continued from the smaller remaining area, whilst an additional set of equivalent samples were taken simultaneously from the burnt area. A second fire, in February 1986, covered most of the site. Sampling from this point was from sites burnt in February 1986 and January 1985. This enabled the recovery of wet savanna from fire under different environmental conditions to be studied, as 1986 was a drier year than 1985 (Figure 4.2c).

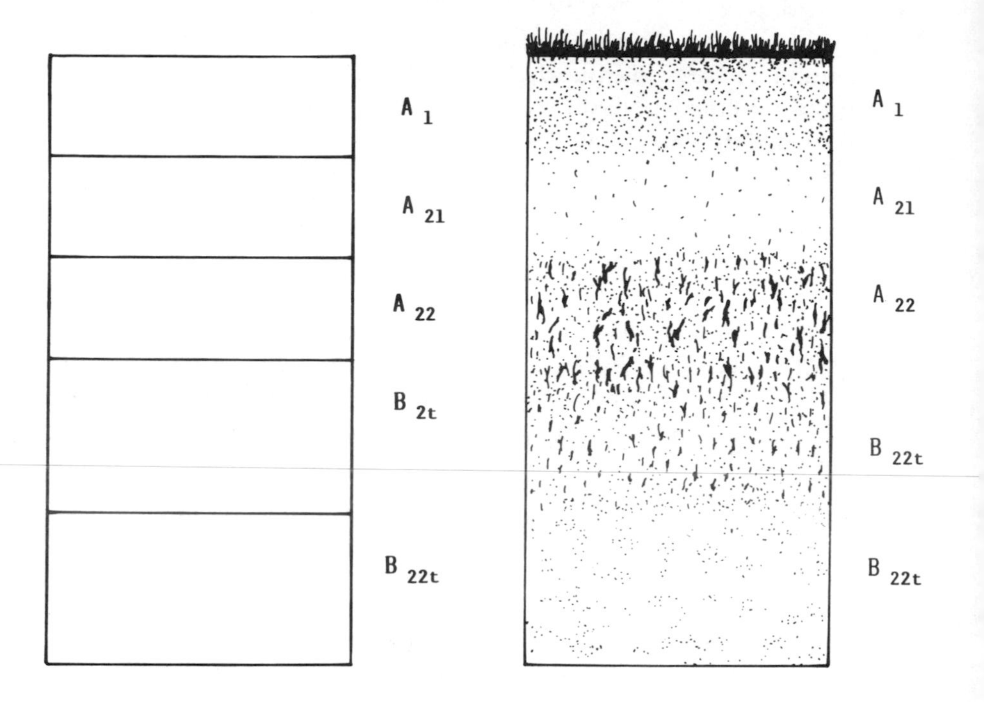

Figure 4.3. Soil profile for the study site. Descriptions of the horizons indicated were as follows. A_1 (0–20 cm), a dark grey-brown (10 YR 3/2) loam; moderate fine to medium subangular blocky structure. Hard when dry, firm moist and plastic when wet. Many very fine and fine roots; acid (pH 5.5) with a clear undulating boundary. A_{21} (20–38 cm), a brown (7.5 YR 5/2) clay loam; massive structure. Extremely firm when moist, sticky and plastic when wet. Some fine roots; acid (pH 5.0) with a clear undulating boundary. A_{22} (38–60 cm), pink-grey (7.5 YR 5/2) clay with many brown (7.5 YR 5/8) and red (2.5 YR 4/6) mottles; massive structure. Very firm when moist, and very sticky and plastic when wet. Some fine roots; acid (pH 5.0) with a clear undulating boundary. B_{2t} (60–90 cm), light grey (10 YR 7/2) clay with prominent red (2.5 YR 5/8) mottles; massive structure. Firm when moist, and very sticky and plastic when wet. Strongly acid (pH 4.5); clear undulating boundary. B_{22t} (90–120 cm), dark brown (7.5 YR 4/4) loamy coarse sand with strong brown (7.5 YR 5/8) mottles; massive structure. Extremely firm when moist, and slightly sticky but non-plastic when wet. Strongly acid (pH 4.5).

4.4 RESULTS

4.4.1 Biomass dynamics

Above-ground biomass

In the unburnt areas, during the period April 1984 to December 1985, no strongly defined periodic cycle in quantities of biomass could be perceived (Figure 4.4a), although lower levels of biomass roughly coincided with the dry period of each year, for example in February to March 1985. Biomass ranged between 250 and 450 g m^{-2} with a mean of 347 g m^{-2}. Averaged over 1984 and 1985, on the unburnt sites, the bulk of the biomass was provided by the two grasses *E. trispicata* (35%) and *L. intermedius* (38%), the other two species of significance were the grass *F. tristachya* (14%) and the dicotyledon *D. hookerii* (15%). Both of the grasses exhibited wide variation in amounts of biomass present in any 1 month; *E. trispicata* ranging between 39 and 250 g m^{-2} and *L. intermedius* between 80 and 225 g m^{-2}. Both species reached peak biomass during the 'wet' seasons, with *E. trispicata* corresponding closely with the amount of water available and *L. intermedius* achieving maximum shoot biomass when not subjected to either excessive or deficient water supply. *L. intermedius* also seemed to be less immediately responsive to drought than *E. trispicata*. As data were not collected continuously throughout 1984, comparison of mean biomass in that year with that in 1985 is limited to the period April to December. Mean above-ground biomass for these months in 1984 was 387.6 g m^{-2} (±18.2), whilst the corresponding figure for 1985 was 323.6 g m^{-2} (±14.2) (Table 4.2). These means were found to be significantly different at $P > 0.05$.

In the burnt areas, regrowth of shoots started soon after the fires (Figure 4.5a). The rate of recovery varied between 1985 and 1986. In 1985, shoot biomass in the burnt area recovered to levels comparable to those in the unburnt area, it exceeded 250 g m^{-2} within 5 months, and exceeded 350 g m^{-2} for 4 months of the year. In 1986, the burnt area showed rapid growth for the first 3 months, but then declined. Biomass only just reached 250 g m^{-2} the following November. Although precipitation was below average in the first half of 1986, the latter half showed above average values (Figure 4.2c). Nevertheless, above-ground biomass failed to recover to the levels observed in previous years on unburnt areas. Above-ground biomass was reduced significantly in the area following the second fire in 1986 (Table 4.2).

Below-ground biomass

In the unburnt areas, below-ground biomass varied between approximately 200 and 600 g m^{-2} during 1984-85 (Figure 4.4a). No real pattern could be discerned from the monthly fluctuations. For the

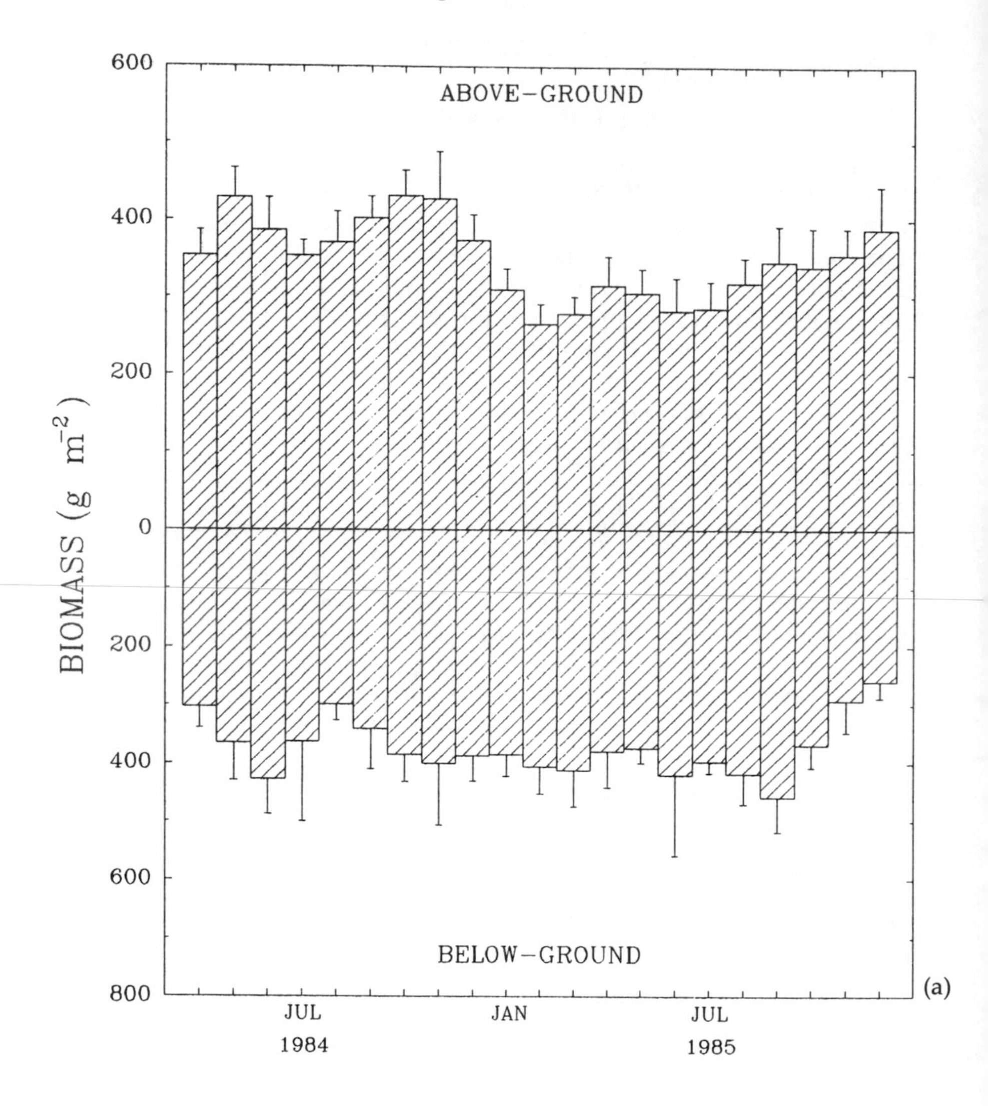

Figure 4.4. Monthly means (±1 se) from April 1984 to December 1985 of (a) above- and below-ground biomass (dry weight) for the areas of the study-site that were not burnt until January 1986; (b) above- and below-ground quantities of the dead vegetation for the same samples; (c) above- and below-ground relative rates of decomposition, i.e. weight loss per unit mass of dead vegetation ove a 1-month interval; and (d) total or absolute quantities of dead vegetation decomposed in each month, both above and below ground.

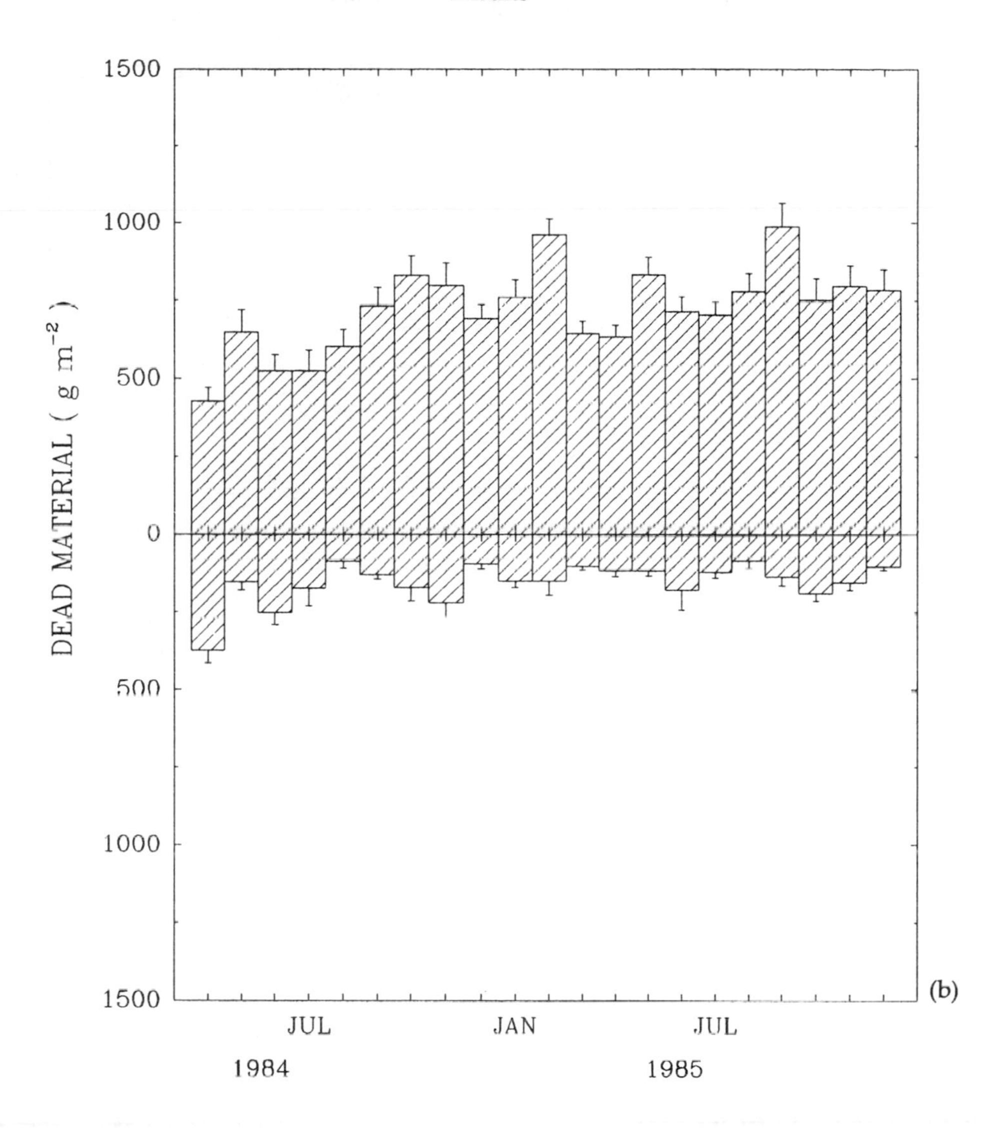

Figure 4.4. *continued*

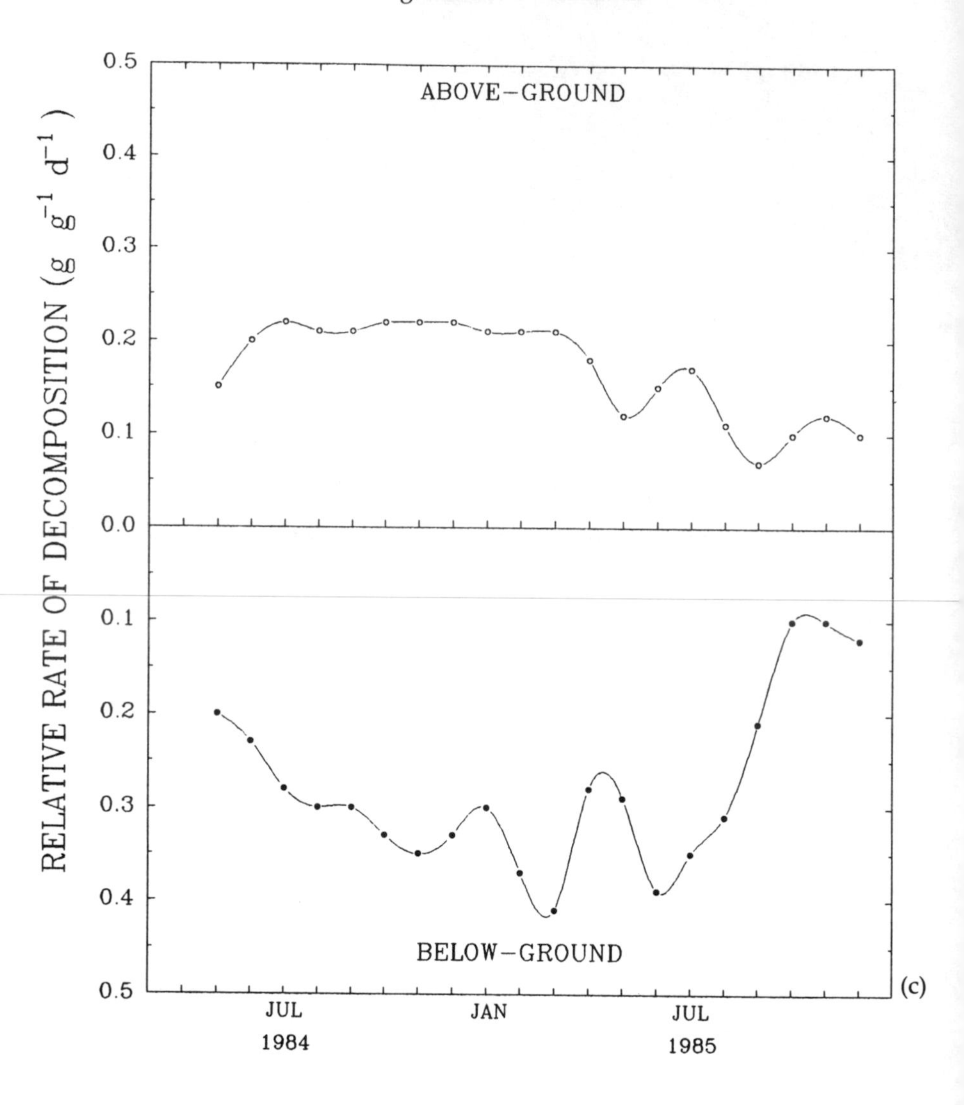

Figure 4.4. *continued*

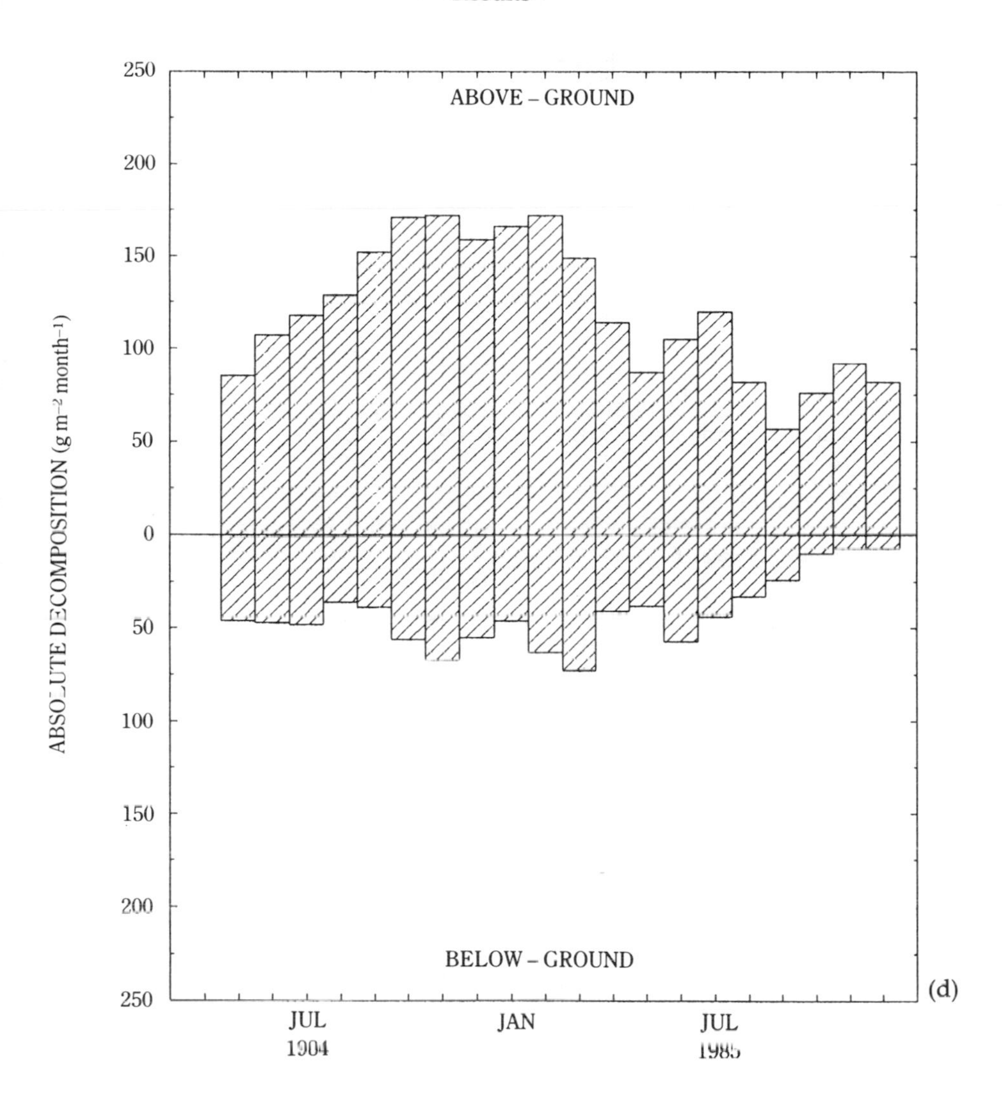

Figure 4.4. *continued*

Table 4.2. Comparison of average amounts of biomass and dead matter during the period April to December of 1984 and 1985

Condition	*Mean mass ($g\ m^{-2}$)*		*Significance*
	1984	*1985*	
Unburnt areas			
Above-ground biomass	387.61	323.55	*
Below-ground biomass	362.23	413.39	NS
Total	749.84	736.94	—
Above-ground dead matter	643.61	792.25	**
Below-ground dead matter	194.99	126.37	*
Total	838.60	918.62	
Burnt areas	1985	1986	
Above-ground biomass	319.48	181.91	**
Below-ground biomass	511.11	349.60	*
Total	830.59	531.51	
Above-ground dead matter	261.66	119.29	**
Below-ground dead matter	136.16	92.43	*
Total	397.82	211.72	

* $P > 0.05$; ** $P > 0.01$.
NS = not significant.

period April to December, the mean value was 362.3 g m^{-2} (±31.0) in 1984, rising to 413.4 g m^{-2} (±38.6) in 1985, but the means were not statistically different. Below-ground biomass as a proportion of total biomass rose from 48% in 1984 to 56% in 1985.

The fire in 1985 appeared to have little or no effect on below-ground biomass, data from burnt and unburnt plots remained remarkably similar (cf. Figures 4.4a and 4.5a). However, a small decline is indicated at the time of regrowth following the fire in 1985 and a much larger decline coinciding with regrowth in 1986 (Figure 4.5a). After the fire in 1986, however, below-ground biomass continued to decline until August, from nearly 600 g m^{-2} in March to only 240 g m^{-2} the following December. Only during the wet period of September to November of this year was any increase in below-ground biomass apparent.

Correlation of biomass with environmental factors

On a month to month basis, little correlation was found between biomass levels in the unburnt plots in 1984 and 1985 and environmen-

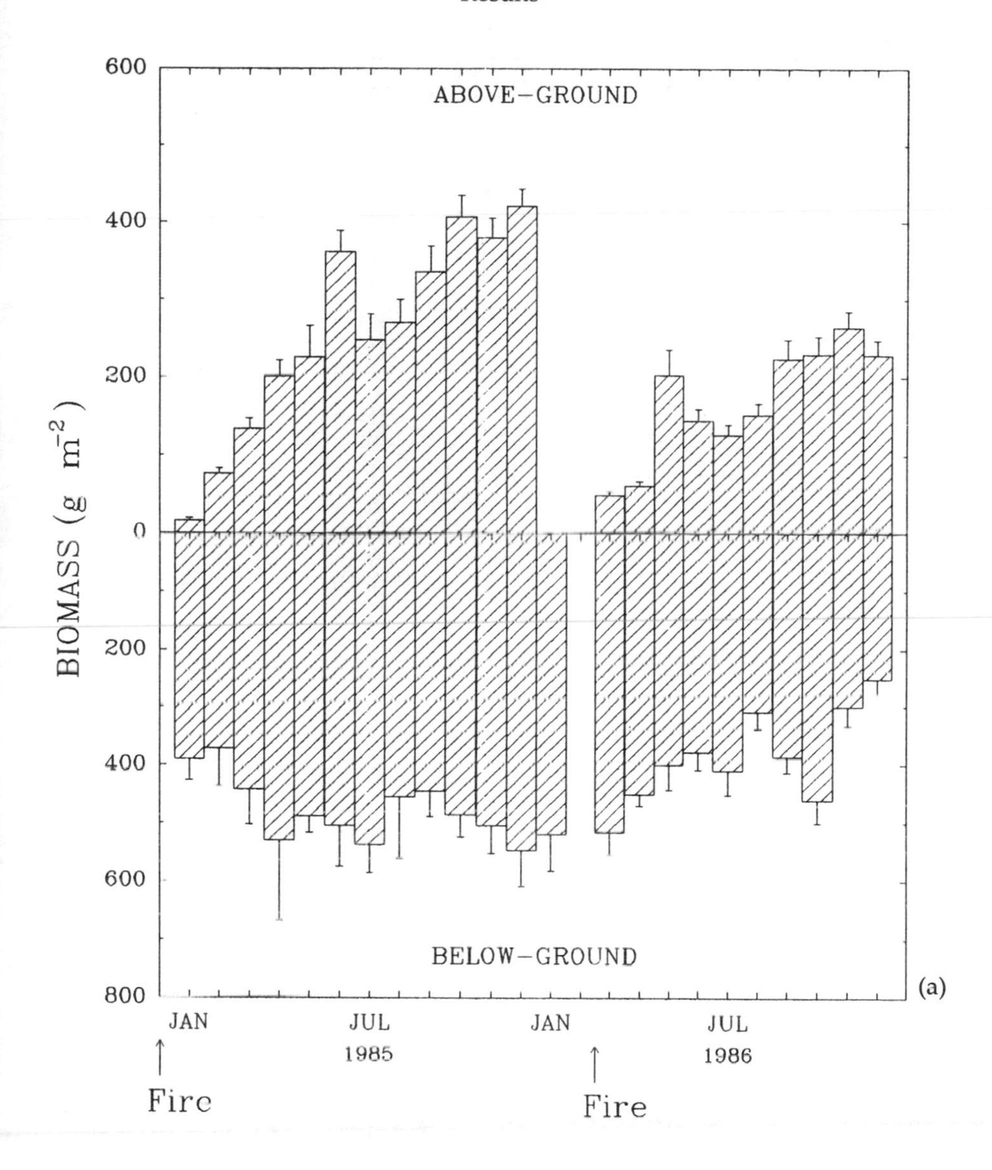

Figure 4.5. Monthly means (±1 se) from January 1985 to December 1986 of (a) above- and below-ground biomass (dry weight) for the areas of the study-site that were burnt in 1985 and again in 1986, and (b) above- and below-ground quantities of the dead vegetation for the same samples.

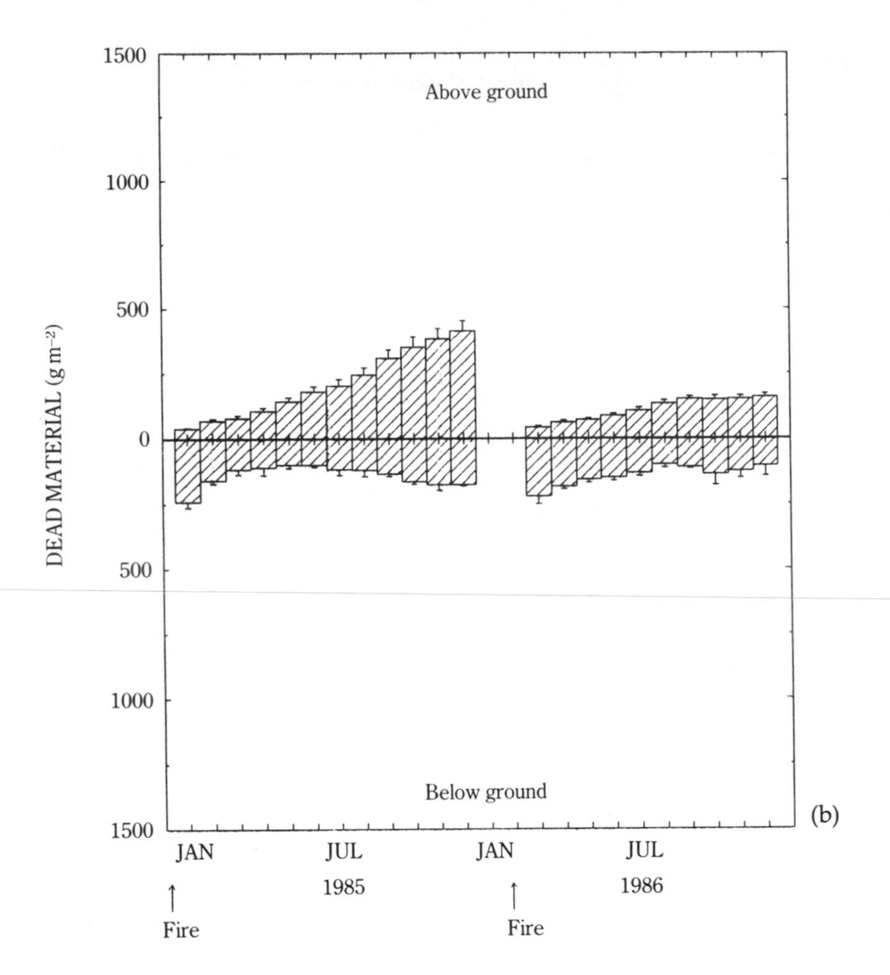

Figure 4.5. *continued*

tal factors such as rainfall, radiation and evaporation. Each year's data can, however, be organized on a seasonal basis, depending on wet or dry periods: first dry period (January to March); first wet period (April to May); second dry period (June to August); second wet period (September to December). When the biomass data for the unburnt plots were so arranged, very little correlation was found even so between underground biomass and the environmental factors referred to above (Table 4.3). However, a correlation was found between above-ground biomass and total rainfall for each season (r^2 = 0.27, df = 6; $P <$ 0.05), whilst there was a strong correlation between total biomass and seasonal rainfall (r^2 = 0.71, df = 6; $P <$ *0.001).*

4.4.2 Dead vegetation

A gradual accumulation of dead vegetation was observed above ground in both burnt and unburnt areas. In the latter case, dead material rose from about 430 g m^{-2} in April 1984 to around 800 g m^{-2} in December 1985 (Figure 4.4b). Averaged over the years values rose from 643.6 g m^{-2} (± 48.2) in 1984 to 792.3 g m^{-2} (± 37.4) in 1985, a highly significant difference at $P > 0.01$ (Table 4.2). Undoubtedly this accumulation was due to the absence of fire. In both burnt areas dead vegetation initially dropped almost to zero (Figure 4.5b). As biomass gradually increased in these areas, so did levels of dead material, although this increase was much slower in 1986 than in the previous year.

Levels of dead vegetation were much lower below-ground, averaging around 160 g m^{-2} (Figure 4.4b). The average of the monthly values was significantly greater in 1984 (195 ± 30 g m^{-2}) than in 1985 (126 ± 12 g m^{-2}) (*t*-test, $P > 0.05$). Following the first fire in 1985 a steady decline in the amount of dead vegetation below-ground is apparent, possibly reflecting an inhibition of turnover of below-ground organs, whilst the canopy is being resynthesized. A similar pattern of decline, again with about 6–8 consecutive months of decline, is apparent following the second fire in 1986 (Figure 4.5b). Comparison of Figures 4.4b and 4.5b shows the overall effect of fire and repeated fire in progressively decreasing quantities of dead vegetation, both above and indirectly below ground.

4.4.3 Decomposition

Throughout 1984 and the early part of 1985, relative rates of decomposition of dead vegetation above ground remained remarkably constant (Figure 4.4c). The above-ground rate was approximately 0.2 g g^{-1} $month^{-1}$ and that below ground around 0.3 g g^{-1} $month^{-1}$. During this period, the litter bags used had a very small mesh size and it would seem that this may have affected the internal micro-climate surrounding the decomposing material. When bags with a larger mesh size were introduced in 1985, relative decomposition rates fluctuated. Above ground, the rate varied from about 0.1 to 0.2 g g^{-1} $month^{-1}$. Below ground, even greater variation was observed, from around 0.1 to nearly 0.4 g g^{-1} $month^{-1}$. The highest rates in 1985 followed the period of heavy rains in April and May, both above- and below-ground. Rates then declined for most of the remainder of the year. Throughout, a more rapid rate of turnover of the below-ground dead material is indicated. Figure 4.4d shows the total predicted quantities

Table 4.3. Correlation between the rainfall on a seasonal basis and the corresponding biomass and net primary production (in normal unburnt condition)

Seasonal period	*Rainfall (mm)*	*Above-ground biomass ($g\ m^{-2}$)*	*Above and below ground ($g\ m^{-2}$)*	*Total AP_n*	*P_n*
Wet 1 (1984)	328.95	378.30	687.65	424.65	406.44
Dry 2 (1984)	279.50	355.20	713.93	218.67	46.13
Wet 2 (1984)	469.00	416.57	807.86	770.49	1 121.09
Dry 1 (1985)	240.60	271.00	679.95	316.98	557.33
Wet 1 (1985)	339.20	324.47	669.74	414.32	350.62
Dry 2 (1985)	245.50	289.97	696.99	299.23	486.95
Wet 2 (1985)	636.80	348.29	800.54	415.74	411.51

of dead vegetation lost though decomposition in each month. Despite lower relative decomposition rates, amounts lost above ground were considerably greater, reflecting the very much greater quantities of dead vegetation above ground (Figure 4.4b). On average, above ground, *ca* 100 g m^{-2} $month^{-1}$ of litter was decomposed, compared to *ca* 40 g m^{-2} $month^{-1}$ below ground.

4.4.4 Net primary production

In the unburnt areas, above-ground net primary production (AP_n) was positive in all months from May 1984 to December 1985 with a minimum of 20 g m^{-2} in October 1985 and a maximum of 290 g m^{-2} in September 1985 (Figure 4.6a). Over the first 12 months of the study (April 1984 to April 1985), net production above and below ground was 2036 $\pm$ 109 g m^{-2} y^{-1}. Below-ground organs contributed 23% of this total. The shoot material turned over on average once every 2.7 months, compared to a mean turnover time of 7.7 months for below-ground plant organs. The sum of monthly AP_n values for the period May to December 1984 was 1438 g m^{-2} compared to 892 g m^{-2} for the same period in 1985. Below-ground net primary production (BP_n) for these areas was negative during 5 months, presumably reflecting translocation to the above-ground tissues. Values ranged from −70 g m^{-2} in November 1985 to 115 g m^{-2} in October 1984 (Figure 4.6a). The sums of monthly BP_n values for May to December were 159.9 g m^{-2} in 1984 and 31.5 g m^{-2} in 1985. Total P_n (above and below ground) can be seen to be generally higher in most months of 1984, by comparison to the equivalent months in 1985, with 3 months with a P_n of over

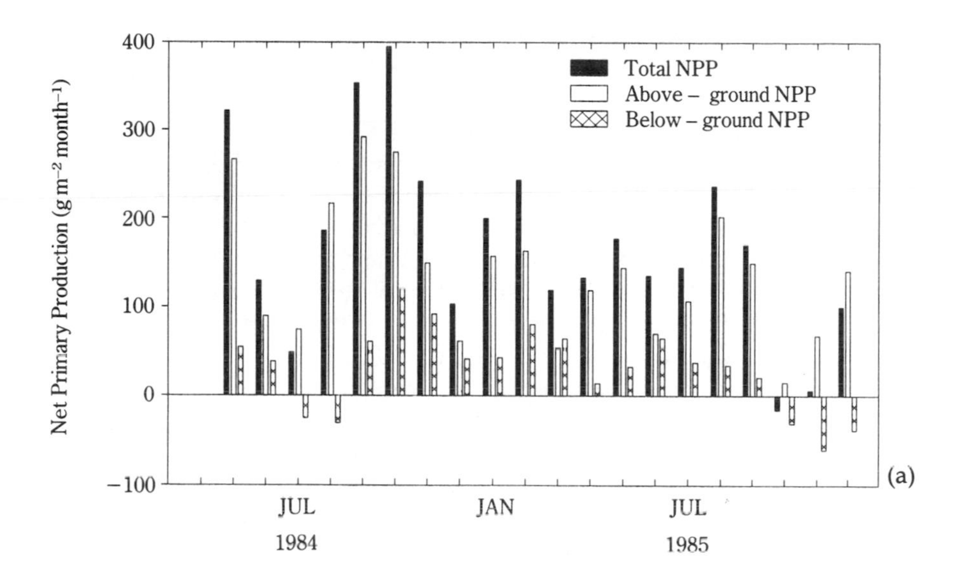

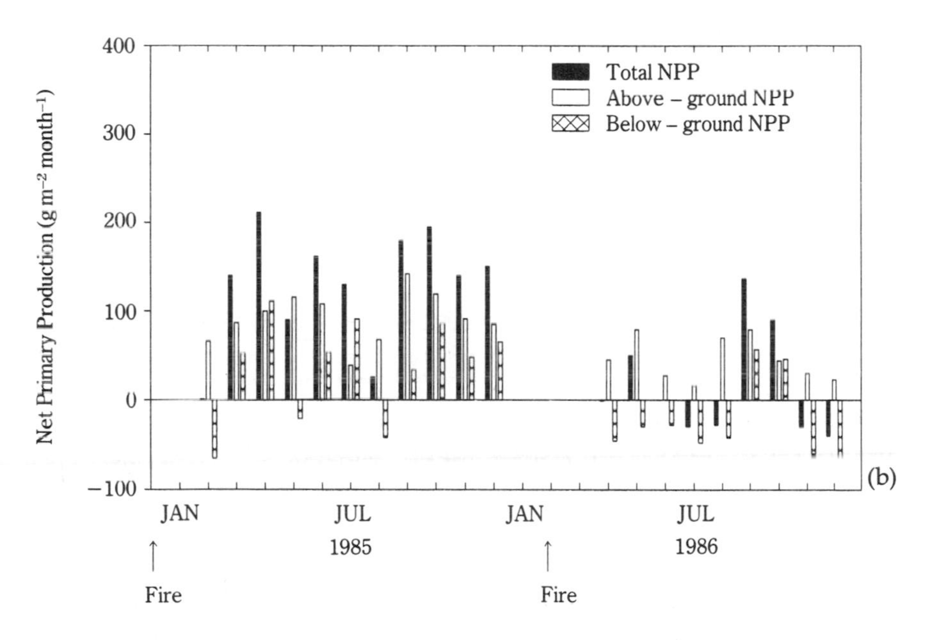

Figure 4.6. Net primary production (above and below ground, and total) (a) from April 1984 to December 1985 for sites which were not burnt until January 1986; and (b) from January 1985 to December 1986 for sites burnt in both 1985 and 1986.

300 g m^{-2} in 1984, whilst in 1985 only 2 months show a P_n above 200 g m^{-2} and none approach 300 g m^{-2} (Figure 4.6a). The total P_n for 1985 was 1676.5 ±135 g m^{-2}.

After both fires, BP_n is seen to be strongly negative (Figure 4.6b). However, whilst BP_n is negative for only the first month following the first fire in 1985, it remained negative for the first 5 months following the fire in 1986. Total P_n was also negative in 4 of the 9 months following the second fire. Total AP_n for May to December 1985 was 756 g m^{-2}, lower than the 896 g m^{-2} for the unburnt area. However, total BP_n for the May to December 1985 period was 185.9 g m^{-2}, greater than in the unburnt areas. The total P_n in 1985 was 1524 ±121 g m^{-2}, only slightly less than in the unburnt area.

By comparison to the burnt and unburnt areas in 1985, P_n showed substantially lower values following the second fire. Even though precipitation in the last 3 months of 1986 exceeded totals for the equivalent period in 1984 and 1985, there was no recovery of P_n. Above-ground net primary production for the May to December period in 1986 was only 406 g m^{-2} and BP_n was negative, −229.5 g m^{-2}, indicating sustained translocation of material from roots to shoots during this drier year, rather than root death which would have been associated with an increase in dead material below ground. The total P_n for 1986 on the site that had been burnt twice was only 134 ±65 g m^{-2}.

4.4.5. Canopy structure and conversion efficiency

Figure 4.7a shows leaf area indices (L) on the unburnt sites during 1984 and 1985. The leaf area index of the community exceeded 1.2 throughout 1984, with a peak of 2 in December. The subsequent decline during January to March 1985 coincided with three consecutive months in which evaporation substantially exceeded precipitation (Figure 4.2c). Leaf area index throughout 1985 remained lower than in 1984 with a mean of *ca* 1. On average, over 60% of L was accounted for by the two grasses *E. trispicata* and *L. intermedius*. In 1984, *L. intermedius* constituted the larger portion of L in all months. However, the relative proportions of *E. trispicata* varied throughout the year. In May it constituted less than 4% of L, but over 20% in December. Generally, *E. trispicata* showed a progressive rise through the course of both years, suggesting a seasonal shift in the relative importances of these two C_4 grasses (Figure 4.7a). The same pattern was apparent after the first fire, although by the middle of the year the L of *E. trispicata* was approximately equal to that of *L. intermedius* (Figure 4.7b).

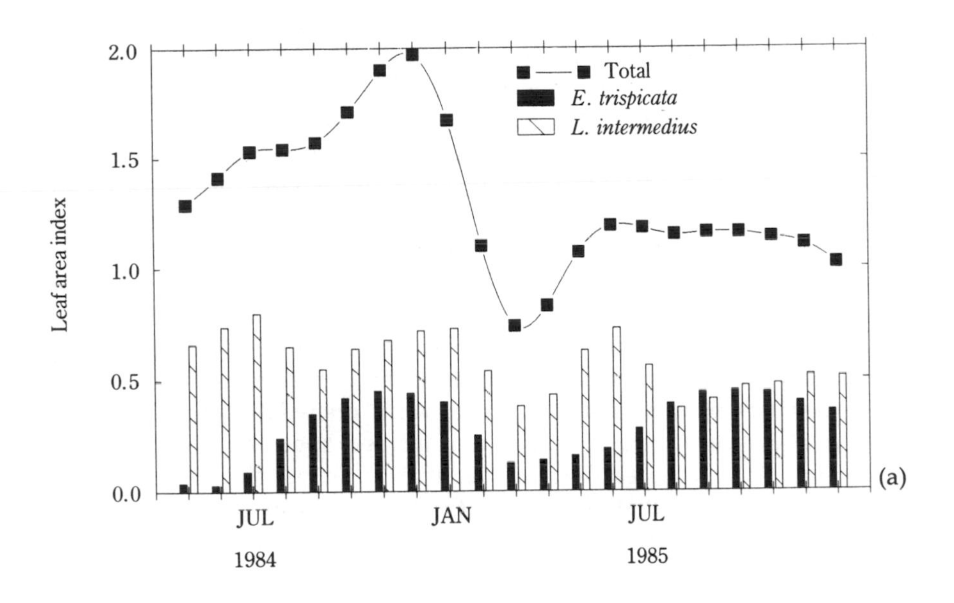

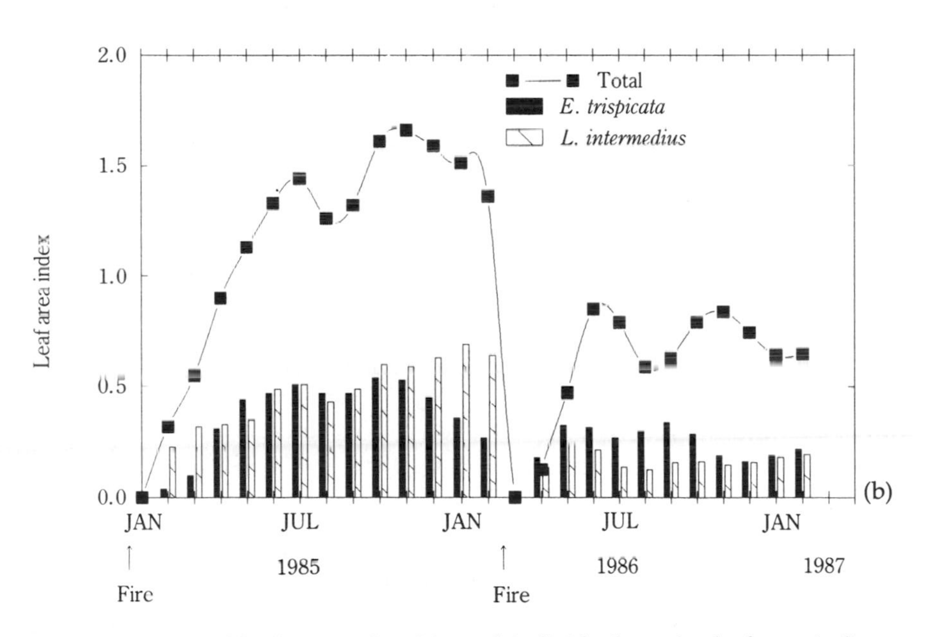

Figure 4.7. Total leaf area index (L), and individual species leaf area indices, (a) from April 1984 to December 1985 for sites which were not burnt until January 1986; and (b) from January 1985 to December 1986 for sites burnt in both 1985 and 1986.

Leaf area index showed a steady recovery following the fire, rising over six consecutive months to 1.5 in July, compared to 1.2 in the unburnt plots (Figure 4.6b). Nevertheless the L in the first 2 months was substantially decreased relative to the unburnt plots and this no doubt accounts for the relatively lower rates of production (Figure 4.6b) and radiation interception (Figure 4.2b). Following the second fire, L again showed an initial rapid rise, to 0.85, but then showed no further increase, such that L was substantially decreased relative to previous years. On this occasion though, *E. trispicata* was the dominant contributor to the combined L for all grasses within the community, whilst non-grass species constituted over 50% of the total L. This pattern suggests that the combination of repeated burning and decreased precipitation may have not only decreased below-ground biomass and total production, but also altered species composition of the community expressed on a leaf area basis. Prior to the fire in 1985, efficiency of solar radiation interception by the canopy was *ca* 90% (Figure 4.2b). As expected the proportion of radiation intercepted dropped markedly following both fires. In 1985, 8 months were required before efficiency appeared to return to 90%. However, calculated interception values were undoubtedly overestimates. Litter continuously accumulated on the solarimeters at the canopy base, and although these were cleared at monthly intervals, would have led to overestimation of interception. Standing litter was also interspersed with live vegetation through much of the canopy, and thus it was not possible to separate interception by live vegetation from that of standing dead, except in the months following the fires when there were no dead shoots.

When cumulative P_n is plotted against the accumulated quantities of solar radiation intercepted by the canopy, approximate linear relationships are apparent (Figure 4.8). For the period September 1984 to November 1985 on the unburnt plots the apparent efficiency of conversion of intercepted radiation into production of new biomass was 0.66 g MJ^{-1}, compared to 0.61 g MJ^{-1} from Jan.–Nov. 1985 on the burnt plots. However, from March 1986 to November 1986, on the sites which had been burnt twice, efficiency was just 0.08 g MJ^{-1} (Figure 4.8).

4.5 DISCUSSION

The variability of biomass and net primary production between years and between unburnt and burnt sites reflects the dominant influence of physical factors on this ecosystem. An important regulator is the

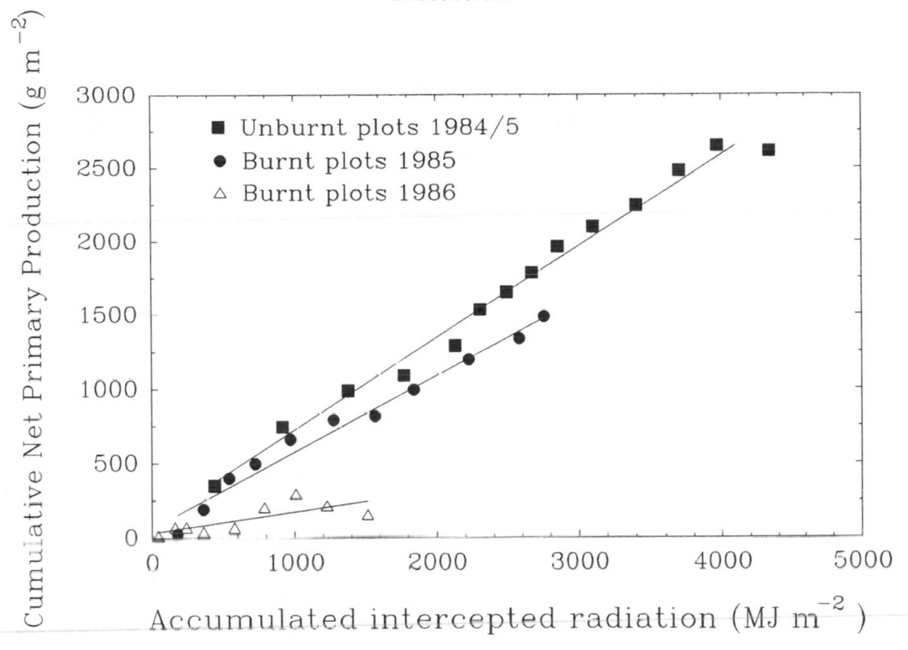

Figure 4.8. Cumulative net primary production (above and below ground) plotted against the accumulated quantity of solar radiation intercepted by the canopy for three periods/groups of samples. Straight lines indicate the best-fitting linear regressions fitted by the least-squares method to the illustrated data points: from September 1984 to November 1985 on the unburnt plots; from January 1985 to November 1985 on the site burnt at the beginning of 1985; from March 1986 to November 1986 on the site burnt for a second time at the beginning of 1986.

amount of rainfall received throughout the year. Comparison of seasonal rainfall revealed a similar distribution pattern for 1984 and 1985, whereas 1986 was much drier during the first three quarters of the year (Table 4.1). On a seasonal basis, total biomass correlated well with the amount of rainfall received (Table 4.3). The fact that similar correlations could not be found between P_n and rainfall indicates the more complex interplay of factors affecting productivity, and the longer term responses of below-ground processes to the aerial environment. The lower levels of biomass attained following the second fire compared to those attained following the first might seem attributable to the lower rainfall of 1986. However, rainfall was higher in the final quarter of 1986, than in either 1984 or 1985 (Table 4.1), yet recovery was still poor. This suggests that the lower production in this year might also result from repeated burning. Possibly, recovery from the first fire depleted reserves such that there was a decreased capacity to recover from a further fire.

Although annual rainfall during the study was lower than the 10-year average, biomass levels within the unburnt area were within the range for Indian sub-humid tropical grasslands quoted by Singh and Joshi (1979). Mean above-ground biomass in the burnt area in 1985 was only slightly less than in the unburnt area. In 1986, however, levels of above-ground biomass in the burnt area were much reduced. The fire appeared to have little effect on below-ground biomass in the burnt area in 1985, but after the second fire, below-ground biomass gradually decreased through the year (Figure 4.5a). It appears that under the drier conditions of 1986, the shoots were unable to recover without sustained translocation from the roots. Recovery of this grassland after fire is therefore greatly dependent on moisture availability, and perhaps frequency of burning. In the unburnt area, above-ground dead material peaked towards the end of the dry season/start of the rainy season. Live shoots rapidly senesce at the end of their life span and this process will be accelerated under dry conditions. Minimal levels of above-ground dead material tended to occur in March or April, following the driest month, February. This is the starting point for the accumulation of dead material for the rest of the year. Over the whole study, and in all areas, dead material accumulated above ground. The community may therefore be seen as a 'fire climax', with occasional fires removing dead material and allowing regeneration of new shoots. However, the results from 1986 suggest that frequent fires or the coincidence of decreased rainfall and fire will not only remove dead vegetation above ground, but will indirectly decrease the rate of input of dead material to the soil, so decreasing below-ground dead material. It was noticeable that after burning, particularly when coupled with drought stress, the rate of recovery differed between the different species categories. *E. trispicata* quickly recovered its former dominance. This species has several attributes which enhance its competitiveness under such conditions. For example, its metaxylem vessels are very large, facilitating rapid water transport. Furthermore, transpiration may be reduced by the thick cuticle of the leaves and by the ability of the leaves to roll. All of these factors may contribute to the higher water use efficiency observed for this species (Kiettilakhagul, 1987), possibly explaining its predominance in 1986. *L. intermedius* was strongly affected by the second fire. Despite the fact that, after the first fire, above-ground biomass fell to zero, P_n was only slightly lower than in the unburnt area. Leaf area index actually increased after the first fire, showing higher values than in the unburnt areas. This may reflect a release of nutrients from the dead vegetation or a smothering effect of dead vegetation within the canopy on live leaves at lower canopy levels.

After the second fire in 1986, however both L and P_n were substantially reduced.

A strong relationship between accumulated P_n and accumulated intercepted solar radiation was apparent (Figure 4.8). The efficiency of conversion of intercepted radiation into new biomass of 0.66 g MJ^{-1} was well below the maximum of 2.3 g MJ^{-1}, recorded for other C_4 species (see Chapter 5). This might be explained, at least in part, by the overestimation of interception due to dead vegetation within the canopy. However, this cannot explain the very low conversion efficiency of 0.08 MJ m^{-2} recorded in 1986. Here, the second fire removed all dead vegetation above ground, and thus interception, at least in the first 6 months, would not have been overestimated. The lower efficiency here might be explained by the large respiratory costs involved in resynthesizing the canopy and mobilizing reserves from the below-ground organs.

The highest total P_n occurred in the unburnt area in the first 12 months of the study and was equivalent to 20.4 t ha^{-1}, figures for 1985 are also high and equivalent to 16.8 t ha^{-1}. These exceed the 10 t ha^{-1} suggested as an average for tropical grasslands by Murphy (1975). This difference is more likely methodological than environmental, since most previous studies have not taken mortality between harvests into account. When this is taken into account, a P_n of 16 t ha^{-1} seems low in comparison to the results of Gupta and Singh (1982) and Menaut and Cesar (1979). Both of these studies took some account of mortality and estimated P_n as more than 30 t ha^{-1} in Indian and African tropical grasslands, respectively. The estimates of the present study do compare favourably with some estimates for production of agricultural crops in Thailand, though. Yabuki *et al.* (1985) calculated the annual production of a paddy field to be 12 t ha^{-1}. However, the same authors estimated the annual production of a sugar cane field to be 34 t ha^{-1}. It is important to remember, however, that this grassland is situated on poor soil and receives none of the fertilizer input of agricultural land. With this is mind, wet savanna may still be considered to be a very productive system.

The pattern of biomass change with time on unburnt plots emphasizes the problems of estimating production from biomass alone. Figure 4.4a shows that total plant biomass of this community shows relatively little variation from month to month, reflecting humid grassland conditions, i.e. conditions conducive to growth for most of the year. This absence of marked seasonal variations in biomass seriously limits any attempts to estimate production from biomass changes, since growth of new biomass and death of old occurs simultaneously in this community. The results also show how widely estimates of production can vary from year to year on the same site. Over 2000 g m^{-2} of production occurred over the first 12 months of the study compared to less than 200 g m^{-2} over the last 12 months, i.e. production over 12-month periods on the same site may vary 10-fold.

4.6 CONCLUSIONS

This study has shown that moist savanna can attain a relatively high production. Both production and levels of biomass are affected by the amount of available water and may be greatly reduced when fire and drought conditions combine. One of the main species in this grassland, *E. trispicata,* was able to recover effectively after fire or drought stress; however, its co-dominant *L. intermedius* showed a significant decline in its relative contribution to the production of the community.

4.7 SUMMARY

1. Production of plant biomass was determined for a semi-natural ungrazed grassland dominated by *Eulalia trispicata* and *Lophopogon intermedius* in southern Thailand. This area has a monsoonal climate and a brief 'dry' season in which soil water reserves become depleted. Monthly changes in biomass and dead vegetation were determined by cutting above-ground vegetation from randomly selected quadrats and by removing below-ground material in soil cores. Production was determined as the change in biomass between harvests corrected for losses through mortality.

2. In the absence of fire, total dry weights of plant biomass varied relatively little between months over a 2-year period, from 640 g m^{-2} to a peak of 810 g m^{-2} in the monsoon period. Leaf area index ranged from 0.6 in the 'dry' season to 2 in the monsoon. Between 38 and 45% of the biomass was below ground. Amounts of dead vegetation above ground increased more-or-less continuously in the absence of fire, from 450 g m^{-2} in April 1984 to 630 g m^{-2} by December 1985. Only *ca* 20% of the total quantity of dead vegetation was below ground on unburnt sites. Decomposition rates below ground averaged 0.30 g g^{-1} $month^{-1}$, compared to less than 0.20 g g^{-1} $month^{-1}$ above ground.

3. Total biomass production was determined at 2036 ± 109 g m^{-2} y^{-1} over the first 12 months of the study, below-ground organs contributing 23% of this total. The period April to December 1984 accounted for

1777 g m^{-2} of this production. The shoot material turned over on average once every 2.7 months, compared to a mean turnover time of 7.7 months for below-ground plant organs. Net production was positive throughout the year with the highest productivities in the 3-month period preceding the monsoon. The lowest productivity was in the 'dry' season, when there was substantial death of shoot material, but substantial input into the below-ground organs. Production in 1985 was substantially lower, the total for the 12 months being 1677 g m^{-2} y^{-1}. This decrease relative to 1984 coincided with a *ca* 35% decrease in mean leaf area index.
4. A fire at the beginning of 1985 removed all above-ground live and dead vegetation from a substantial proportion of the study area and allowed comparison of production on burnt and unburnt areas. The vegetation recovered rapidly and after 6 months the leaf area index on the burnt site was significantly higher than on the unburnt areas. After an initial period of negative net primary production, total production for the year was only 10% less than on the unburnt site. A second fire at the beginning of 1986 was followed by a drought. During this year, net primary production was decreased by over 80% and leaf area index remained below 1. The second fire was followed by substantial decreases in below-ground biomass and dead vegetation. Substantial changes in species contributions to total biomass occurred following this fire, including a decreased contribution by the grass species.

REFERENCES

Bourlière, F. and Hadley, M. (1970) The ecology of tropical savannas. *Annual Review of Ecological Systematics*, **1**, 125–52.

Caldwell, M.M. (1975) Primary production of grazing lands, in *Photosynthesis and Productivity in Different Environments*, IBP Vol. 3 (ed. J.P. Cooper), Cambridge University Press, Cambridge, pp. 41–74.

Dobby, E.H.G. (1973) *Southeast Asia*, University of London Press, London.

Evenson, J.P. (1983) Climate of the Songkla Basin. *Songklanakarin Journal of Science and Technology*, **5**, 175–7.

Gilliland, H.B. (1971) *A Revised Flora of Malaya*, Vol. 3: *Grasses of Malaya*, Government Printing Office, Singapore.

Gupta, S.R. and Singh, J.S. (1982) Influence of floristic composition on the net primary production and dry matter turnover in a tropical grassland. *Australian Journal of Ecology*, **7**, 363–74.

Kiettilakhagul, N. (1987) *Studies on Adaptive Responses to Water Stress of Plant Species in a Natural Ecosystems*. M.Sc. Thesis, Prince of Songkla University.

Menaut, J.C. and Cesar, J. (1979) Structure and primary productivity of Lamto savannas, Ivory Coast. *Ecology*, **60**, 1197–210.

Misra, K.C. (1979) Introduction to tropical grasslands, in *Grassland Ecosystems of the World: Analysis of Grasslands and their Uses*, IBP Vol 18 (ed. R.T. Coupland), Cambridge University Press, Cambridge, pp. 189–95.

Murphy, P.G. (1975) Net primary production in tropical terrestrial ecosystems, in *Primary Productivity of the Biosphere* (eds H. Lieth and R.H. Whittaker), Springer-Verlag, Berlin. pp. 217–31.

Myers, N. (1980) *Conversion of Tropical Moist Forests*. A report for the Committee of Research Activities in Tropical Biology of Natural Resources, National Academy of Sciences, Washington.

Singh, J.S. and Joshi, M.C. (1979) Tropical grasslands: primary production, in *Grassland Ecosystems of the World: Analysis of Grasslands and their Uses*, IBP Vol 18 (ed. R.T. Coupland), Cambridge University Press, Cambridge, pp. 197–218.

Smitinand, T. and Larsen, K. (1972) *Flora of Thailand*, ASRCT Press, Bangkok.

Whitmore, T.C. (1984) *Tropical Rain Forests of the Far East*, 2nd edn, Clarendon Press, Oxford.

Yabuki K., Aoki, M., Kiyota, M., Chunkao, K. and Stienswat, W. (1985) Primary production of natural and cultivated plants in Thailand, in *Proceedings of the International Seminar on Environmental Factors in Agriculture Production*, Prince of Songkhla University, Hat Yai, Songkla, Thailand.

5

A floodplain grassland of the central Amazon

M.T.F. PIEDADE, W.J. JUNK, and J.A.N. DE MELLO

5.1 INTRODUCTION

The Amazon basin covers about 7 million km^2 north and south of the equator. Precipitation in the central part varies from 2000 to 3000 mm a year. On the slopes of the Andes, maximum precipitation can exceed 5000 mm per year, declining to less than 1500 mm annually on the basin's northern and southern borders. Rains are not evenly distributed throughout the year, resulting in more or less pronounced 'dry ' and 'rainy' seasons. Consequently, river discharge varies strongly.

The large annual amplitude between high and low water of 10 m or more in the Amazon River and its main tributaries results in the periodic flooding of about 300 000 km^2 of Amazonian lowlands (Junk, 1991). Within the tropics, river floodplains are considered to be particularly productive regions (Lieth and Whittaker, 1975).

Productivity is, however, related to the nutritional status of water and sediments. According to the geochemistry of their catchment areas, Amazonian rivers vary in water quality and sediment load. Sioli (1967) and Junk and Furch (1985) differentiate between white-water, black-water and clear-water rivers. White-water rivers drain the Andes and their foothills. They transport large amounts of suspended solids, giving the water a greyish colour. The sediment load consists of fine sand and differing amounts of montmorillonite, illite, kaolinite and, in some tributaries chlorite. Accordingly, the content of dissolved minerals in the water is relatively high, the pH varying between 6 and 7. White-water floodplain soils are very fertile.

Black-water rivers, such as the Rio Negro, have dark, translucent waters due to the large amounts of dissolved humic substances, leached from giant spodosols. They are highly acid, with pH between 4 and 5, and contain very low amounts of dissolved inorganic material. Fertility of both the water and the floodplain soils is low.

Clear-water rivers drain the archaic shields of the Guianas and central Brazil. Their water is translucent with a greenish colour, and the sediment load is small. Amounts of dissolved inorganics, levels of fertility and pH values vary, but are normally intermediate between black- and white-water rivers.

Two types of floodplains are distinguished according to their plant and animal communities and their productivity: the Varzea of the white-water rivers and the Igapo of the black- and clear-water rivers (Sioli, 1951, 1964; Irmler, 1977; Prance, 1980; Junk and Furch,1985). Covering about 200 000 km^2 of the total floodplain area (Junk, 1991), the Varzea is associated with high productivity, since neither water nor nutrients are likely to be limiting.

An important biological driving force in floodplains is the annual flood pulse (Junk *et al.*, 1989). It can vary in frequency, amplitude and timing. However, a predictable, prolonged and monomodal flood pulse is typical of large tropical rivers, and particularly so of the Amazon. The biota colonizing the floodplain must adapt to the change between a terrestrial and an aquatic phase, with its attendant physiological stresses. Plant species have become adapted to specific habitats as a result of the flood regime. They form a gradient from the higher permanently terrestrial positions to lower permanently aquatic conditions. The stable habitats at the top of this gradient are colonized by floodplain forest. Those at the other extreme support annual and perennial herbs at times of low water. At high water, a luxuriant community of aquatic and semi-aquatic vegetation develops. Some of the most frequent free-floating species are *Eichhornia crassipes*, *Pistia stratiotes* and *Salvinia* spp. Several aquatic and semi-aquatic grasses occupy large areas, for example *Hymenachne amplexicaulis*, *Oryza perennis*, and the C_4 species *Paspalum fasciculatum*, *Paspalum repens* and *Echinochloa polystachya* (Hedges *et al.*, 1986). Together with *P. repens*, *E. polystachya* constitutes about 90% of the 'floating meadows' which drift down the Amazon (Junk, 1970). Species occupy certain habitats according to their particular ecological requirements and relative competitive ability. Waves and strong currents can damage and even destroy these aquatic communities.

Studies on the primary production of the Varzea indicate that phytoplankton production is about 6 t ha^{-1} y^{-1} (Schmidt, 1973). Total production of the forest is estimated to be about 33 t ha^{-1} y^{-1} (Junk, 1985). Production of terrestrial and aquatic herbaceous plants has been determined to be between 10 and 45 t ha^{-1} y^{-1}, depending on species, community and length of growing season (Junk and Howard-Williams, 1984; Junk, 1985). It has been suggested that the productivity of *E. polystachya* approaches 50 t ha^{-1} y^{-1} (Junk, 1986). These herbaceous plants therefore play an important role in the carbon cycle of the area.

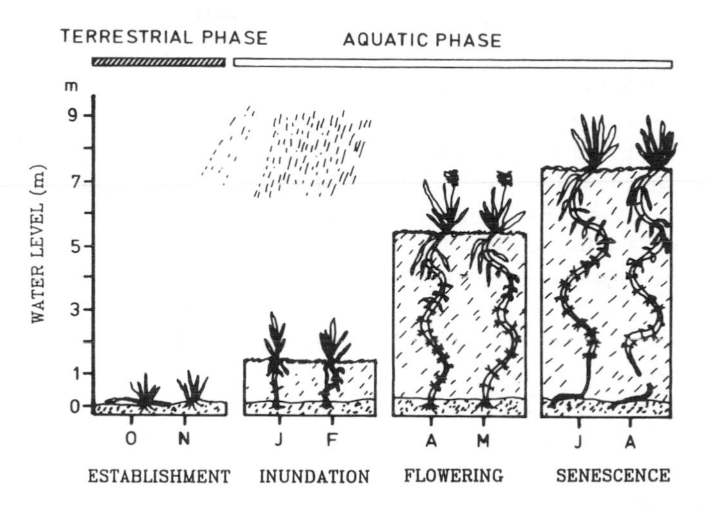

Figure 5.1. An illustration of the annual growth cycle of *Echinochloa polystachya* in relation to the annual cycles of water level on the floodplains of the central Amazon.

5.1.1 Growth pattern of *Echinochloa polystachya*

E. polystachya is a perennial, tall C_4 grass that occurs from Mexico to Argentina (Hitchcock, 1936). In the central Amazon basin it is most frequent on shores and mudflats of river channels and floodplain lakes with a high nutrient status. There it forms large, monospecific stands. In areas of low fertility, its growth becomes inhibited and it is absent in the black-water floodplains.

The plant's life cycle is regulated by the annual cycle of water levels and can be divided into a terrestrial and an aquatic phase (Figure 5.1). The terrestrial phase starts when the water level drops to expose the sediments during October and November in the Manaus region. During the dry phase, new shoots are formed at the nodes of old stems as the latter die. Shoots in contact with the sediment vigorously set root and form individual, unbranched stems. The large amount of decaying material from the preceding generation covers the area, inhibiting the growth of competing species. Thus, each year during low water, a single cohort of new plants is produced with little overlap between the previous year's plants. The young plants start to grow vertically, typically reaching 1–2 m in height by the start of the aquatic phase.

Flowering commences during the aquatic phase and extends from March to September. A large amount of seed is formed, but vegetative propagation is more frequent (Junk, 1970). Seedlings are mainly important for the formation of new colonies, or in the regeneration of stands which have been destroyed by strong currents or fire.

After flowering, as the water column starts to fall in July, individuals continue to grow but at a reduced rate. Often the basal portion of the plant starts to rot. With the lower water level, the upper parts of the stems are exposed to air and adventitious roots become desiccated. On exposure of the sediments again in October/November a new generation starts vegetatively from the dying, old stems. In years when the water level does not drop sufficiently to expose the sediment, the old stems may persist for another flood period, although new stems may still be formed from their nodes.

The species provides a food source for capybaras (*Hydrocherus hydrochoeris*), manatees (*Trichechus inunguis*), some turtles and herbivorous fish. The leaves are consumed by some terrestrial invertebrates, while the roots provide a substrate for aquatic invertebrates and a habitat for young fish (Junk, 1973). In the central Amazon region, the plant is also grazed by cattle and buffalo (*Bubalus bubalis*) (Ohly, 1987).

5.1.2 Objectives

Although some previous estimates have been made of the productivity of *E. polystachya* they have not taken into account the seasonal turnover of the canopy nor quantified the seasonal cycle of production. This study combines regular destructive measurements of stem and leaf biomass with monthly determinations of plant density to calculate total net primary production (P_n) of an *E. polystachya* stand in the Amazon. In addition, leaf area development and canopy radiation interception were used to assess the stand's efficiency of light conversion into dry matter. The photosynthetic response of *E. polystachya* to changes in water level over an annual cycle was also investigated in the field.

5.2 STUDY SITE

5.2.1 Location

The study site is located at Ilha de Marchantaria, Manaus, Amazonas state at latitude 3°15′S and 60°00′W. It is the first island upstream of the confluence of the Amazon and the Rio Negro, standing 30 m above sea level (Figure 5.2a). The island lies almost at the centre of the

Amazon basin, some 20 km south-west of the city of Manaus, 200 km south of equator and about 1500 km from the mouth of the river. According to Radam/Brasil (1972), Ilha de Marchantaria has an area of 32 km^2. However, during the dry period (i.e. the terrestrial phase), the island's area can double. At low water, about one-third of the island is occupied by floodplain forest. The remainder is periodically covered by lakes and terrestrial, semi-aquatic and aquatic herbs. There is no quantitative information about the total island area colonized by *E. polystachya*. It is, however one of the most common species, forming large, monospecific stands. At a conservative estimate, it covers at least 5–6 km^2.

Fieldwork was conducted from July 1985 at Lago Camaleão (Figure 5.2b), one of the largest lakes of the Ilha de Marchantaria. The lake basin is about 300–500 m wide and 7 km long. The depth varies depending upon the water level of the Amazon. All measurements were taken within a representative study area measuring 100 m by 35 m, marked out with lines attached to tethered floats.

5.2.2 Climate and hydrology

The climate at Manaus has been described in detail by Ribeiro and Adis (1984). The mean monthly temperature ranges from 22.9 to 23.8°C, with a maximum between 30.2 and 33.2°C. Monthly evapotranspiration is estimated at 38.5–85.9 mm, wind speed at 3 m above ground level between 2.3 and 2.6 m s^{-1} and mean monthly insolation ranges from 105.3 to 248.5 h. The relative humidity of the air at 1 m above the ground lies between 75.6 and 86.7%. Mean annual precipitation is 2100 mm with about 75% of this falling within the rainy season from December to May. However, large differences can occur between years in the local pattern of precipitation.

During the study period, precipitation (Figure 5.3c) and sunshine hours were recorded at the climatic station of INPA at Ilha do Careiro, about 14 km from the study site (Figure 5.2a). Precipitation was concentrated within the period December 1985 to March 1986, with the highest rainfall in February (373.3 mm). The driest period was in July and August 1986. In 1985, October had the least rain with only 57.2 mm. Sunshine hours were least in December in both years, 52.6 h in 1985 and 45.9 h in 1986. The months with the most sunshine were September 1985 (231.3 h) and August 1986 (301.5 h). Solar radiation receipts per month varied from 250 MJ m^{-2} in February 1986 to 580 MJ m^{-2} in August 1986, these also being the wettest and driest months, respectively, of the period of study. Relative humidity and temperature were measured during the study period at Reserva Ducke

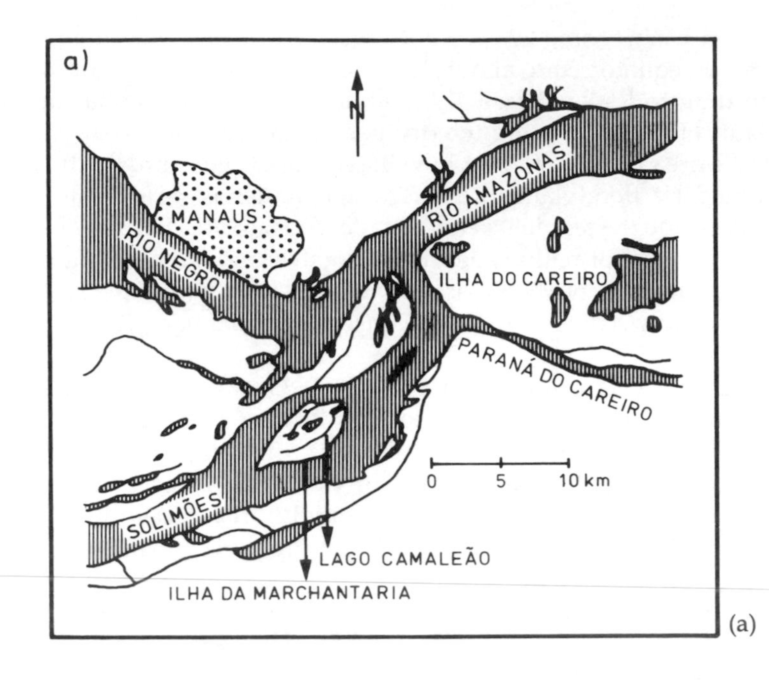

(a)

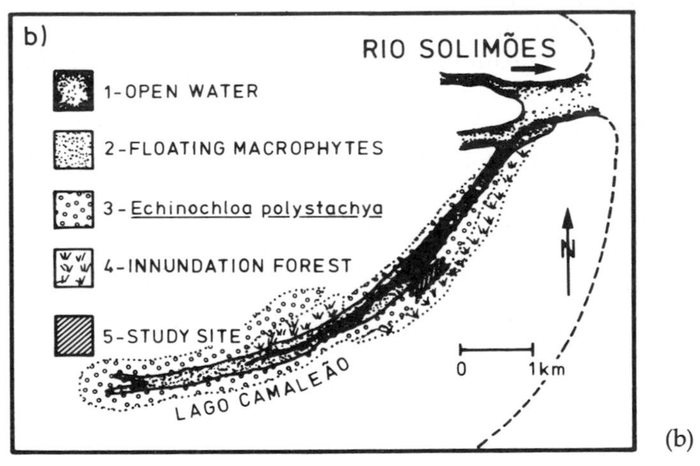

(b)

Figure 5.2. (a) Location of the island, Ilha da Marchantaria, within the Manaus area of the Central Amazon (03°15′S, 60°00′W) (b) detail of the lake, Laqgo Camaleã. The location of the study site within the *Echinochloa polystachya* community on the southern side of the lake is indicated.

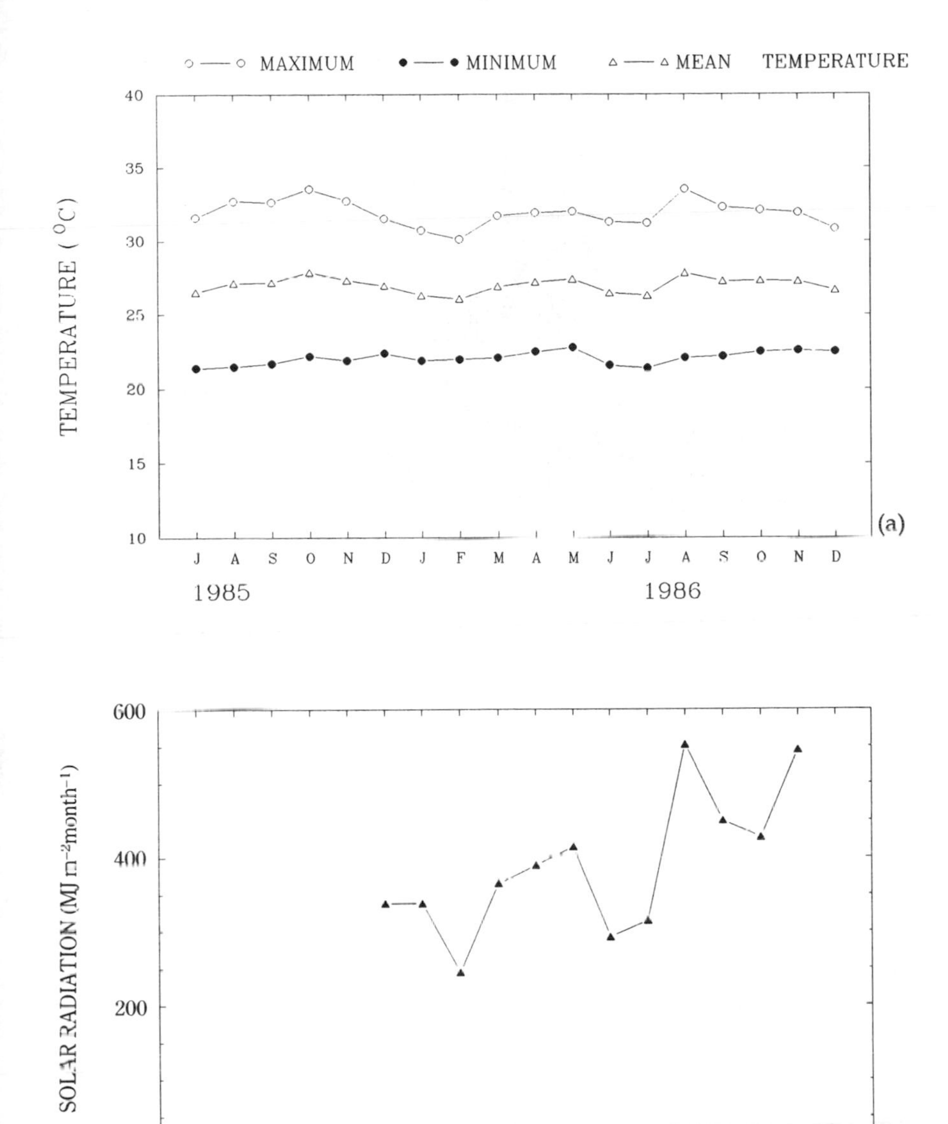

Figure 5.3. Mean monthly weather records at the study site for the period July 1985 to December 1986. (a) Average monthly maximum, mean and minimum air temperatures recorded at Reserve Ducke, on the northern outskirts of Manaus; (b) incident solar radiation recorded at the study site.

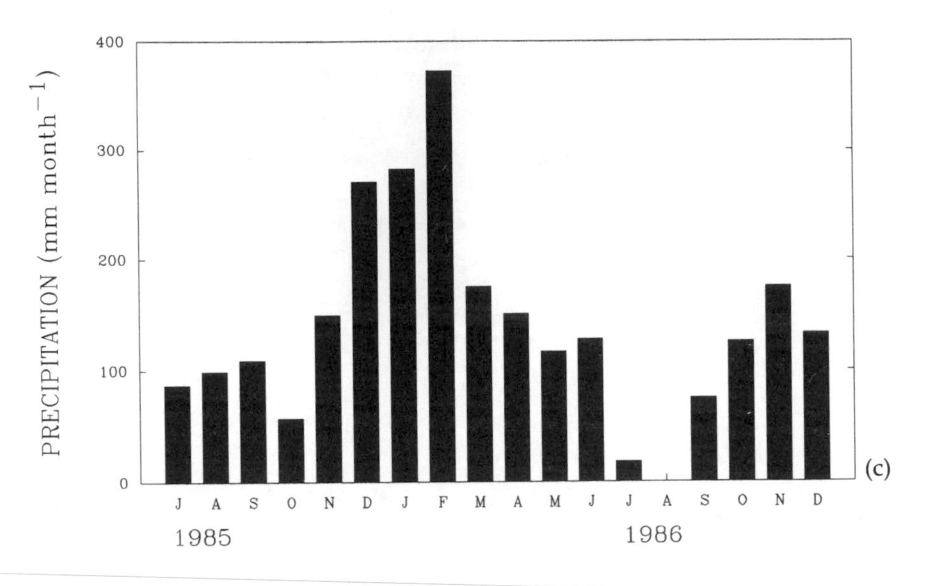

Figure 5.3. *continued* (c) Monthly precipitation recorded at Ilha do Careiro, *ca* 15 km north-west of the study site.

(road AM-O10), a Biological Reserve, approximately 40 km from the study site. Relative humidity was high throughout with a maximum of 86.8% in February 1986 and a minimum of 73.8% in August 1986. Temperature was quite stable (Figure 5.3a), with the mean maximum monthly temperature ranging from 30.1°C in February 1986 to 33.5°C the following August. The mean minimum monthly temperature varied even less, from 21.4°C in July 1985 to 22.8°C in May 1986. Mean temperature varied by only 1.8°C, from 26.1°C in February 1986 to 27.9°C in October 1985.

The hydrograph of the Amazon River is quite regular, exhibiting a monomodal, sinusoidal flood curve. The 80 years mean annual oscillation is about 10 m, with a maximum amplitude of 16 m noted since records were kept (from 1902). At very low water the lake dries out, except for some small shallow pools. During very high floods, water depth may reach almost 8 m (Soares *et al.*, 1986). The water level in the rivers and lakes connected with the Amazon at all sites examined, within 100 km of Manaus, corresponds within a few centimetres to the water level regularly recorded in the harbour at Manaus (Schmidt, 1973). Therefore the water-level records of the harbour at Manaus were used to describe the hydrology of Lago Camaleão.

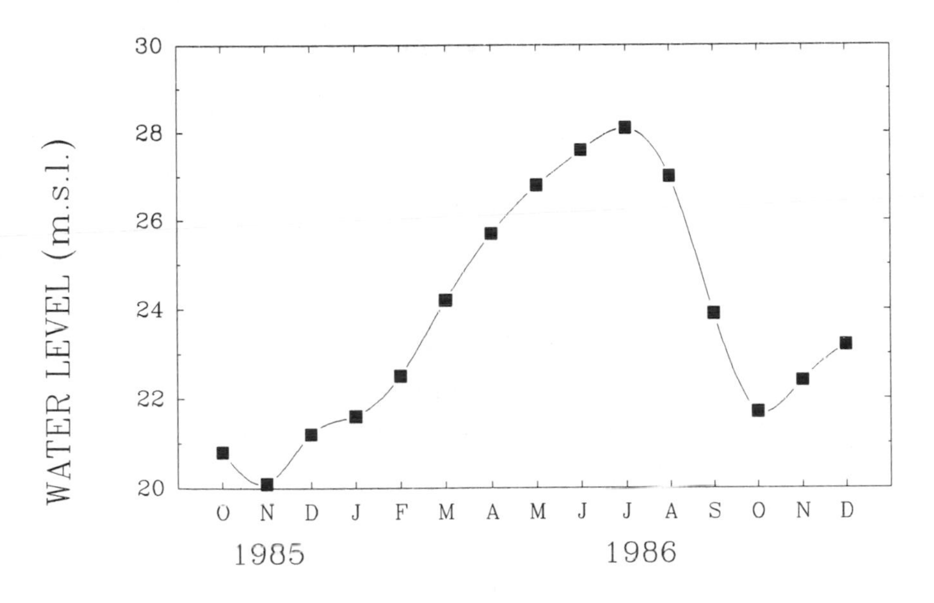

Figure 5.4. Changes in water level during the study period, recorded at Manaus harbour on the Rio Negro. Water level is indicated in relation to the mean sea level (m.s.l.).

The maximum water levels at Lago Camaleão in 1985 and 1986 were 26.27 and 28.14 m above sea level, respectively (Figure 5.4). The corresponding minima were 19.74 and 21.40 m. The amplitude was therefore 6.53 m in 1985 and 6.74 m in 1986. The low water level was 1.66 m higher in 1986 than in 1985 and much of the *E. polystachya* stands remained inundated in the second year, including those of the study area.

5.2.3 Soil structure and hydrochemistry

The soils of Ilha de Marchantaria are alluvial in origin. Carbon (^{14}C) dating of sediments from the central part of the island indicates them to be *ca* 5000 years old (Irion and Junk, unpublished data). Large amounts of sediments are continually being deposited immediately downstream of the island, where there is appreciable accretion.

Hydrochemical conditions in Lago Camaleão vary with season. The highest concentrations of elements are found at low water, when total dissolved minerals may be 15 times higher than in the Amazon River. With rising water level, the lake becomes increasingly dilute, although it retains a higher concentration of electrolytes than the river (Furch *et al.*, 1983; Furch 1984a, b).

Phosphate levels are usually more than 10 $\mu g\ l^{-1}$, reaching nearly 300 $\mu g\ l^{-1}$ in the residual pools at low water. The concentration of nitrate is normally low (0.1–8.0 $\mu g\ l^{-1}$), which is in part due to the near-anoxic lake conditions. Maximum values of around 100 $\mu g\ l^{-1}$ are reached only during the highest water levels, when the river actually passes over the island. Thus nitrogen supply to the island ecosystem is highly dependent on the annual water cycle. Levels of ammonium ions vary between 15 and 100 $\mu g\ l^{-1}$ at rising and high water, increasing to nearly 3000 $\mu g\ l^{-1}$ in the low water pools.

The great variation in the chemical composition of the lake water is caused by its internal biogenic processes, which are in turn related to the large annual amounts of production and decomposition of herbaceous plants (Furch *et al.*, 1983; Furch,1984a).

5.3 METHODS

5.3.1 Biomass dynamics

Due to the depth of water quantitative harvesting in defined quadrats was not possible. Further, *E. polystachya* stands are monotypic and are made up of single unbranched large stems presenting the possibility of a demographic approach to production estimation. Thus, an alternative to the method outlined for terrestrial grassland sites was necessary and appropriate (cf. Chapter 1).

Individual shoots of *E. polystachya* were used as the unit for biomass determination. Fifteen plants were selected for destructive harvesting by a randomized block design. During the terrestrial phase, the plant root system was excavated; during the aquatic phase it became necessary to remove the roots from the sediment by uprooting the plants. Since the sediment was soft and semi-fluid this could be done with a minimum of damage to the roots. The numbers of internodes on each of the harvested plants were counted and the shoot was divided into the following categories: live stem, dead stem, live sheaths and lamina, dead sheaths and lamina, flowers, basal root and adventitious roots. Dead material was considered to be all necrotic (leaves) or rotting material (stems). All material was dried to constant weight at 95°C in a forced draught oven.

Plant density was determined each month in 20, 1 m^2 quadrats, positioned at random along a 100 m transect. The biomass of each category of material and total biomass were calculated by multiplying the mean values obtained from the 15 plants by the mean plant density.

5.3.2 Demography

In June 1985, 20 individual plants were randomly chosen and the youngest visible node above the water level marked with a coloured strip. At weekly intervals thereafter, the growth in length of each tagged plant was recorded, together with the number of new leaves and flowers. When strips or plants were lost new plants were selected to maintain the number under study at 20.

5.3.3 Calculation of net primary production

Monthly net primary production (P_n) was calculated from:

$$P_{ni} = N \,.\, (n_i \,.\, W_{a,i} + n_i \,.\, W_{b,i} + n_i \,.\, W_{c,i}) \qquad [1]$$

where: P_{ni} = net primary production during month i (g m^{-2} mo^{-1})
N = mean density of stems during month i
n_i = mean number of new internodes during month i
$W_{a,i}$ = mean biomass of new internodes at the end of month i (g m^{-2})
$W_{b,i}$ = mean biomass of new leaves and sheaths at the end of month i (g m^{-2})
$W_{c,i}$ = mean biomass of new roots at the end of month i (g m^{-2})

Annual P_n was calculated as the yearly sum of the monthly values. By combining data from the harvested plants with data from the individual growth measurements, it was possible to calculate P_n, even when the distal parts of the plants decomposed and total biomass declined (corrected values). Root losses by grazing and decomposition could not be evaluated under water. Total biomass was therefore calculated using the maximum values from individual measurements. Thus the P_n estimated for the roots is probably an underestimate (cf. Chapter 1).

5.3.4 Leaf area index

To determine leaf area index (L), 10 quadrats of 1 m^2 were selected at random along the 100 m transect each month and all the living leaves within them collected. Total leaf area was determined for each quadrat with a leaf area meter (type AM1, Delta-T Devices, Cambridge, UK) and the mean L calculated.

5.3.5 Light interception

Paired tube solarimeters (type TSL, Delta-T Devices) were sited above and below the canopy. The output from each solarimeter was recorded on a millivolt integrator (type MVl, Delta-T Devices) and read at weekly intervals. During the aquatic phase, the solarimeters were kept above the water on floats.

5.3.6 Photosynthetic characteristics

Leaf photosynthetic CO_2 assimilation and stomatal conductance were measured with a portable open gas exchange system. Air was supplied at a measured and controlled rate (ASU, ADC Ltd., Hoddesdon, UK) to a leaf cuvette (PLC, ADC Ltd) and the water vapour and CO_2 concentration changes across the cuvette were measured with a capacitance humidity sensor and an infra-red gas analyser, respectively (LCA, ADC Ltd). The diurnal responses of photosynthesis and stomatal conductance were determined by measurements at hourly intervals on 10 expanded leaves of the upper canopy. Leaves chosen were those accessible from a boat, in a representative area of the stand and measurements were made in the middle of the leaves. These measurements were repeated on a monthly basis. In addition, light response curves for the upper and lower canopy leaves were determined at photon flux densities ranging from 50 to 2000 $\mu mol\ m^{-2}\ s^{-1}$. Using natural sunlight, around the middle of the day, neutral density filters were placed above the chamber window to vary the light level at the leaf surface. These measurements were repeated at monthly intervals.

5.3.7 Chemical analysis

Plants for chemical analysis were collected from April 1986 to March 1987. A representative selection of leaves, floating roots and the upper, middle and lower parts of stems were collected. The material was dried at 95°C. Material was then transferred to the Max-Planck-Institut, Plön, Germany for analysis.

Immediately prior to analysis the material was ground and dried to constant weight. Ash contents were determined by ignition at 450°C. The ash was wetted with distilled water, dissolved in 30% HCl and evaporated to dryness before the procedure was repeated. The resulting ash was taken up in 1% HCl and made up to volume.

Sodium, potassium, calcium and magnesium were determined by atomic absorption spectrophotometry. Phosphorus was determined spectrophotometrically (vanadium molybdate method; Anon., 1971) and nitrogen by Kjeldahl digestion. Crude protein contents were estimated by multiplying the nitrogen content by 6.25 (Boyd, 1970).

5.4 RESULTS

5.4.1 Demography; leaf area index and radiation interception

Plant density increased from the establishment of the new population to a maximum of 30–35 stems m^{-2} throughout April to October 1986. The average number of leaves per plant rose from 6.5 in December 1985 to 9.9 the following August, then declined towards the end of the growing season (Table 5.1).

Stem extension rates were high throughout the study, but greater rates were observed at the start of the aquatic phase (1.17 m $month^{-1}$). After August 1986, stems rotted from the distal end. From individual extension rate data, total stem length was calculated to be about 15 m at the end of the growth period (corrected values, Table 5.1). Each month, about 7 new internodes were formed.

With rising water level an increasing proportion of the stem became submerged. This proportion accounted for about 90% of total biomass at high water. Only the youngest internodes remained above water level and leaves quickly decomposed when submerged. The mean leaf lifespan was 34 days, indicating a rapid turnover of photosynthetic tissue.

Leaf area index ranged from 2.3 in November 1985 to 4.75 the following August. Light interception by the canopy was high, varying between 72 and 89%. Leaf number and light interception reached maximum values in August, as the water level began to subside (Table 5.2).

5.4.2 Biomass and productivity

Following the formation of the new shoots in November 1985 and until September 1986, biomass increased steadily to a peak of 68.6 t ha^{-1}, or 80.1 t ha^{-1} if attached dead material is included. In the 9-month period from the December harvest until the September harvest, total dry weight increased by 70.0 ± 0.35 t ha^{-1} giving a mean daily growth rate of 25.9 ± 0.1 g m^{-2} d^{-1} (Figure 5.5). At the beginning of

Table 5.1. Demographic data (mean values; n = 15) during the annual growth cycle of *Echinochloa polystachya* at Ilha de Marchantaria (1985–86)

Month/ year	No. of plants/ m^{-2}	Mean stem length (m)	Mean stem length with correction* (m)	Accum. no. of internodes without correction	Accum. no. of internodes with correction*	Number of submersed internodes	% Internodes submersed	New internodes formed	No. of live leaves per plant	No. of plants with flowers/m^2 (n = 20 m^2)
D/85	25	1.5	1.5	10.0	10.0	1.0	10.0	n.d.	6.5	0
J/86	24	3.2	3.2	18.3	18.3	14.1	77.3	8.3	7.5	0
F/86	26	5.1	5.1	28.7	28.7	23.1	80.7	10.4	8.9	0
M/86	28	5.6	5.6	30.1	30.1	26.3	87.2	1.5	7.3	3
A/86	33	6.8	6.8	37.5	37.5	32.6	87.0	7.3	8.7	10
M/86	30	7.5	7.5	42.1	42.1	38.0	90.2	4.7	8.2	7
J/86	35	8.4	8.4	47.9	47.9	43.2	90.2	5.7	8.9	7
J/86	33	9.6	9.6	54.7	54.7	49.8	91.1	6.8	9.0	4
A/86	30	11.4	11.4	63.4	63.4	57.7	92.5	8.7	9.9	4
S/86	31	10.2	12.3	56.6	68.2	52.3	92.3	4.8	8.6	3
O/86	31	8.6	13.0	48.9	73.0	44.8	91.7	4.4	7.9	0
N/86	38	9.5	14.1	51.5	78.3	48.5	94.2	5.7	7.1	0
D/86	31	8.0	15.2	49.1	84.3	46.1	93.8	6.0	7.1	0

* After September 1986, the basal part of the plants decomposed and total biomass declined. At this moment, the corrected biomass was calculated by summing to the biomass of August, the biomass of the internodes formed per month by the individual plants (n = 20, weekly measured).

Table 5.2. Leaf area index, incident and intercepted radiation and percentage absorbed by the canopy of *Echinochloa polystachya* at Ilha de Marchantaria from November 1985 to December 1986

Month/ year	*Leaf area index* (n = *10*)	*Incident radiation* ($MJ\ m^{-2}$)	*Intercepted radiation* ($MJ\ m^{-2}$)	*Intercepted radiation* (%)
11/85	2.28 ± 0.128	613.3	507.0	82.7
12/85	2.05 ± 0.190	385.7	318.9	82.7
01/86	3.05 ± 0.171	408.7	337.6	82.6
02/86	2.51 ± 0.230	295.5	244.3	82.7
03/86	2.45 ± 0.134	440.3	364.0	82.7
04/86	2.56 ± 0.168	576.7	413.0	71.6
05/86	1.79 ± 0.127	570.0	413.4	72.5
06/86	3.17 ± 0.201	381.4	292.2	76.6
07/86	3.93 ± 0.336	434.5	313.9	72.2
08/86	4.75 ± 0.448	680.2	551.8	81.1
09/86	3.19 ± 0.431	508.3	448.9	88.3
10/86	2.35 ± 0.180	477.6	426.4	89.3
11/86	1.50 ± 0.116	611.0	544.4	89.0
12.86	1.50 ± 0.233	426.3	349.7	82.0

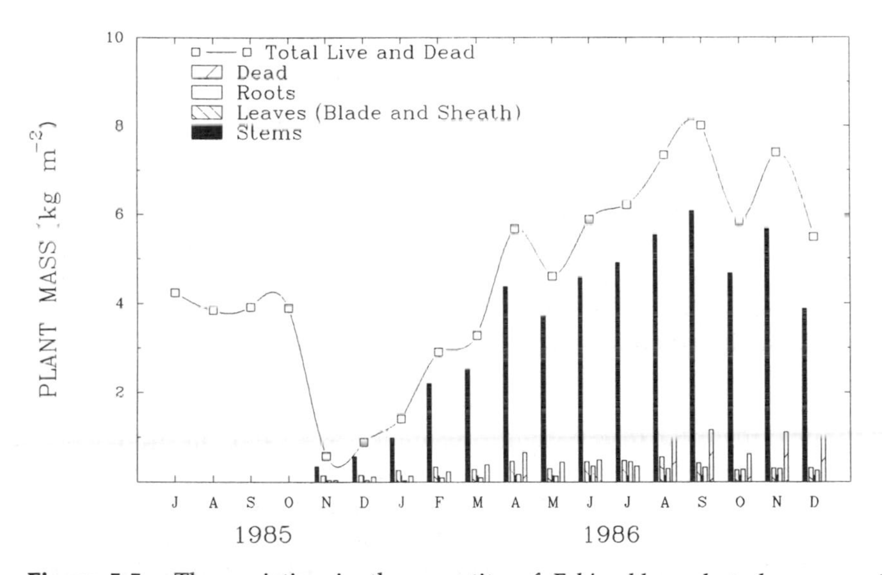

Figure 5.5. The variation in the quantity of *Echinochloa polystachya* per unit ground area on a dry weight basis. The continuous line indicates the total dry weight of both live (biomass) and dead vegetation. The vertical bars indicate the quantities of live stems, leaves (including sheaths) and roots, together with all categories of dead vegetation at Ilha da Marchantaria for each month from July 1985 to December 1986.

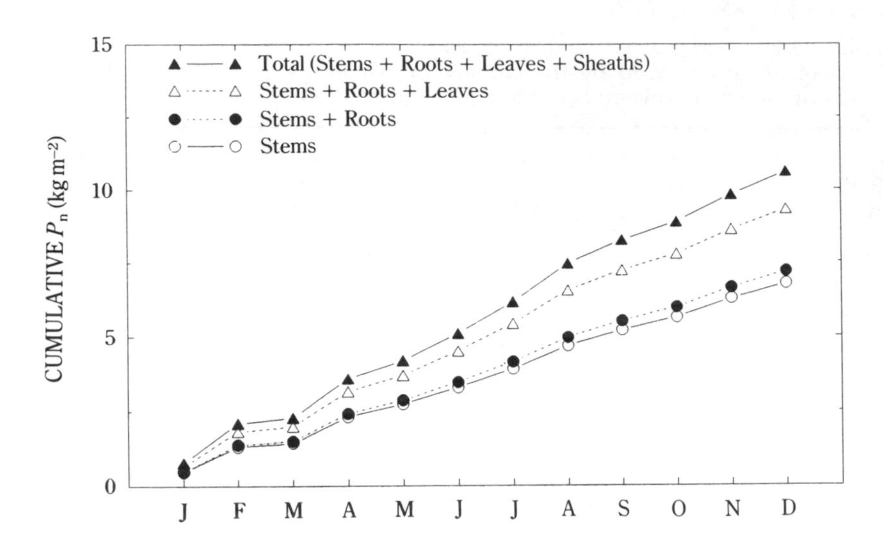

Figure 5.6. Accumulated monthly values of net primary production (P_n) for *Echinochloa polystachya* at Ilha da Marchantaria during 1986. The contributions of individual organs to total P_n are plotted in a cumulative manner, e.g. stems + roots, to illustrate the fractions of the total contributed by different plant parts.

the growth period, stems contributed around 60% of total biomass. This proportion increased to nearly 90% from September until the end of the growing season. Leaves and sheaths contributed from 3 to 20% to total biomass and roots from 2.3 to 7.8% (Figure 5.5).

Considerable differences were observed in biomass levels between 1985 and 1986. In 1985 the study area dried out completely and the old population of stems died and decayed during the dry phase. In November 1985 the total biomass was 5.8 t ha^{-1} and the population was composed entirely of new plants, indicated by the low number of internodes at this time. In 1986 the lowest water level was about 2 m higher than in the previous year and the study area remained under water. The old plants continued to grow, producing additional shoots from the nodes. Maximum biomass levels were accordingly higher at 68.8 t ha^{-1}.

Monthly net primary production was calculated using equation (1). Taking into account losses of decaying leaves and stems, annual P_n in

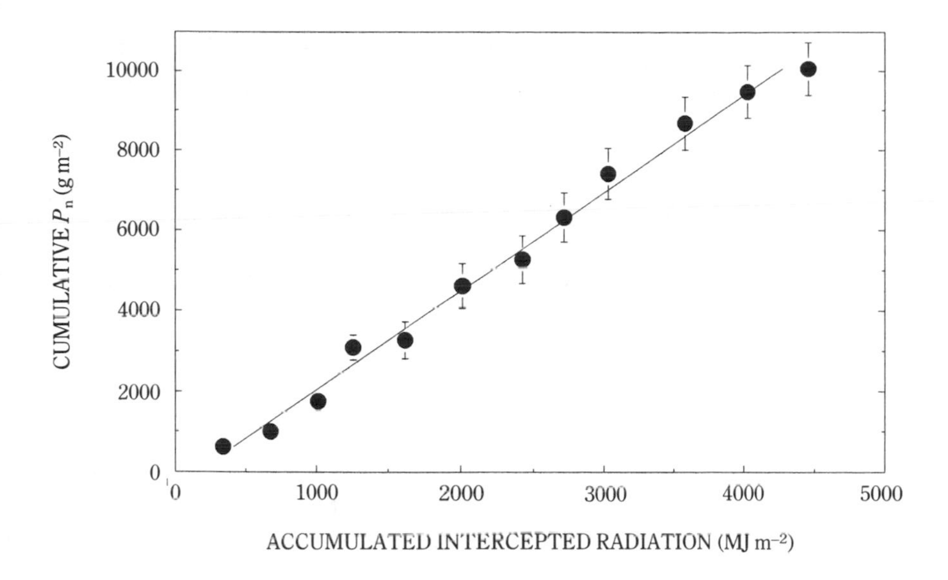

Figure 5.7. Net primary production (P_n ± 1 se) of *Echinochloa polystachya* accumulated month by month through 1986 in relation to the accumulated quantities of solar radiation intercepted by the canopy in each month. The best-fit straight line for the illustrated data points determined by the least squares method is illustrated. The slope of this line is 2.3 g MJ^{-1}.

1986 was 99.3 t ha^{-1} (Figure 5.6). Of this total, stems contributed 63.8%, leaves and sheaths 30.7% and roots 5.5%. There was a linear relationship between P_n and light interception by the canopy, indicating a constant conversion efficiency (ε_c) of 2.3 g of plant material per MJ of intercepted radiation (Figure 5.7).

5.4.3 Chemical analysis

Generally, N, P, and K contents on a dry weight basis of all plant parts were relatively high throughout the year. The phosphorus content of the leaves varied little and were 1500–2000 μg g^{-1} throughout the study period. Levels of nitrogen rose to a maximum of 25 000 μg g^{-1} in July 1986, but were not found to decline below 15 000 μg g^{-1} (1.5%) in any month. Potassium in the leaves was also high varying between 13 000 and 21 000 μg g^{-1} over the period of study. The ash contents of the leaves varied between *ca* 5 and 10%, whilst the stems showed a near constant ash content of *ca* 5%. The ash content of the roots was extremely high in June (*ca* 23%) and this is attributed to the inorganic

Table 5.3. Quantities of elements stored in a stand of *Echinochloa polystachya* with biomass of 70 t (dry weight) ha^{-1}, in comparison with the nutrient concentration of the Amazon River water (Furch, 1984a), considering a water column of 5 m depth

	Stems (*kg ha*$^{-1}$)	*Leaves* (*kg ha*$^{-1}$)	*Roots* (*kg ha*$^{-1}$)	*Total* (*kg ha*$^{-1}$)	*Water concent.* (*kg ha*$^{-1}$)	*Plant: water ratio*
Dry matter	62 000	4 600	3 400	70 000	—	—
N (total)	161.5	86.9	50.3	298.4	ND	—
$N(NO_2 + NO_3)$	ND	ND	ND	ND	7.7	38.70
P (total)	59.5	7.4	4.4	71.3	5.3	13.50
$P(PO_4)$	ND	ND	ND	ND	1.3	54.80
K	1159.4	64.6	25.4	1248.5	44	28.40
Ca	62.6	17.8	9.6	90.0	486	0.19
Mg	41.5	8.6	4.4	54.5	68	0.80
Na	43.4	0.4	2.9	46.7	150	0.31
Protein	1007.5	543.1	314.4	1865.0	—	—
Ash	3372.8	435.6	355.0	4163.4	—	—

ND = not determined.

sediments swept in by the high water. Of the various plant organs, the leaves possessed the highest levels of N, P, Ca, Mg and ash on average. Potassium was slightly enriched in the stems and sodium in the roots and stems. The greatest differences were observed in the nitrogen content, which was on average 8 times higher in the leaves than in the stems, and the sodium content, which was 9–10 times lower in the leaves than other plant organs (Table 5.3).

From the observed levels of nitrogen, protein content is highest in the leaves; nearly 12% on average. By contrast, the protein content of the stems is very low, only 1.6%. As the stems provide the overwhelming bulk of total biomass, the protein content of whole plants was correspondingly low, only 2.7% at high water.

5.4.4 Photosynthetic characteristics

The highest midday assimilation rates, under clear sky conditions, were observed in July 1986, during high water: 31 μmol m^{-2} s^{-1}. The lowest rates (13 μmol m^{-2} s^{-1}) were recorded in October/November 1985, the driest period. At that time the study area was completely dry and the *E. polystachya* stand was senescing. Although the sediment is

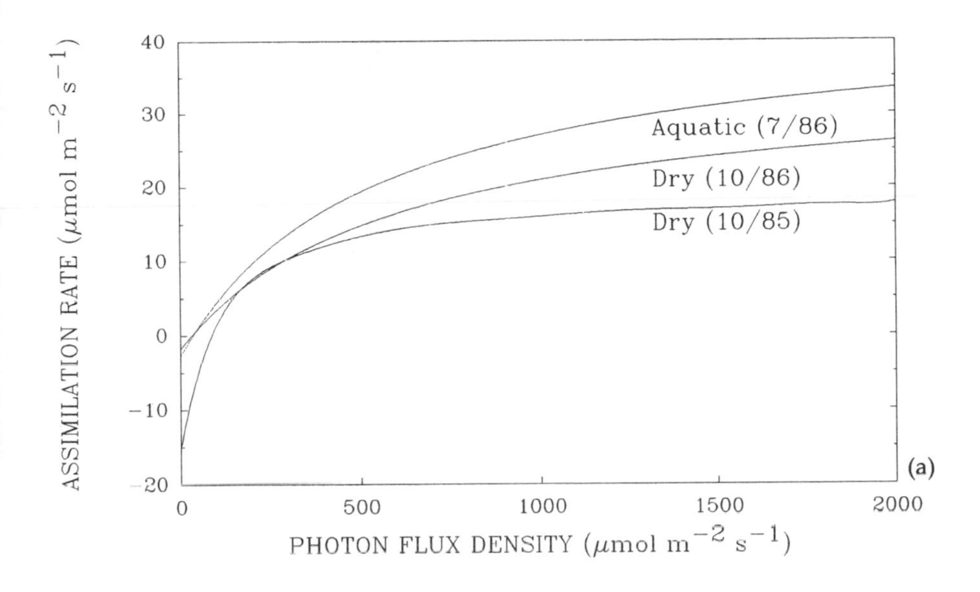

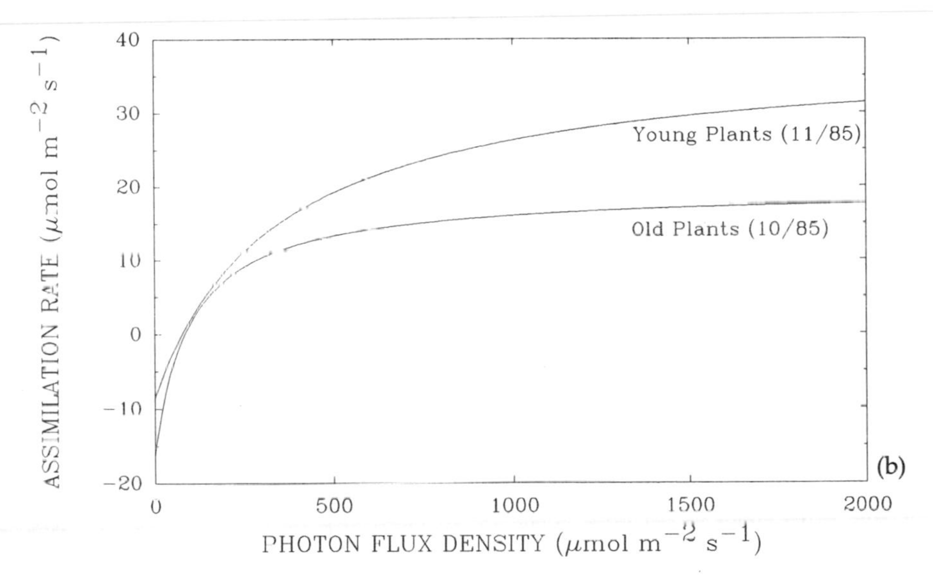

Figure 5.8. Fitted rectangular hyperbolas of the response of CO_2 uptake per unit leaf area to photon flux for leaves within the canopy of *Echinochloa polystachya* at the study site. The illustrated curves are fitted to >120 individual measurements of leaves selected at random within the upper canopy and made throughout the day for (a) and (b); here variation of photon flux is provided by normal diurnal changes. In (c) and (d) curves were fitted to CO_2 uptake rates of single leaves (10 replicates) measured around midday; here photon flux was varied by placing neutral density filters over the chamber window. (a) Photosynthetic rates during the aquatic phase (July 1986) and the terrestrial phase (October 1985 and October 1986); (b) photosynthetic rates for old (October 1985) and young plants (November 1985) during the terrestrial phase.

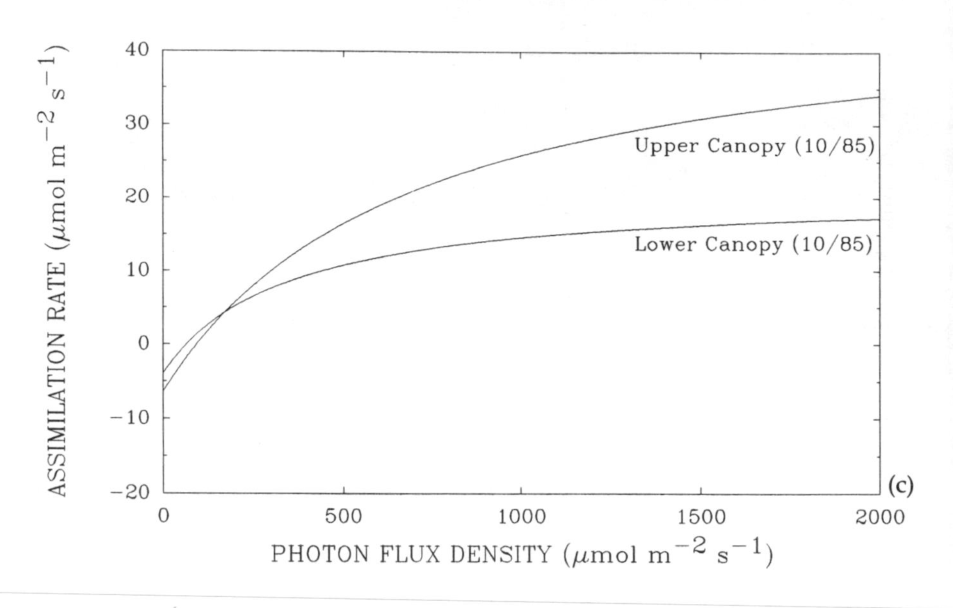

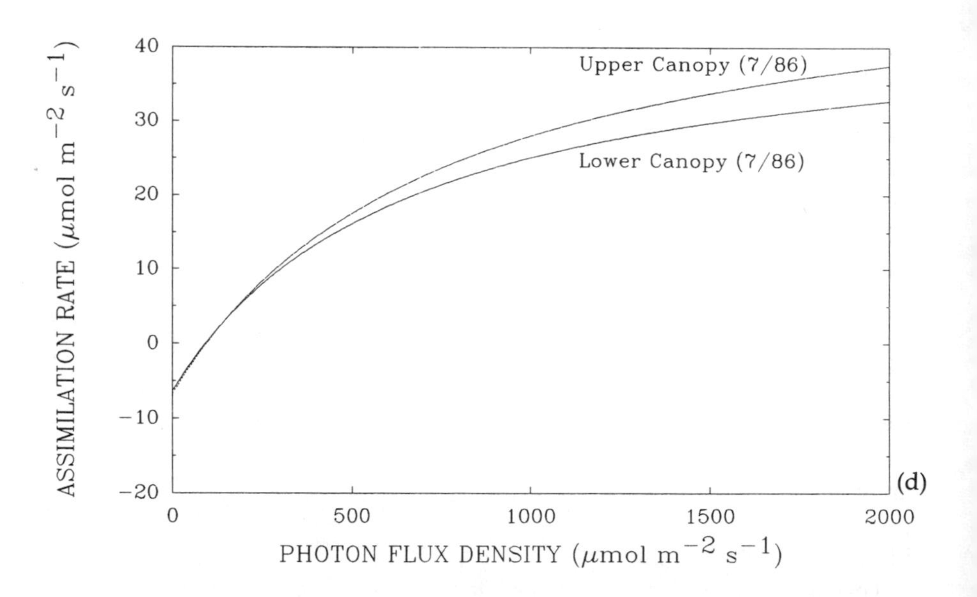

Figure 5.8. *continued* (c) Light response curves during the terrestrial phase (October 1985) for the upper and lower canopy; and (d) light response curves during the aquatic phase (July 1986), for the upper and lower canopy.

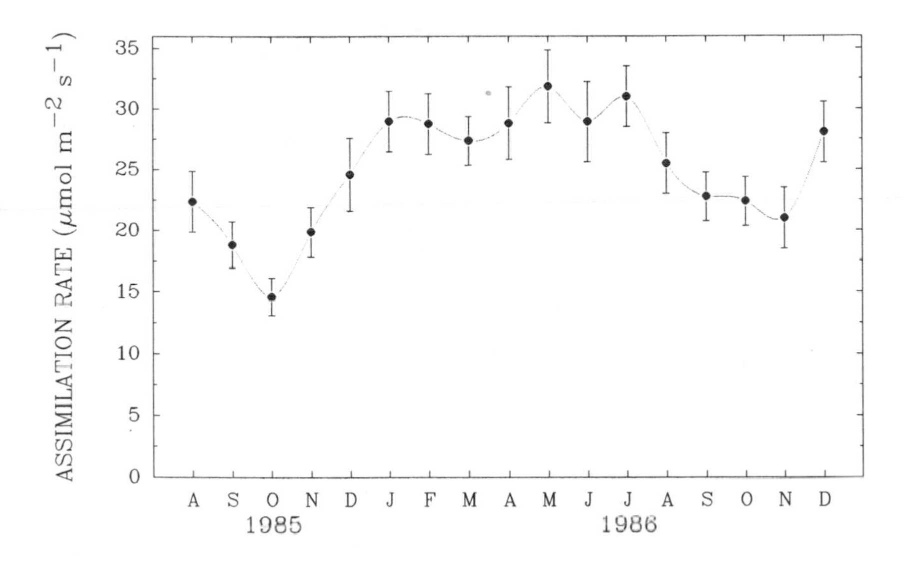

Figure 5.9. The average leaf photosynthetic rates of CO_2 uptake (±1 se) of *ca* 20–30 leaves of the upper canopy sampled within ±1 h of solar noon, on one day of each month from August 1985 to December 1986 at Ilha da Marchantaria.

usually uncovered during October/November, at this time in 1986 the water level was abnormally high and many plants of the older generation continued to grow. These plants exhibited intermediate assimilation rates (Figure 5.8a). The lowest photosynthetic rates were observed in the old plants during the low water period, i.e. October and November (Figure 5.8b). However, the new plants at this time showed rates of 18 μmol m^{-2} s^{-1}, *ca* 30% more than the old plants (Figure 5.8b).

The light-saturated rates of assimilation (A_{sat}) were always higher for leaves from the upper canopy. At low water in 1985, A_{sat} values for upper canopy leaves were 16 μmol m^{-2} s^{-1} higher than lower canopy leaves (Figure 5.8c). However, at high water, the difference in A_{sat} values was only about 4 μmol m^{-2} s^{-1} (Figure 5.8d). During the dry period of 1986, differences in light response between various canopy layers were much less pronounced than during the same period of the previous year (Figure 5.8d). Presumably this was due to the unusually high water level at that time in 1986.

The diurnal pattern of assimilation showed a significant midday reduction during days in October and November each year, dropping to about 13 μmol m^{-2} s^{-1}. Higher midday rates of about 30 μmol m^{-2} s^{-1} were observed during days in April to August (Figure 5.9).

5.5 DISCUSSION

5.5.1 Biomass and production

The life cycle of *E. polystachya* is closely correlated with the annual cycle of water level in the Amazon River. Thus maximum biomass is observed at high water and minimum at low water. When the habitat does not dry out, as in 1986, biomass and plant density remain high. In such circumstances, it is probable that total production in the following high water period would be reduced, as complete vegetative rejuvenation of the population would have been impeded. In fact, *E. polystachya* does not normally occur in permanently flooded habitats.

In terms of biomass and production, the subterranean portion of the plant is negligible. Root growth in the sediment is obviously restricted to the short terrestrial phase. After flooding, free floating adventitious roots, grouped at the nodes, take over the role of the basal roots. The latter provide only 10% of total root biomass.

The submerged proportion of the biomass increases with the rising water level, until at high water it constitutes 80–90% of total biomass. Leaves must be maintained above the water surface which may rise by 2 m a month. This is achieved by very rapid growth of new internodes rather than by any lengthening of the existing internodes (Table 5.1).

The annual P_n of 99.3 t ha^{-1} y^{-1} compares with the highest values ever recorded for natural communities (Table 5.4). Westlake (1963) estimated the annual P_n of aquatic macrophytes to lie between 8 and 60 t ha^{-1} y^{-1}. For comparison, Lieth and Whittaker (1975) estimated the P_n of tropical rain forests to be around 25 t ha^{-1} y^{-1}. There is some evidence from the literature that *Cyperus papyrus* may exceed *E. polystachya* in productivity. Thompson (1976) suggested that under optimal conditions *C. papyrus* may produce more than 100 t ha^{-1} y^{-1}. However, there have been no long-term measurements made to support this suggestion and there is further evidence that productivity may decrease to less than 30 t ha^{-1} y^{-1} when nutrients are deficient (Thompson, 1976; Thompson *et al.*, 1979).

In crop plantations, biomass levels of 88 t ha^{-1} have been achieved by elephant grass (*Pennisetum purpureum*) under management (Beadle *et al.*, 1985). This compares with a figure of 80.1 t ha^{-1} for *E. polystachya* (Figure 5.7). In studies in Venezuela on the use of *E. polystachya* as a crop, Melendez (1978) showed that the species could produce 25–30 t ha^{-1} fresh weight over 60 days. At Lago Camaleão, *E. polystachya* averaged growth rates of 25.9 ± 0.1 g m^{-2} d^{-1} in 9 separate

Table 5.4. Biomass and productivity (P_n) (dry weight) of some tropical herbaceous species

Species	*Place*	*Maximum biomass* ($t\ ha^{-1}$)	*Annual* P_n ($t\ ha^{-1}$)	*Author*
Oryza sativa	New South Wales, Australia	40	40	Westlake (1963)
Eichhornia crassipes	Mississippi, USA	15	15–44	Westlake (1963)
Cyperus papyrus	Africa	—	48–143	Thompson *et al.* (1979)
Paspalum fasciculatum	Amazon, Brazil	28	39	Junk and Howard-Williams (1984)
Saccharum officinarum	Hawaii, Java	73–85	108	Westlake (1963)
Echinochloa polystachya	Amazon, Brazil	70	108	Present study

months. Cultivated C_4 plants may surpass this for short periods only. For example, *Zea mays* may produce 51 g m^{-2} d^{-1} for 2–3 weeks only (Beadle *et al.*, 1985).

The high productivity of *E. polystachya* is especially remarkable when considering the environmental conditions under which it is achieved. The habitats are unstable, the change from a terrestrial to an aquatic phase produces high physiological stress and communities suffer great losses of biomass every year (although these are compensated by a rapid growth rate). The species can quickly colonize suitable new habitats. However, when conditions stabilize, *E. polystachya* communities are replaced by floodplain forest. Thus in spite of its considerable adaption to floodplain conditions, *E. polystachya* tends to show many characteristics of r-strategy (Pianka, 1970). By contrast, *C. papyrus* might be considered more of a K-strategist in that it colonizes relatively stable habitats. Its below-ground organs constitute 35–45% of its total biomass, but much of this may be dead and slowly decomposing rhizomes (Muthuri *et al.*, 1989). Well-developed stands of *C. papyrus* are considered the final stage in the hydrosere (Thompson *et al.*, 1979).

Prerequisites for r-strategy, based on quick growth and high productivity, are a fertile substrate and a highly efficient photosynthetic system. C_4 plants are better adapted to habitats which are nitrogen deficient than are C_3 plants, except legumes, due to their capacity to achieve high photosynthetic rates with low concentrations of RuBP carboxylase/oxygenase where more than half the soluble protein of a leaf is located. Therefore, the photosynthetic rate per unit nitrogen of a leaf is higher in C_4 plants than in C_3 plants (Brown, 1978). In spite of a generally good nutrient supply in the Varzea, nitrogen may become at least periodically limiting since the huge mass of plant biomass requires large amounts of this element. Furthermore, strongly hypoxic conditions in water and waterlogged soils favour denitrification. Thus the high concentrations of nitrogen and phosphorus in the emergent parts of the plants can be interpreted as a strategy to concentrate these elements in the photosynthetically active parts.

K-adapted *C. papyrus* maintains its nutrient levels by internal recycling between rhizomes and shoots. It is therefore quickly at a disadvantage when harvested, and under nutrient stress production declines greatly. However, r-adapted *E. polystachya* absorbs nutrients mainly from the soil and water. It has been successfully cultivated for cattle fodder, producing up to 30 t ha^{-1} fresh weight in crops of 60 days (Melendez, 1978). The plants become stunted in habitats of low nutrient status (e.g. clear-water rivers) and the species is completely absent in most of the nutrient-poor black-water rivers (Black, 1950; Junk and Howard-Williams, 1984).

5.5.2 Chemical composition

Concentrations of elements in *E. polystachya* generally fall into the range observed for other aquatic macrophytes (Cowgill, 1974; Howard-Williams and Junk, 1977). An exception is the very low level of nitrogen and correspondingly reduced amounts of crude proteins in the stems. Extrapolating from the nitrogen levels, protein content is highest in the leaves, 11.8% on average. The concentration in the stems is 1.6%. Due to the disproportionately larger amount of stems, the average protein content for the whole plant at high water is only 2.7%. Howard-Williams and Junk (1977) and Ohly (1987) quote total protein levels in *E. polystachya* of between 6.6 and 13.6%, possibly because they analysed young plants or plants with a higher proportion of leaf material.

Considering the large biomass of *E. polystachya* stands, the total amount of elements stored by them is significant. Nitrogen, phosphorus and potassium accumulate in much higher quantities than in a 5 m deep column of Amazon water, i.e. the mean depth of water covering the floodplain at high water.

New plants grow vegetatively from the nodes of old plants, utilizing stored elements. When the plants become rooted, the nutrient pool in the sediments becomes available. Furthermore, large amounts of nutrients are recycled from decaying organic material during the dry period and from the Amazon River as the water begins to inundate the floodplain.

Furch (1984a) pointed to the fact that the water of Lago Camaleão has a much higher concentration of potassium and other elements than the adjacent Amazon River. This is attributed to nutrient enrichment and internal recycling by aquatic and terrestrial macrophytes, together with the input of litter from the inundation forest. Furch (1984a) calculated that the quantity of macrophyte biomass necessary to allow the observed surplus of potassium corresponds to 75 t (dry matter) ha^{-1}.

Changes in the availability of phosphorus and nitrogen in the system, may explain the depletion of these elements in the stems of *E. polystachya* . As the water level falls, there is no additional input into the system either by the river or from recently flooded and decaying terrestrial vegetation. Since the plants continue to grow normally (Piedade *et al.*, 1991), there may have been an accumulation of phosphorus and nitrogen in the roots and leaves to avoid any deficiency of these elements in new growth.

Of interest is the fact that element concentration changes little along the length of the stems. Only sodium is slightly enriched towards the base. This element is, however, considered to be of little importance for plant growth (Long and Baker, 1986). Considering the reproductive strategy of *E. polystachya*, it is obvious, that the stems not only function for vegetative rejuvenation of established populations but also as propagules for vegetative propagation. Stems are vulnerable to damage by water currents and small pieces drift in great numbers in the river. Even small stem portions with one node are successful in vegetative propagation on exposed sediments at low water. The uniform distribution of elements within the stem therefore assists in vegetative propagation as each portion of stem possesses an adequate store of nutrients.

5.5.3 Leaf area index and photosynthetic characteristics

The progressive immersion of the stems as the water rises results in the permanent submergence and loss of leaves. Old leaves are therefore continuously replaced. Thus the average life span of a leaf is only 34 days. Leaf area index was lowest during the dry period as the new generation became established. Maximum values occurred at high water when growth rates were greatest. Leaves of *E. polystachya* are typically about 80 × 5 cm, but are smaller in young and flowering plants. This explains the low leaf area index (L) in November/December each year, and in the main flowering period during May 1986.

In comparison with other highly productive species L is smaller in *E. polystachya*. High yielding paddy rice may attain an L of 7 to 8 (Zelitch, 1971). The lower L of *E. polystachya* is probably the consequence of the rapid turnover of the canopy necessitated by the rising water. Nevertheless, canopy architecture is such that most incident light is absorbed. The leaves are alternate and erect, reclining at the end due to their size. Light interception is fairly constant throughout the year, varying only between 72 and 89% of incident solar radiation. Further, in high light CO_2 assimilation rates are relatively high throughout the year. At high water level, maximum photosynthetic rate (A_{sat}) shows only a slight difference between upper and lower canopy leaves (Figure 5.8d). With inundation, as the leaves become submerged and die, they would act as a net sink for photosynthate and nutrients from the plant. Thus the rapid turnover of leaves and a canopy highly efficient at CO_2 assimilation allows the high productivity shown by this species. The maximum efficiency of conversion of intercepted light into biomass in C_4 plants is considered to be about 30% greater than in C_3 plants (Monteith, 1977). The maximum assimilation rate of *E. polystachya* at Marchantaria island reached 31 μmol m^{-2} s^{-1}, whilst the equivalent rates of *Eichhornia crassipes* and *Hymenachne amplexicaulis*, both C_3 species, reached only about 20 μmol m^{-2} s^{-1} (Piedade, unpublished data).

5.5.4 C_4 Photosynthesis and the distribution of C_4 species

The C_4 pathway is considered especially effective under drought stress (Teeri and Stowe, 1976). The number and frequency of C_4 plants increases from the humid to the semi-arid and arid tropics and sub-tropics (Teeri and Stowe, 1976; Long, 1983). At first glance, C_4 plants with high water use efficiency do not have competitive advantages over C_3 plants in moist habitats such as the Amazonian Varzea. Highly productive C_3 plants like *Eichhornia crassipes* may even benefit from a high water loss as this would cool the leaves more efficiently and allow an increased uptake of dissolved nutrients (Thompson, 1985). Water supply may, however, be temporarily limiting for plants in the Varzea during the dry phase. To produce 1 kg of organic matter under identical conditions, C_4 plants require 0.3–0.4 m^3 of water whereas C_3 plants need 0.6–0.8 m^3 (Medina *et al.*, 1976). Efficient use of available soil moisture during the critical phase of establishing the new generation may give a C_4 species a competitive advantage. Furthermore, the C_4 species would be quicker to produce plants sufficiently robust to withstand the heavy mechanical stresses of waves and currents during the aquatic phase. Thus the C_4 *E. polystachya* possesses a definite advantage over its C_3 competitors.

Nitrogen use efficiency may also be of competitive advantage under certain conditions, e.g. in floodplain lakes, where nitrogen instead of phosphorus may become the limiting factor for plant growth (Junk, unpublished data). The low content of nitrogen in the submersed stems of *E. polystachya* in comparison with the high nitrogen level in the photosynthetically active emergent parts may be interpreted as a strategy to optimize the use of nitrogen. The importance of a sufficient nitrogen level is also emphasized by the observation by Melendez (1978) that *E. polystachya* responds markedly to addition of nitrogen under cultivation.

Measurements of photosynthetic activity of *E. polystachya* during the terrestrial phase show a significant midday reduction in CO_2 uptake which was not observed during the aquatic phase. A similar phenomenon can be seen in *Paspalum fasciculatum*, another frequent

C_4 species of the Varzea. This is presumably due to mild drought when the floodplain surface is exposed.

5.6 CONCLUSIONS

The vigorous growth and high productivity of *E. polystachya* allow it to occupy throughout the year habitats which other species are unable to invade. In areas which are flooded to a depth of 8 m or more, growth rates of *E. polystachya* are sufficient to maintain the canopy above the water level and produce stems strong enough to withstand physical damage by wind, waves and current. *E. polystachya* may therefore be one of the rare cases where the high productivity potential of C_4 species has decisive importance for the colonization of a specific habitat.

The high net primary production of *E. polystachya* is of significance when the role of the Amazon basin in the global cycling of carbon is considered. Although the Amazonian floodplain covers a far smaller area of the region than the rain forest, this difference may be compensated for by the former's higher productivity. Assuming that *E. polystachya* covers an area of 5000 km^2, the species can consume 75 million tonnes of carbon per annum. A large proportion of this would be rapidly recycled as stems and leaves die *in situ*, but a part would be exported in the waters of the Amazon. Another part would be incorporated into lake and river sediments. *E. polystachya* stands may therefore be important sinks for atmospheric carbon in this region.

5.7 SUMMARY

1. Biomass, demography, production, gas exchange and chemical composition were monitored at monthly intervals from July 1985 to December 1986 for an *Echinochloa polystachya* dominated community near Manaus. This species is one of the most abundant C_4 grasses of the Amazon region and inhabits floodplain areas which receive a large, annual flood pulse. Conditions vary from a terrestrial to an aquatic phase.
2. In one annual growth cycle, plant mass increased from the formation of new shoots during the dry season to a peak at high water level of 80.1 t (dry matter) ha^{-1}. The bulk (85.5%) was provided by living material, with stems contributing 75.9%, leaves (laminae and sheaths) 5.5%, roots 4.1% and dead material 14.5%.

3. The mean light saturated rate of photosynthesis, measured at solar noon, decreased from approximately 35 μmol m^{-2} s^{-1} during the aquatic phase to 15 μmol m^{-2} s^{-1} during the terrestrial phase.
4. The annual net primary production of 99.3 t ha^{-1} is among the highest ever observed for a natural community and approaches the maximum yield for C_4 crops.
5. The stand intercepted between 72 and 89% of the annual average of 486.4 MJ m^{-2} of radiation received. Conversion efficiency was calculated to be 4.3% of photosynthetically active radiation. As they are so abundant in the Amazon floodplain, *E. polystachya* communities are important sinks for atmospheric carbon.

REFERENCES

Anon. (1971) Referenzmethoden für die Bestimmung der Mineralstoffe in den Pflanzen. I. Stickstoff, Phosphor, Kalium, Natrium, Calzium, Magnesium, in *Monatliche Mitteilungen des Internationalen Kali-Institutes*, Internationalen Kali-Institutes, Bern (Schweiz), pp. 1–18.

Beadle, C.L., Long, S.P., Imbamba, S.K., Hall, D.O. and Olembo, R.J. (1985) *Photosynthesis in Relation to Plant Production in Terrestrial Environments*, UNEP/Tycooly, Oxford.

Black, G.A. (1950) Os capíns aquticos da Amazônia. Inst. Agronômico do Norte, Belém-PA. *Boletin Técnico do Instituto Agronômico do Norte*, **19**, 54–94.

Boyd, C.E. (1970) Amino acid, protein and caloric contents of vascular aquatic macrophytes. *Ecology*, **51**, 902–6.

Brown, R.H. (1978) A difference in N use efficiency in C_3 and C_4 plants and its implications in adaptation and evolution. *Crop Science*, **18**, 93–8.

Cowgill, U.M. (1974) The hydrochemistry of Linsley Pond, North Branford, Connecticut. II. The chemical composition of the aquatic macrophytes. *Archiv fur Hydrobiologie Supplement*, **45**, 1–119.

Furch, K. (1984a) Seasonal variations of the major cation content of the Varzea-lake Lago Camaleão, middle Amazon, Brazil, in 1981 and 1982. *Verhandlungen der Internationalen Vereinigung für Theoretische und Angewandte Limnologie*, **22**, 1288–93.

Furch, K. (1984b) Water chemistry of the Amazon basin. The distribution of chemical elements among freshwaters, in *The Amazon: Limnology and Landscape Ecology of a Mighty Tropical River and its Basin*. (ed. H. Sioli), Dr. W. Junk Publishers, Dordrecht, pp. 167–99

Furch, K., Junk, W.J., Dieterich, J. and Kochert, N. (1983) Seasonal variation in the major cation (Na, K, Mg, and Ca) content of the water of Lago Camaleão, an Amazonian floodplain-lake near Manaus, Brazil. *Amazoniana*, **8**, 75–89.

Hedges, J.I., Clark, W.A., Quay, J.P.D., Richey, J., Devol, A.H. and Santos de M.U. (1986) Compositions and fluxes of particulate organic material in the Amazon River. *Limnology and Oceanography* **31**, 717–38.

Hitchcock, A.S. (1936) *Manual of the Grasses of the West Indies*. Miscellaneous Publ. 243., US Department of Agriculture, Washington, DC.

Howard-Williams, C. and Junk, W.J. (1977) The chemical composition of Central Amazonian aquatic macrophytes with special reference to their role in the ecosystem. *Archiv für Hydrobiologie*, **79**, 446–64.

Irmler, U. (1977) Inundation forest types in the vicinity of Manaus. *Biogeographica*, **8**, 17–29.

Junk. W.J. (1970) Investigations on the ecology and production-biology of the 'floating meadows' (Paspalo-Echinochloetum) of the Middle Amazon. I. The floating vegetation and its ecology. *Amazoniana*, **2**, 449–95.

Junk, W.J. (1973) Investigations on the ecology and production-biology of the 'floating meadows' (Paspalo-Echinochloetum) on the Middle Amazon. II. The aquatic fauna in the root zone of floating vegetation. *Amazoniana*, **4**, 9–102.

Junk, W.J. (1985) The Amazon floodplain – a sink or source of organic carbon? in, *Transport of Carbon and Minerals in Major World Rivers*, Part 3 (eds E.T.H. Degens, S. Kempe and R. Herrera), Mitteilung Geologisch und Palaeontologisch Institut Universität Hamburg. SCOPE/UNEP Sonderbd., Hamburg. pp. 267–83.

Junk, W.J. (1986) Aquatic plants of the Amazon system, in *The Ecology of River Systems* (eds B.R. Davis and K.F. Walker), Dr W. Junk Publishers, Dordrecht, pp. 319–37.

Junk, W.J. (1991) Wetlands of northern South-America, in (eds D. Whigham, S. Hejny and D. Dykyjova), Dr W. Junk Publishers, Dordrecht, (in press).

Junk, W.J. and Howard-Williams, C. (1984) Ecology of aquatic macrophytes in Amazonia, in *The Amazon: Limnology and Landscape Ecology of a Mighty Tropical River and its Basin*. (ed. H. Sioli), Dr. W. Junk Publishers, Dordrecht, pp. 269–93.

Junk, W.J. and Furch, K. (1985) The physical and chemical properties of Amazonian waters and their relationship with the biota. in *Key Environments: Amazonia* (eds G.T. Prance and T.E. Lovejoy), Pergamon Press, Oxford, pp.3–17.

Junk, W.J., Bayley, P.B. and Sparks, R.E. (1989) The flood pulse concept in river-floodplain systems. in *Proceedings of the International Large River Symposium (LARS)* (ed. D.P. Dodge), Canadian Special Publications of Fisheries and Aquatic Sciences, no. 106, Ottawa, pp. 110–27.

Lieth, H. and Whittaker, R.H. (1975) *The Primary Production of the Biosphere*. Springer-Verlag, Berlin.

Long, S.P. (1983) C_4 photosynthesis at low temperatures. *Plant Cell and Environment*, **6**, 345–63.

Long, S.P. and Baker, N.R. (1986) Saline terrestrial environments. in *Photosynthesis in Contrasting Environments* (eds N.R. Baker and S.P. Long) Elsevier, Amsterdam, pp. 63–102.

Medina, E., Bifano de, T. and Delgado, M. (1976) Differenciacion fotossintetica en plantas superiores. *Interciencia*, **1**, 96–104.

Melendez, M.E. (1978) Aspectos agronomicos del pasto aleman (*Echinochloa polystachya*), in *Aniversario del Fondo Nacional de Investigaciones Agropecuarias*, Estacion Experimental Calabozo, Venezuela, pp. 1-11.

Monteith, J. (1977) Climate and the efficiency of crop production in Britain. *Philosophical Transactions of the Royal Society of London, Series B*, **281**, 277–94.

Muthuri, F.M., Jones, M.B. and Imbamba, S.K. (1989) Primary productivity of papyrus (*Cyperus papyrus*) in a tropical swamp; Lake Naivasha, Kenya. *Biomass*, **18** 1–14.

Ohly, J. (1987) Untersuchungen uber die Eignung der natürlichen Pflanzenbestande auf den Uberschwemmungsgebieten (Varzea) am mittleren Amazonas, Brasilien, als Weide für den Wasser-buffel (*Babalus bubalis*) wahrend der terrestrischen Phase Ökosystems. *Gottinger Beitrage zur Land- und Forstwirtschaft in den Tropen und Subtropen*, **24**, 1–200.

Pianka, E.R. (1970) On r and K selection. *American Naturalist*, **104**, 592–7.

Piedade, M.T.F., Junk, W.J. and Long, S.P. (1991) The productivity of the C_4 grass *Echinochloa polystachya* on the flood-plain of the Amazon. *Ecology* (in press).

Prance, G.T. (1980) A terminologia dos tipos de florestas Amâzonicas sujeitas a inundacão. *Acta Amazonica*, **10**, 495–504.

Radam/Brasil. (1972) *Mosaico Semi-Controlado de Radar*, fola Sa 20-Z-D, escala 1.250:000, Ministério das Minas e Energia Departamento Nacional de Producão Mineral, Rio de Janeiro.

Ribeiro, M. de N.G. and Adis, J. (1984) Local rainfall variability – a potential bias for bioecological studies in the central Amazon. *Acta Amazonica*, **14**, 159–74.

Schmidt, G.W. (1973) Primary production of phytoplankton in the three types of Amazonian waters. II. The limnology of a tropical flood-plain lake in Central Amazonia (Lago do Castanho). *Amazoniana*, **4**, 139–203.

Sioli, H. (1951) Algunas resultados e problemas da limnologia Amazonica. *Boletin Técnico do Instituto Agronômico do Norte*, **24**, 2–44.

Sioli, H. (1964) General features of the limnology of Amazonia. *Verhandlungen der Internationalen Vereinigung für Theoretische und Angewandte Limnologie*, **15**, 1053–8.

Sioli, H. (1967) Studies in Amazonian waters. *Atas do Simposio sobre a Biota Amazonica. Rio de Janeiro*, **3**, 9–50.

Soares, M.G.M., Almeida, R.G. and Junk, W.J. (1986) The trophic status of the fish fauna in Lago Camaleão, a macrophyte dominated floodplain lake in the Middle Amazon. *Amazoniana*, **9**, 511–26.

Teeri, J.A. and Stowe, L.G. (1976) Climatic patterns and the distribution of C_4 grasses in N. America. *Oecologia*, **23**, 1–12.

Thompson, K. (1976) Swamp development in the headwaters of the White Nile, in *The Nile* (ed. J. Rzoska), Dr W. Junk Publishers, The Hague, pp. 177-96.

Thompson, K. (1985) Emergent plants of permanent and seasonally flooded wetlands, in *The Ecology and Management of African Wetland Vegetation*. (ed. P. Denny), Dr W. Junk Publisher, The Hague, pp. 43–107.

Thompson, K., Shewry, P.R. and Woolhouse, H.W. (1979) Papyrus swamp development in the Upemba Basin, Zaire: studies on population structure in *Cyperus papyrus* stands. *Botanical Journal of the Linnean Society*, **78**, 299–316.

Westlake, D.F. (1963) Comparisons of plant productivity. *Biological Reviews*, **38**, 385–425.

Zelitch, I. (1971) *Photosynthesis, Photorespiration, and Plant Productivity*, Academic Press, New York.

6

Bamboo in sub-tropical eastern China

G.-X. QIU, Y.-K. SHEN, D.-Y. LI, Z.-W. WANG, Q.-M. HUANG, D.-D. YANG and A.-X. GAO

6.1 INTRODUCTION

Bamboo plants have multiple uses. They provide the raw materials for pulp, construction work, furniture and personal items such as baskets and hats (Hong, 1988). In addition, young bamboo shoots are edible and are used as vegetables (Liese, 1989). Bamboos are woody perennials but belong to the Gramineae. They are amongst the most 'advanced' monocotyledonae, and exhibit an extreme range of growth form, from less than 10 cm in height to 15–20 m. Bamboo-dominated communities can therefore be classified both as grasslands and forests (Hsiung, 1987; Liese, 1989).

There are more than 1 300 species of bamboo, mostly distributed in the tropics and sub-tropics (Bahadur, 1979; Keng, 1984). About 1000 species are found in Asia, covering over 180 000 km^2. Within China there are 300 species, grouped into 44 genera and spread over 33 000 km^2, approximately 3% of the country's total forest area, although new species and areas of bamboo are being described annually (Keng, 1987; Yi, 1988; Hsiung, 1987; Ma, 1989). Much of this cover (about 2 × 10^6 ha) consists of a single species, *Phyllostachys pubescens* Mazel ex H. delehaire, which is indigenous to China. The natural habitat of *P. pubescens* is the sub-tropical region of the middle and lower reaches of the Yangtze River, mainly in Hunan, Jiangsi, Fujian and Zhejiang provinces (Figure 6.1). It is favoured by a climate which has a yearly mean temperature between 12.6 and 20°C, a minimum monthly mean temperature greater than 0°C in all months, typically 3–10°C in January, with an annual precipitation of 1200–1800 mm and mean

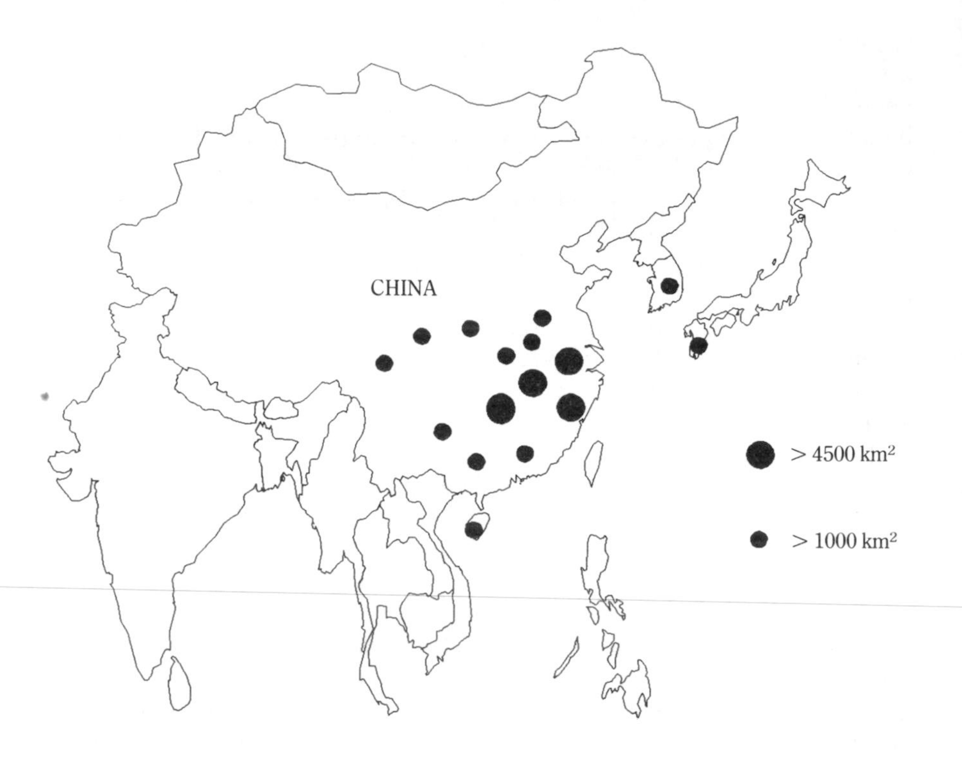

Figure 6.1. The distribution of *Phyllostachys pubescens* in Asia.

relative humidity over 80% (Chang, 1982; Qiu, 1982; Liu, 1987; Yuan and Xu, 1989).

P. pubescens forms monospecific stands and is deciduous, with leaf fall occurring in spring. In common with most bamboo species flowering and seed set are very rare, occurring on average once every 50–60 years (Janzen, 1976; Liese, 1989). Reproduction is therefore predominantly vegetative, by rhizome extension. Young shoots emerge in spring from rhizomal buds. Shoot diameters are then already nearly at their maximum dimensions. Shoot elongation proceeds for 50–60 days until a maximum height of around 15–20 m is reached, in late May or early June. Once this maximum height is reached shoot diameter remains constant, but the density of the wood increases in subsequent years. The maximum shoot density is 3000 to 4500 ha^{-1}, depending on the depth of the soil layer, soil fertility and hydrological state and the degree of slope. Leaf emergence takes place shortly after the shoots reach maximum height and the canopy is fully expanded by early June. New shoots renew their leaves in the second spring; thereafter shoots renew their leaves every 2 years. Rhizomes grow horizontally underground, and have a life span of approximately 20 years. *P. pubescens* exhibits annual alternation in growth of above- and below-ground parts, i.e. more new shoots are formed in one year and more new rhizomes in the next (Cheng, 1983; Anon., 1986; Liao, 1988).

P. pubescens tends to shade out other species, despite the fact that its leaflets are small (usually <10 cm^2) and its stands are not very dense. In summer and autumn only about 5% of solar radiation incident at the top of the canopy reaches the floor of the stand. The result is that only a reduced ground flora can exist for a short time in early summer, when around 10–15% of incident solar radiation penetrates through the canopy. Accordingly, under favourable climatic conditions, the area colonized by the species is limited principally by edaphic factors and water supply (Chen, 1982; Chang, 1982).

Ecosystems dominated by *P. pubescens* may exert a major influence on the local environment, particularly with respect to the carbon and water cycles, since the species grows fast and forms forest very quickly. It also has a much longer life span than other grasses.

Little work has previously been published on the biomass dynamics or productivity of *P. pubescens*, or indeed on bamboo in general. Veblen *et al.* (1980) estimated above-ground biomass for *Chusquea culeou* and *C. tenuiflora* in south-central Chile at an altitude of over 700 m. They used a previously calculated relationship between culm (stem) diameter at 1 m height and culm dry weight. Above-ground biomass was found to be approximately 15.6 16.2 kg m^{-2} for *C. culeou* and about 1.3 kg m^{-2} for *C. tenuiflora*. Net primary production (P_n) for *C. culeou* was calculated from the yearly leaf fall and total above-ground biomass of all 1-year-old culms, and found to be around 1.0–1.4 kg m^{-2} y^{-1}. Suwannapinunt (1983) calculated the biomass and net primary production of *Thyrostachys siamensis* at Kanchanaburi, Thailand, from equations relating total height and breast height of stems to their dry weight. Biomass ranged from 1.08 kg m^{-2} on a poor site to 5.38 kg m^{-2} on a good site. P_n similarly varied from 0.16 to 0.81 kg m^{-2} y^{-1}. From analysis of culm dynamics, Taylor and Qin (1987) estimated the above-ground biomass and annual P_n of three bamboo species at Sichuan province, China. The mean biomass of *Fargesia spathacea* was 2.37 kg m^{-2}, while *F. scabrida* and *Sinarundinaria fangiana* averaged 0.93 and 0.74 kg m^{-2} respectively. P_n was found to be 0.36 kg m^{-2} y^{-1} for *F. spathacea* and 0.15 kg m^{-2} y^{-1} for *S. fangiana*. P_n for *F. scabrida* was suggested to be 0.3 kg m^{-2} y^{-1}. However, these figures did not include biomass losses to grazers such as the giant panda. If so corrected, Taylor and Qin (1987) assumed that P_n could average 0.45 kg m^{-2} y^{-1} for *F. spathacea* and 0.17 kg m^{-2} y^{-1} for *F. scabrida*. None of these studies has simultaneously examined below-ground production and organ turnover.

This study investigated the P_n of a *Phyllostachys pubescens* forest in China by a combination of non-destructive and harvest techniques, taking into account losses due to the death above and below ground. The efficiency of photosynthetic conversion of solar radiation was also calculated. In addition, the diurnal and seasonal photosynthetic characteristics of *P. pubescens* were also determined in the field.

6.2 THE STUDY SITE

The *P. pubescens* forest under investigation is located at Miao Shan Valley, Zhejiang province, at latitude 30°01′N, longitude 120°15′E and with an elevation of about 100 m above sea level. It is situated approximately 10 km north-east of the Sub-tropical Forest Institute, Chinese Academy of Forest Science, at Fuyang county and a little more than 30 km south-west of the city of Hongzhou (Figure 6.2). The *P. pubescens* stand spreads along the south-east slope of a hill, on an incline of around 20°. The soil is an acidic red earth with a pH of 6.0 and an average depth of 60 cm, overlying sandstone bedrock.

The forest is semi-natural and subjected to low intensity management, which consists chiefly of the cropping of mature shoots (i.e. those over 8 years old) for wood at the end of each year. As the forest originated from many different rhizomes, there is some heterogeneity with respect to the typical biennial growth cycle of bamboo (Anon., 1986). Over the forest as a whole, then, there is less pronounced fluctuation in total above-ground biomass from year to year than is usually the case. An area of 2 ha was selected within the forest as the study site and 16 plots, measuring 10 × 10 m each, were marked out in May 1984. An 18 m high tower was also erected within the plots, so that photosynthetic measurements could be made within the bamboo canopy (Figure 6.3).

A meteorological station is situated about 100 m away from the study plots (Figure 6.3). In general, there is a warm season from April to September, during which the average temperature is higher than 15°C and two-thirds of the yearly precipitation is received. Annually, temperatures range from a minimum of −7.6°C to a maximum of 37.2°C, with a mean of 16°C. In January, the average temperature is 3.4°C, whilst in July it rises to 27.1°C. Total annual precipitation is approximately 1800 mm, with a distribution pattern reflecting more or less that of temperature. Snowfall occurs in some winters. The area is typical of the *P. pubescens* areas of Zhejiang province.

Detailed site meteorological records were kept from January 1985 to

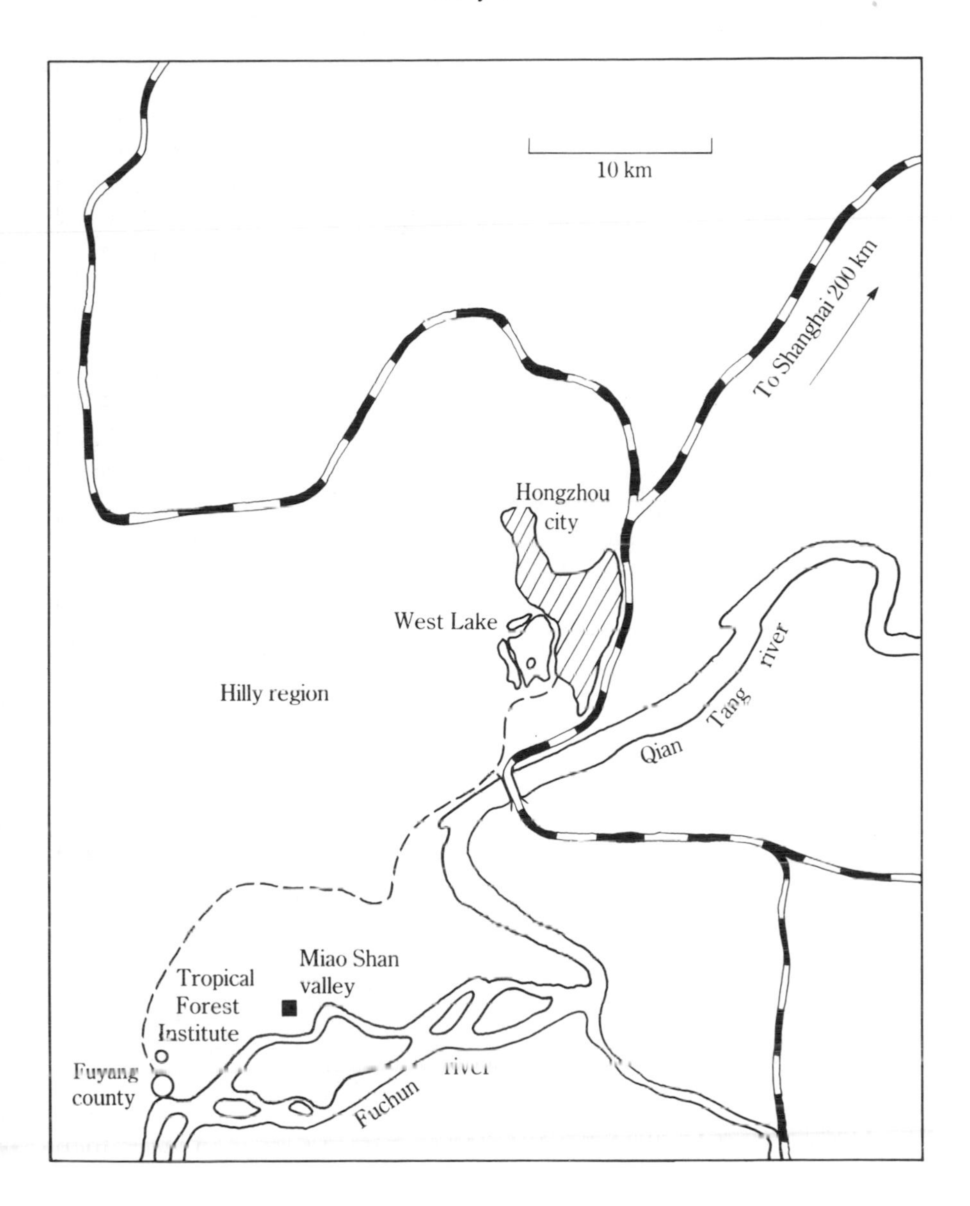

Figure 6.2. The location of the Miao Shan Valley, which contains the study site, in relation to Hongzhou City and its surroundings.

December 1987. The yearly temperature cycle was broadly similar for each of the years (Figure 6.4a). In the open, the lowest monthly minimum was −9.9°C, recorded in March 1987, while the highest was

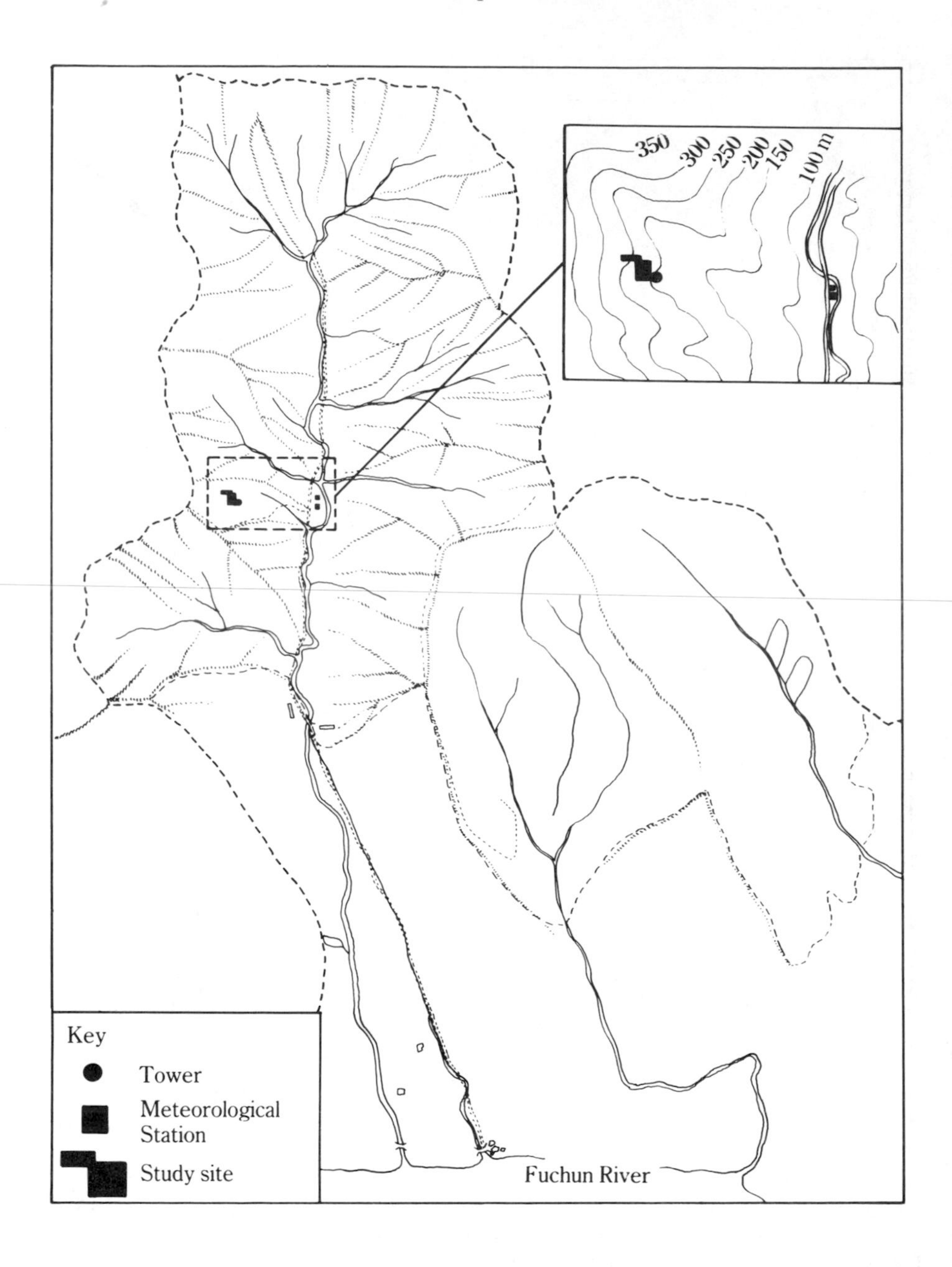

Figure 6.3. Location of the study sites, 18 m sampling tower and meteorological station within the Miao Shan Valley.

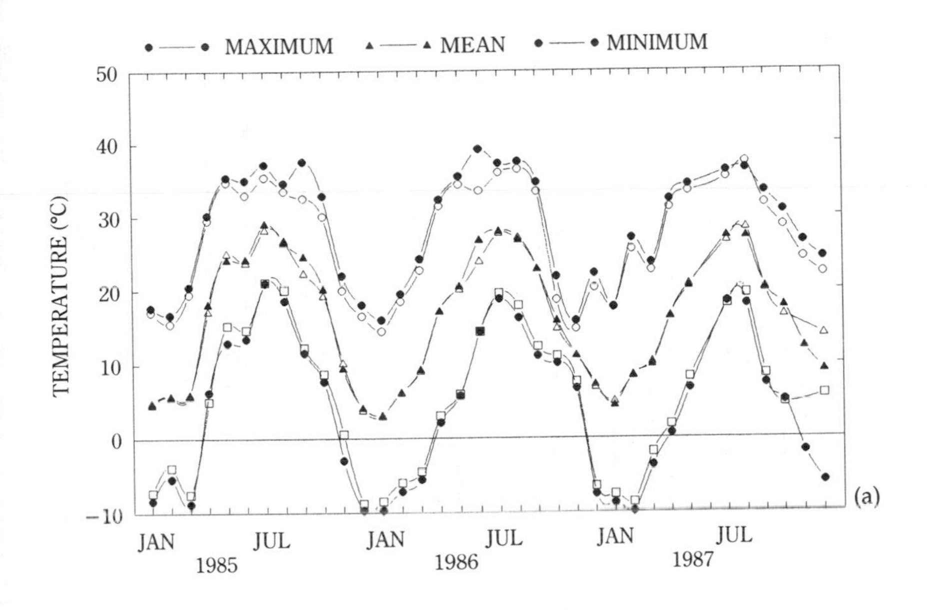

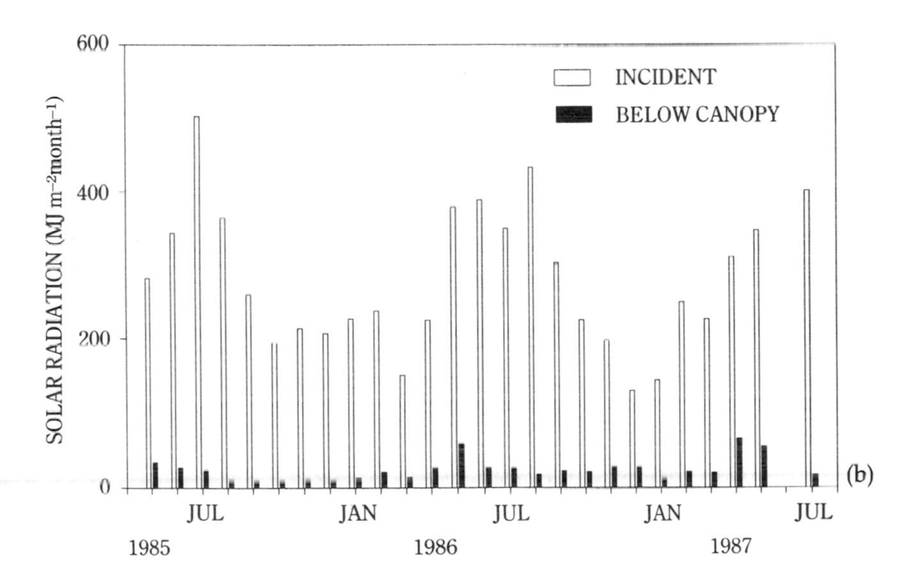

Figure 6.4. (a) The average monthly maximum, mean and minimum temperatures at the meteorological station in the open, adjacent to the forest. The open symbols indicate the corresponding temperatures under the forest canopy in the study site; (b) monthly totals of solar radiation incident and below the canopy.

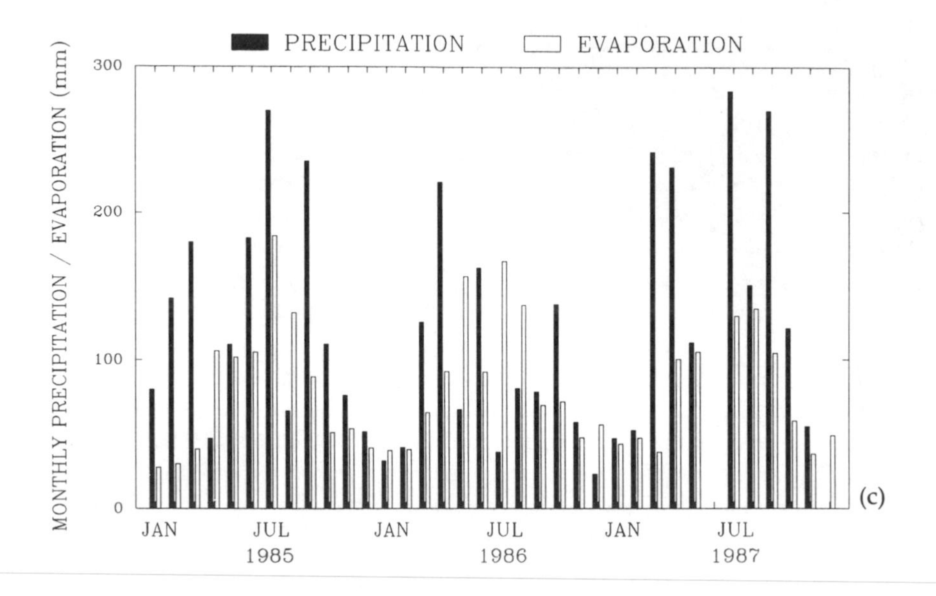

Figure 6.4. (c) Monthly totals of precipitation and pan evaporation. All measurements were made at the study site in the Miao Shan Valley.

27.1°C in July 1985. Temperature fluctuation within the forest was somewhat less pronounced (Figure 6.4a). In 1986, the annual receipt of incident solar radiation was 3342 MJ m^{-2}, of which only 301 MJ m^{-2} reached the floor of the stand (Figure 6.4b). The driest year was 1986, with 1064 mm of precipitation, and it also had the highest evaporation rate of 1033 mm. The year 1985 received 1557 mm of precipitation, and experienced an evaporation rate of 962 mm, whilst the respective figures for 1987 were 1565 and 848 mm (Figure 6.4c). Humidity was high throughout with an average relative humidity of 80%.

6.3 METHODS AND CALCULATIONS

6.3.1 General biomass methods

The techniques used in this study were, due to the biology of *P. pubescens*, somewhat different from those described for grasslands in Chapter 1. Monthly harvesting of shoots was impracticable, but the characteristics of bamboo provide an alternative method of estimating biomass. New shoots of *P. pubescens* emerge each spring and grow to their final size by the summer. This contributes the major increment in biomass each year. The number of shoots and their circumferences can be easily counted and measured. Shoot dry weights can be calculated at specific time intervals from derived relationships between their circumference, volume, age and wood density, as described in section 6.3.2. Older shoots contribute small additions to total above-ground biomass by increases in their wood density, which can be easily determined (section 6.3.2). Thus changes in above-ground biomass over time could be monitored non-destructively. Changes in below-ground biomass were determined directly from the excavation of pits. Since large experimental variation occurs in successive excavations, increments of below-ground biomass were also estimated from above-ground biomass increments. This utilized the ratio of above- to below-ground biomass obtained from excavation of 4 × 4 m plots. Leaf area index (L) was estimated from some destructive harvests and from the collection of fallen leaves.

6.3.2 Estimation of above-ground biomass

The above-ground biomass of *P. pubescens* comprises leaves, branches, stems and shoot bases. Within the forest, each shoot was marked with its year of emergence. Within the plots, the plot number and shoot number were marked on each stem and the stem circumference at a height of 1.6 m recorded.

Determination of wood density

Stems of bamboo form segmented, hollow tubes. Blocks, with an approximate volume of 5 × 2 cm^2, were cut from one side of the tube section with a saw. These were immediately placed into plastic bags to prevent water loss. Ten samples of each age group of shoots were taken every few months during 1985 to 1987. The volume of each block was determined from its weight change when immersed in water. The density of each block was then calculated by dividing its dry weight by its volume.

The relationship between wood density and shoot age

The age of each shoot (x) was calculated from the time its branches became fully expanded, i.e. the July of its year of emergence. The relationship between wood density (y) and shoot age was found to fit a hyperbolic curve of the form:

$$y = x/(ax + b) \qquad [1]$$

where $a = 1.58$ and $b = 0.253$ with a correlation co-efficient of $r = 0.92$ (Figure 6.5a).

The relationship between stem circumference and wood volume
Destructive harvests of shoots, cut at the connection between base and rhizome in three $4 \times 4\ m^2$ plots, were carried out in the spring of 1986 and 1987. Each harvested shoot was then cut into several sections and the total above-ground dry weight determined from its fresh weight and water content. Stem circumferences and wood densities were also measured.

The volume of each shoot was calculated by dividing its dry weight (DW) by its wood density (i.e. volume $= DW/y$). A good linear relationship was then found between shoot circumference (z) and shoot volume, which fitted the equation:

$$\frac{DW}{y} = c.z - d \qquad [2]$$

where $c = 1.904$ and $d = 30.07$ with a correlation co-efficient of $r = 0.95$ (Figure 6.5b).

Calculation of total above-ground biomass
By combining equations (1) and (2) with the ages of the shoots in the 16 plots, the total dry weight of shoots of year i (DW_i) in each plot is given by:

$$DW_i = n_i.x_i\ (c.z_i - D)/(a.x_i + b) \qquad [3]$$

where n_i is the number of i-year-old shoots in the plot, x_i is the age of year i shoots and z_i is the shoot circumference. Total above-ground biomass of all the *P. pubescens* plants in a plot was then given by:

$$\sum_{i=1979}^{i=1986} DW_i \qquad [4]$$

and the annual increment in biomass (ΔDW) during 1986 was given by:

$$\Delta DW = \sum_{i=1979}^{i=1986} DW_i - \sum_{i=1979}^{i=1985} DW_i \qquad [5]$$

6.3.3 Estimation of below-ground biomass

In 1985, below-ground biomass was determined by direct excavation of six pits, each $1\ m^2$ in area and 60–100 cm deep, at the beginning and in the middle of the year. For the subsequent 2 years, three

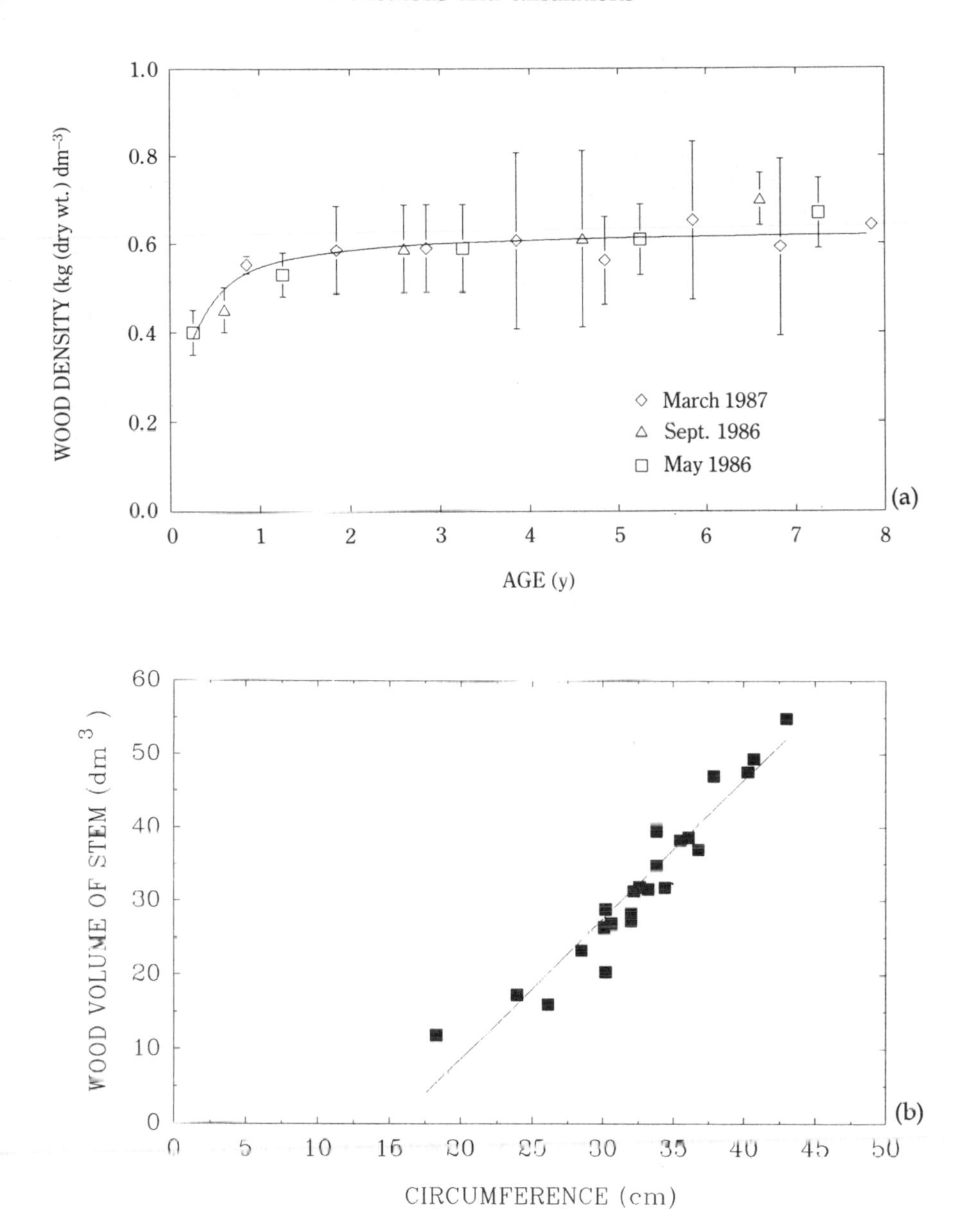

Figure 6.5. (a) Wood density in relation to age of stem. Each point is a mean (±1 se) value for samples from one age group. The line is the best-fit hyperbolic curve fitted by the least squares method; and (b) the relationship between stem circumference and stem wood volume at 1.6 m. Volume = 1.904 × circumference − 30.07.

excavations were carried out each spring and the area of each pit excavated was increased to 16 m^2. This increase in area helped reduce statistical variation within the data and also enabled longer rhizomes to be harvested for age classification. In addition, six 20 × 20 cm pits were also excavated every 2 months in order to provide data more representative of the site as a whole.

Sampling of underground biomass and dead material

Excavated samples of below-ground biomass and dead material, together with attached substrate, were passed through a 1 cm^2 mesh sieve to retrieve the larger roots. These were then washed, air dried and weighed. Subsamples of these roots and rhizomes were oven dried at 80°C for 30 h in order to obtain the ratio of air dry to oven dry weight. This ratio was then used to calculate the oven dry weight of the original sample. All of the material which passed through the 1 cm^2 sieve (i.e. the short, fine, broken roots and attached soil) was weighed and a representative 1% subsample taken. The subsample was placed in a 1 mm^2 mesh nylon bag and washed until the attached soil was removed. The root material was oven-dried as above and weighed. This weight was multiplied by 100 and added to the total underground biomass.

Distribution profile of underground biomass

At approximately 2 month intervals starting in May 1986, six 20 × 20 cm pits were excavated at 10 cm intervals down to a depth of 60–100 cm, in order to study the distribution profile of below-ground biomass. Excavated material was subsampled (to 10%). Subsamples were placed in 1 mm^2 mesh nylon bags, the attached substrate washed away, and the sample oven-dried as above and weighed.

Age distribution of rhizomes

Rhizomes excavated from the 4 × 4 m^2 pits were long enough for their ages to be determined from the presence and colour of the rhizome sheath, the colour of the rhizome and the appearance of the buds on the internodes after the method of Liao (1988).

6.3.4 Estimation of dead material

Above-ground dead material of *P. pubescens* consists of fallen leaves, dead shoots, dropped sheaths of young shoots and a few dead branches. Some young shoots died soon after their emergence in spring. These were gathered in May and their dry weight determined after oven drying as above. Dead sheaths were collected in June and fallen leaves gathered in the litter trays (section 6.3.6). Below-ground, dead material is composed of dead rhizomes and roots. This material was sampled in the pit excavations (section 6.3.3).

6.3.5 Decomposition

From May 1986, relative rates of decomposition were determined by enclosing samples of dead underground vegetation from the pit excavations into fine mesh (2 × 2 mm) litter bags. These were then buried at depths of 10, 20 and 30 cm in the study plots. Initially, litter bags were retrieved after 2 months, but this was extended to 3–4 months during the winter.

6.3.6 Calculation of net primary production

Net primary production (P_n) was calculated from the increments in above- and below-ground biomass, collected fallen leaves, dropped sheaths and dead young shoots, plus the absolute rate of decomposition, i.e. the sum of the increase in biomass and correcting for losses through death.

6.3.7 Canopy leaf area index

Canopy leaf area index (L) was determined directly from the leaves of shoots cut when making the 4 × 4 m pits for investigation of underground biomass. Additionally, L was estimated indirectly from the fallen leaves which were collected on a monthly basis.

Direct determination

All of the leaves on the harvested shoots in the 4 × 4 m pits were collected according to canopy strata. Canopy L was obtained by summation of the leaf areas of all the shoots cut from the 4 m^2 areas on which the pits for root and rhizome sampling were dug.

Indirect determination

Ten leaf collecting trays, each 1 m^2 in area and 15 cm deep, were randomly positioned and levelled in the sample site in 1984. Fallen leaves were collected and their area measured each month. Canopy L could be estimated from these data assuming that:

1. The leaf area of each shoot was the average for the forest as a whole;

2. New shoots renew their leaves in their second year, while mature shoots produce a new canopy, of the same leaf area, every two years;
3. Leaf area remains unchanged from the time it is fully formed until it drops from the shoot. New leaves expanded 2 months after the old leaves had dropped.

6.3.8 Physiological measurements

Measurements of leaf photosynthetic CO_2 uptake and transpiration were made with a portable open gas exchange system, comprising: an infra-red gas analyser (type LCA2), a leaf chamber (type PLC), and an air supply unit (type ASU, Analytical Development Co., Hoddesdon, UK). During each month the following measurements were made over a 3- to 4-day period: (1) leaf photosynthetic rates and stomatal conductances; (2) leaf light response curves, constructed by varying the photon flux at the leaf surface by placing neutral density filters over the chamber window; (3) the diurnal course of leaf photosynthesis; (4) comparison of photosynthetic rates of leaves in different canopy strata; (5) leaf and stem respiration rates.

Calculation of canopy photosynthesis

Canopy photosynthetic CO_2 uptake (A_c) was calculated according to the following equation:

$$A_c = \left[\frac{\alpha . I_o . A_{sat}}{\alpha . K . I_o + A_{sat}} - \frac{R_d}{k} \right] . \quad (1 - e^{-KF}) \qquad [6]$$

Where α is the apparent maximum quantum yield, A_{sat} is the light saturated rate of leaf photosynthetic CO_2 uptake, R_d is the leaf dark respiration rate, k is the canopy extinction co-efficient, L is the canopy leaf area index and I_o is the photon flux incident on the top of the canopy.

Values for A_{sat} were calculated from light response curves after fitting non-rectangular hyperbolic curves (Zhang, 1988). Daily canopy photosynthesis (A_d) was calculated using:

$$A_d = \left[\frac{\alpha . h . A_{sat} . I_a}{\alpha . k . I_a + h . A_{sat}} - \frac{R_d}{k} \right] . \quad (1 - (I_a/I_t)) \qquad [7]$$

Where I_a is the total photon flux intercepted by the canopy, I_t is the total photon flux incident on the canopy and h is the day length.

Yearly canopy photosynthesis (A_y) was estimated from:

$$A_y = \sum_{i=1}^{i=12} \left[\frac{\alpha_i.A_{sati}.h_i.Ia_i}{\alpha_i.k.Ia_i \pm h_i.A_{sat}} \quad \frac{R_{di}}{k} \right] . \left[1 - \frac{I_{a,i}}{I_{t,i}} \right] .24.n_i \qquad [8]$$

Where n_i is the number of days in the $_i$th month.

Stem respiration rates

Stem respiration was measured with an assimilation chamber attached to the stem *in situ*, covering an area of approximately 100 cm^2 (Beadle, 1986).

6.4 RESULTS

6.4.1 Above-ground biomass

At the end of 1986, the total above-ground biomass of the forest was estimated to be 5.578 kg m^{-2}, an increase of 0.877 kg m^{-2} since the end of 1985 (Table 6.1; Figure 6.6). The high standard errors associated with the data in Figure 6.6 arose from the uneven distribution of *P. pubescens* within the 16 permanent plots. During 1987, the yearly increase was 0.656 kg m^{-2}. The lower figure for 1987 is due to the reduced number of new shoots which emerged in that year (60, as compared to 71 in 1986).

6.4.2 Below-ground biomass

The vertical distribution of below-ground biomass is shown in Figure 6.7. Approximately 90% of the total underground biomass was located within 60 cm of the surface. Changes in below-ground biomass for one set of pits is illustrated in Figure 6.8. In February 1987 the mean amount of below-ground biomass was found to be 5.69 kg m^{-2}, an increase of 0.6 kg m^{-2} over that year (Table 6.1). Thus at the end of 1986, below-ground biomass constituted almost 50% of the total biomass. Because rhizomes may live for many years, rhizomes and rhizome segments formed in the 8 years constituted over 90% of the total mass.

Due to the uneven coverage of soil along the hillside and the nature of the root system, it was difficult to maintain low sample variability.

Table 6.1. Summary of data for the Miao Shen Valley *Phyllostachys pubescens* stand

Biomass ($kg\ m^{-2}$)	
Total at the end of 1986	11.268
Above ground	5.578
Below ground	5.690
Increment in 1986	1.477
Above ground	0.877
Below ground	0.600
Harvested at the end of 1985	0.415
Dead materials	0.439
Fallen leaves	0.232
Dead shoots	0.027
Dropped sheaths	0.020
Dead roots and rhizomes	0.160
Decomposition rate (% per month)	6.58
Vegetation	
No. of shoots per hectare	3437 ± 975
Weight (kg) of single shoot (aver.)	17.2
Stem circumference (cm) at 1.6 m	29.7 ± 5.7
Wood density (kg dm^{-3}at 1.6 m	
New shoot	0.40
Old shoot	0.65
Canopy leaf area index	8.02
Canopy extinction coefficient	0.30
Weight % of leaves in total biomass	3.2
Specific leaf weight (g m^{-2})	50
Biomass ratio of above:below ground	58:42
Respiration (kg m^{-2})	
Total amount in 1986	
Maintenance respiration loss	0.106
Growth respiration loss	0.396
Primary production (kg m^{-2} y^{-1})	
Net production from biomass data	
1.477 + 0.232 + 0.020 + 0.027	1.756 kg m^{-2}
Net production plus respiration loss	
1.756 + 0.396 + 0.106	2.258 kg m^{-2}
Calculated photosynthetic gain	
69 mol m^{-2} × 0.030 kg mol^{-1}	2.070 kg m^{-2}
Total amount of solar radiation in 1986	3342 MJ m^{-2}
Efficiency of solar radiation: biomass conversion	
1.756g m^{-2} ÷ 3342 MJ m^{-2}	
or solar energy utilization	0.525 g MJ^{-1}
0.525 g MJ^{-1} × 18.8 kJ g^{-1}	1.0%
Development (years)	
Life span of forest	
Flowering at the age of then whole forest died	60–80
Life span of rhizomes	20
Seedling develops into mature forest	8–10

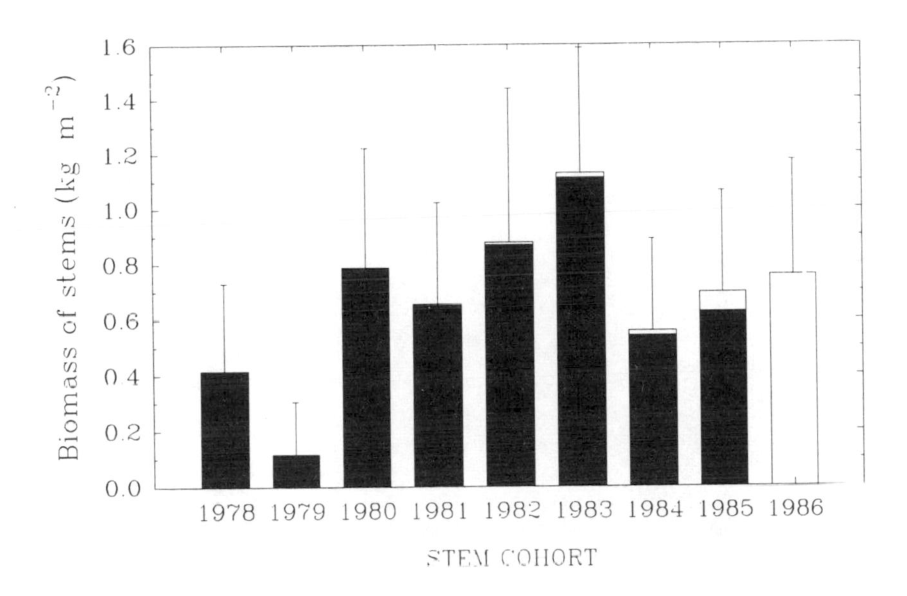

Figure 6.6. The biomass of stems (+ 1 se) plotted against stem age. The solid bar illustrates all of the live shoot biomass (dry weight per unit area) present at the end of 1985, the open bars indicate the additional shoot biomass added to each age class in 1986.

However, a reliable measure of below-ground biomass increments could be obtained from the ratio of below- to above-ground biomass (Huang, 1982), determined from $4 \times 4\,m^2$ excavations and the increments of above-ground biomass. This gave, for 1986, a figure of $0.877 \times (42/58) = 0.635$ kg m^{-2}, as compared to the estimate of 0.600 kg m^{-2} calculated above.

6.4.3 Dead material and decomposition

In 1986, dead shoot material increased by 0.027 kg m^{-2} and the amount of dead sheath material rose by 0.02 kg m^{-2}. Over the same period, fallen leaves were found to accumulate by 0.232 kg m^{-2}.

Since the life span of rhizomes is about 20 years, and this bamboo forest is only a little more than 20 years old, it is not surprising that few dead rhizomes were found in the samples. The dry weight of dead rhizomes and roots was 0.16 kg m^{-2} in February 1986 and 0.11 kg m^{-2} in February 1987. The amount of dead roots was low, perhaps the dead fine roots decayed very quickly and were therefore undetectable by the sampling methods used.

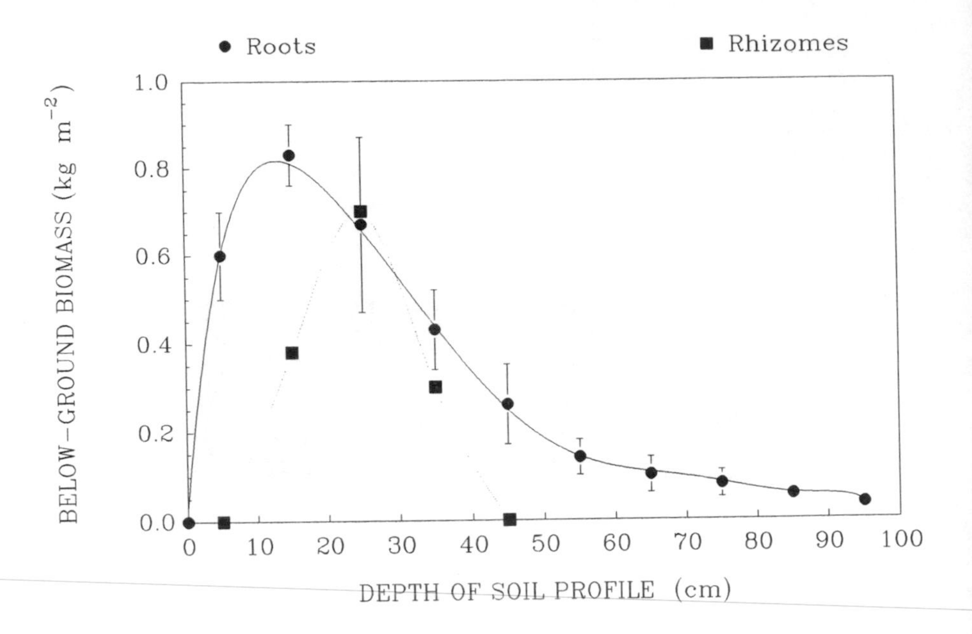

Figure 6.7. The distribution profile of live roots and rhizomes with depth of soil, expressed as dry weight per unit ground area in each 10 cm slice of soil.

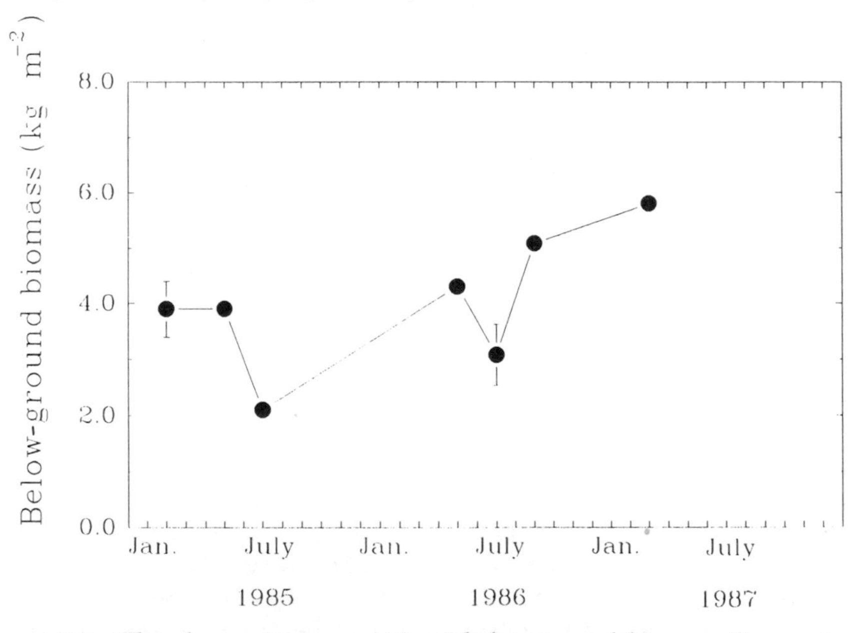

Figure 6.8. The change in mean (±1 se) below-ground biomass (live roots and rhizomes), illustrated for one set of pits.

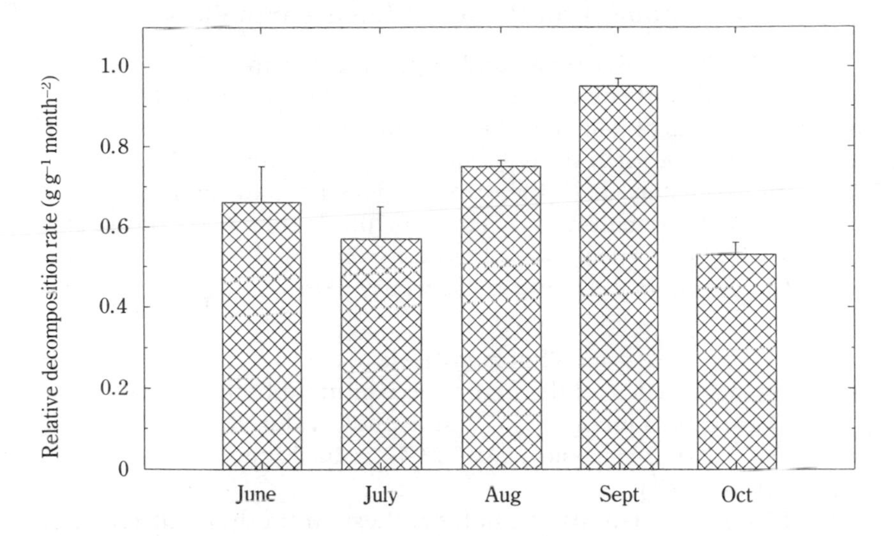

Figure 6.9. The relative rate of decomposition (+1se) of dead vegetation below ground during the summer and autumn of 1986. Values are the means of measurements made at 10, 30 and 50 cm depth.

Relative rates of decomposition from June 1986 are shown in Figure 6.9. There was no significant difference in the rate of decomposition in relation to depth from 10 to 50 cm ($P>0.05$). The lowest rates recorded were 0.04 g g^{-1} $month^{-1}$ at 50 cm in July 1986 and the highest 0.1 g g^{-1} $month^{-1}$, also at 50 cm, in September 1986. The mean relative rate of decomposition for 1986 was 0.07 g g^{-1} $month^{-1}$, and the mean amount of dead material below ground was 0.17 kg m^{-2}. This gave a mean absolute rate of decomposition of 0.012 kg m^{-2} $month^{-1}$ and an annual total of 0.144 kg m^{-2}.

6.4.4 Net primary production and photosynthetic energy conversion efficiency

In 1986, the total increase in biomass was 1.477 kg m^{-2}. Dead material accumulated by 0.279 kg m^{-2} and absolute decomposition was 0.144 kg m^{-2}. Summed together, this gives a total annual P_n of 1.90 kg m^{-2}. Turnover, expressed as P_n/mean biomass, was 1.9/10.28 = 0.18. The annual absorption of solar radiation by the *P. pubescens* canopy in 1986 was 3342 MJ m^{-2}. Assuming that the energy content of organic material is between 17 and 20 MJ kg^{-1} (Roberts *et al.*, 1985), the efficiency of photosynthetic energy conversion was approximately 0.01.

6.4.5 Photosynthesis and related parameters

Leaf area index and light interception

The leaf area index of the *P. pubescens* canopy was calculated from the monthly collection of fallen leaves (Figure 6.10). The main leaf fall occurred between April and June. Although leaves live for 2 years, this loss of approximately half the canopy is sufficient to allow large increases in light penetration (Figure 6.10). Figure 6.11 shows the vertical distribution of leaf area obtained from the 4 × 4 m quadrats, measured in February 1986 and in February 1987.

Photosynthesis

New leaves achieved slightly higher maximum photosynthetic rates (A_{sat}) of 10.24 ± 0.43 μmol m^{-2} s^{-1} at a photon flux density (I) of 833 μmol m^{-2} s^{-1} than 2-year-old leaves (9.40 ± 0.63 μmol m^{-2} s^{-1} at a I of 651 μmol m^{-2} s^{-1}).

The light response curves of photosynthesis from different layers of the canopy are shown in Figure 6.12. A_{sat} was highest for the upper leaves (18.1 μmol m^{-2} s^{-1}) and declined to 7.7 μmol m^{-2} s^{-1} for leaves at the canopy base. Diurnal changes in photosynthetic capacity (Figure 6.13) showed a significant midday depression, particularly during the summer. Both A_{sat} and the apparent maximum quantum yield appear depressed in leaves by mid-afternoon. On a seasonal basis, the highest photosynthetic rates occurred in the autumn (Table 6.2).

6.4.6 Respiration and respiratory losses

Stem respiration rates measured in September 1985 at 20°C were 0.46 μmol m^{-2} s^{-1} for new (1985) shoots, 0.21 μmol m^{-2} s^{-1} for 1-year-old (1984) shoots and 0.11 μmol m^{-2} s^{-1} for 2-year-old (1983) shoots.

Maintenance respiration

The mean maintenance respiration rate of *P. pubescens* was calculated assuming the following:

1. The data in section 6.3.6 are representative of shoots of average size, that is with a circumference of 31.3 cm and total dry weight of 18.00 kg.
2. Average stems are cone shaped with a height of 15 m and total surface area of 0.313 × 15/2 = 2.35 m^2
3. Shoots which are older than two years can be regarded as mature and their respiration taken to be at maintenance level.

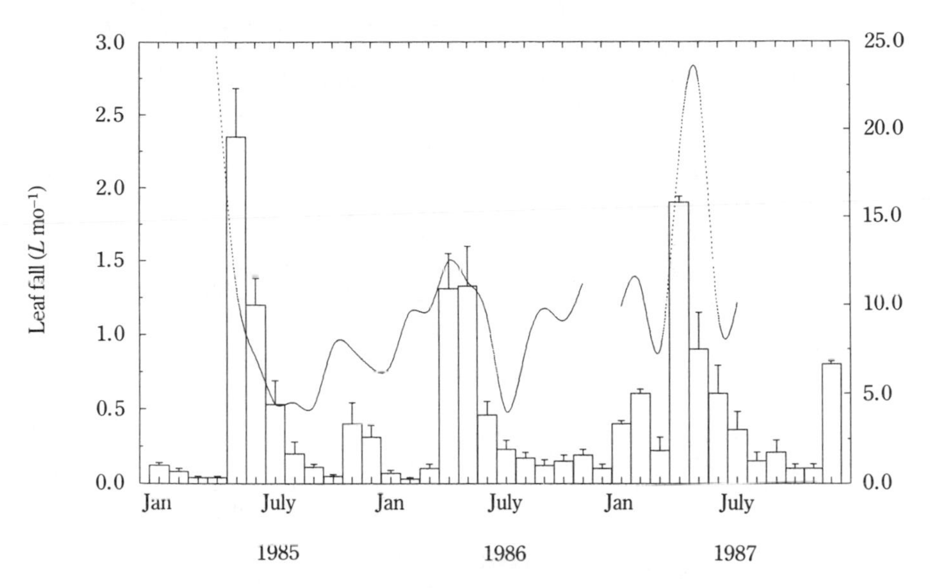

Figure 6.10. Mean (+1 se) leaf fall (vertical bars) per month, expressed as fallen leaf area index, i.e. m^2 of leaf area fallen per m^2 of ground. The dotted line indicates the percentage of light incident at the top of the canopy, which penetrates to the base.

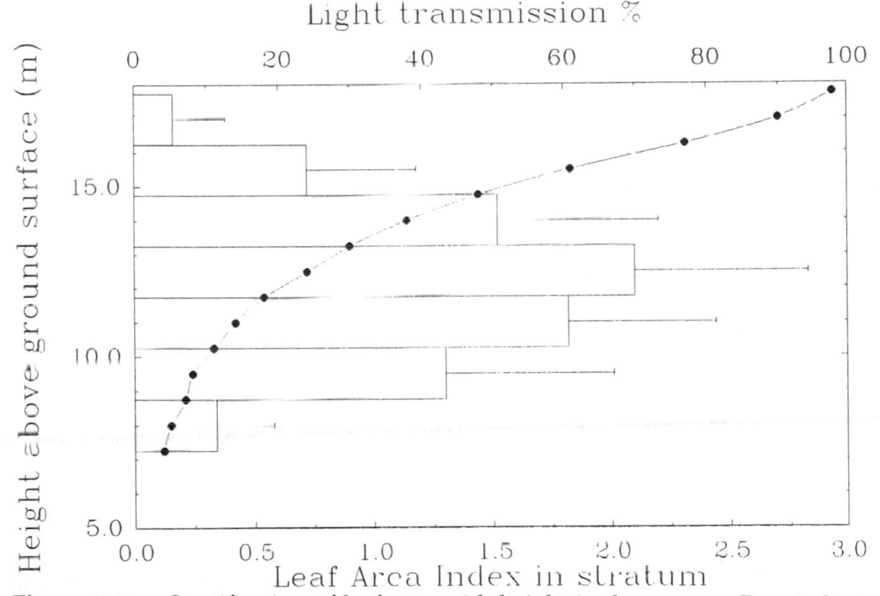

Figure 6.11. Stratification of leaf area with height in the canopy. Bars indicate the leaf area index (+1 se) in each slice of 1.5 m of vertical distance. The continuous line indicates the percentage of light present at the top of the canopy which penetrates to different depths within the canopy.

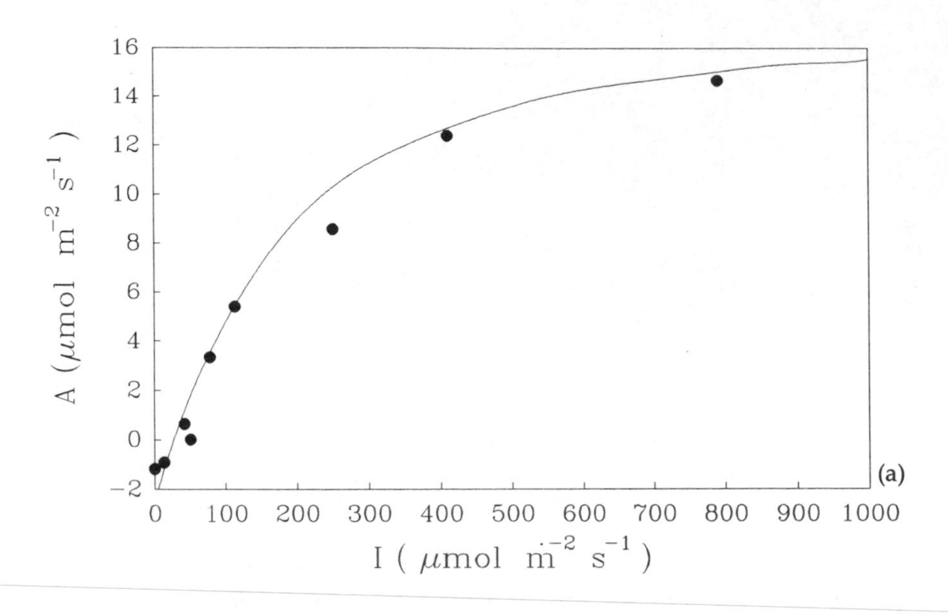

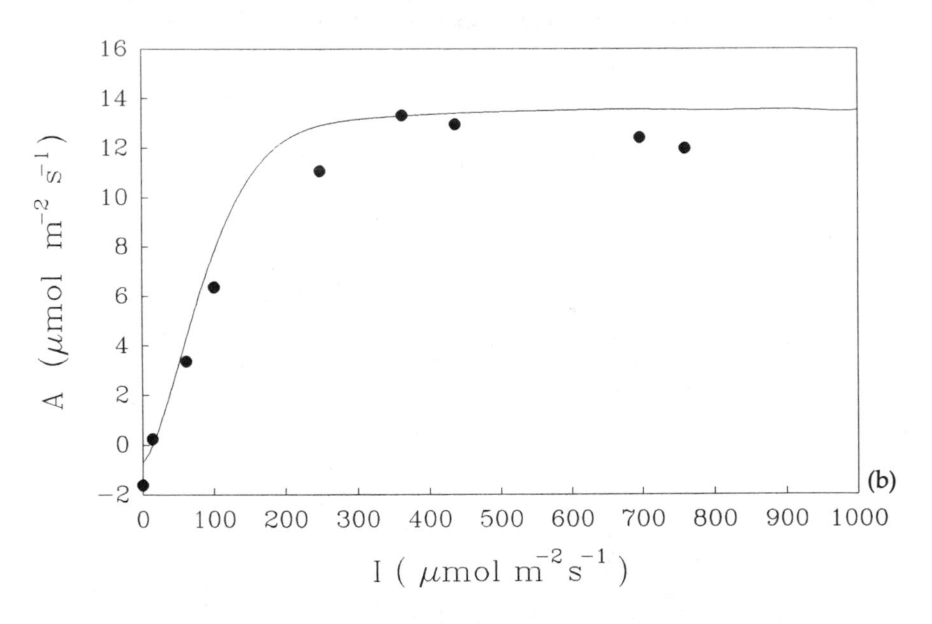

Figure 6.12. (a) An example response of CO_2 uptake rate per unit area of leaf (A) to photon flux (I) for a leaf from the top of the canopy sampled in the early morning; (b) as for (a), but a leaf from the middle of the canopy.

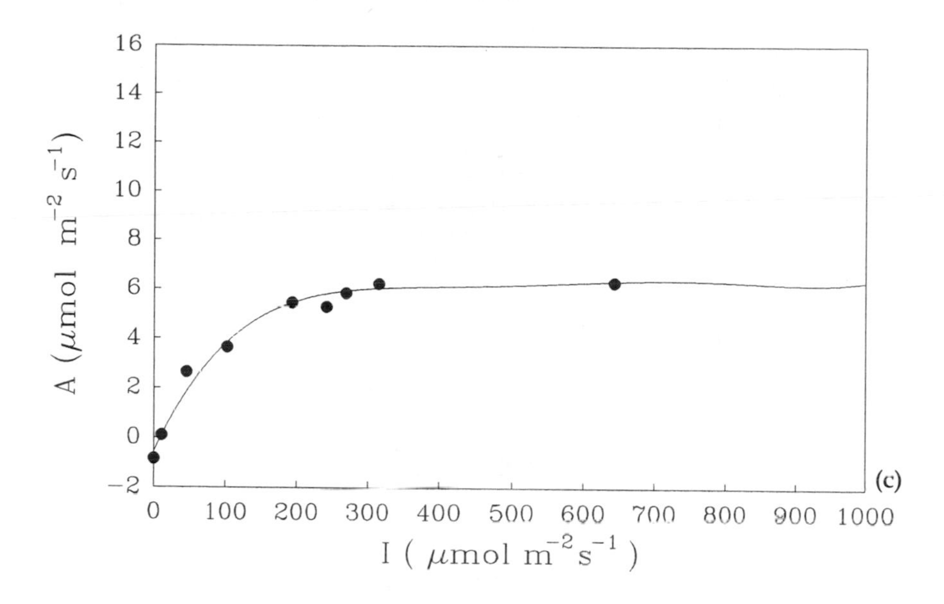

Figure 6.12. *continued* (c) As for (a), but a leaf from the base of the canopy.

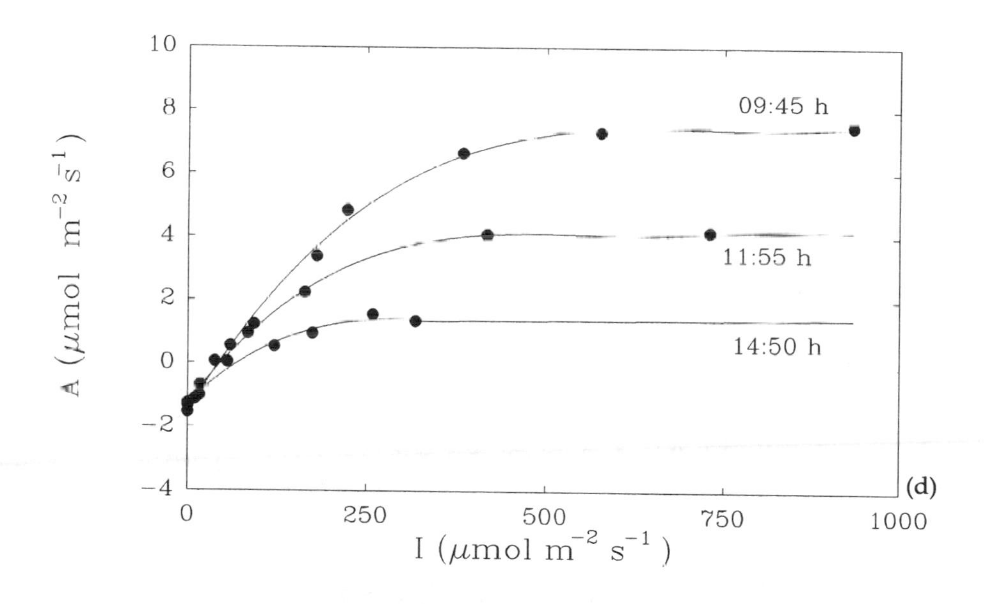

Figure 6.13. Examples responses of CO_2 uptake rate per unit area of leaf (*A*) to photon flux (*I*) for a leaf from the top of the canopy at three different times through the course of 1 day in July 1986.

Table 6.2. Seasonal variation in bamboo leaf photosynthetic rates and related parameters for the upper canopy. Values are means of all measurements made between 11:00 and 13:00 h

Date	T_a (°C)	Q (μmol m^{-2} s^{-1})	g_s (mol m^{-2} s^{-1})	c_i (μmol mol^{-1})	A (μmol m^{-2} s^{-1})	n
29 Sept. 1985	29.4 ± 0.9	1 280 ± 431	0.260 ± 0.043	223 ± 6	9.51 ± 0.98	7
8 Oct. 1985	27.5 ± 1.6	1 197 ± 175	0.348 ± 0.048	243 ± 7	11.13 ± 0.83	9
11 Nov. 1985	24.7 ± 1.2	864 ± 78	0.078 ± 0.022	236 ± 12	3.45 ± 1.14	9
5 Jan. 1986	6.7 ± 1.1	1 265 ± 150	0.031 ± 0.008	330 ± 4	0.31 ± 0.10	6
22 Feb. 1986	14.4 ± 0.6	766 ± 167	0.044 ± 0.090	284 ± 19	1.22 ± 0.54	9
31 Mar. 1986	18.0 ± 1.4	1 553 ± 135	0.101 ± 0.022	283 ± 5	2.56 ± 0.57	7
19 May 1986	33.4 ± 0.7	1 416 ± 235	0.197 ± 0.040	277 ± 13	5.17 ± 1.91	5
27 June 1986	31.5 ± 1.4	887 ± 250	0.217 ± 0.087	222 ± 13	8.80 ± 2.01	4
24 July 1986	35.3 ± 1.1	669 ± 62	0.197 ± 0.036	229 ± 21	6.58 ± 1.36	4
5 Aug. 1986	35.1 ± 1.3	1 108 ± 98	0.277 ± 0.036	221 ± 4	8.56 ± 0.91	4
9 Sept. 1986	31.8	760 ± 193	—	—	11.00 ± 2.64	4
10 Oct. 1986	26.7 ± 1.4	832 ± 232	0.273 ± 0.065	212 ± 17	12.15 ± 2.72	9
6 Nov. 1986	25.1 ± 0.5	1 163 ± 26	0.383 ± 0.116	203 ± 12	15.54 ± 1.50	6

A, leaf photosynthetic CO_2 uptake rate; C_i, intercellular CO_2 concentration; g_s, stomatal conductance; *Q*, photon flux; T_a, air temperature.

The respiration (i.e. maintenance) rate of a two-year-old shoot would therefore be $0.11 \times (2.35/18.00) = 0.0144\ \mu\text{mol kg}^{-1}\ \text{s}^{-1}$. On a monthly basis, the rate at 20°C would be $0.0144 \times 60 \times 60 \times 24 \times 30 = 0.0373\ \text{mol kg}^{-1}\ \text{month}^{-1}$.

Calculation of annual respiratory losses

Assuming that a temperature coefficient of $Q_{10} = 2$ could be applied throughout the year for biomass respiration, then the correction factor (q_i) for the rate of respiration at an average monthly temperature (T_i°C) other than at 20°C will be:

$$q_i = \frac{2.T_i - 20}{10} \qquad [9]$$

where T_i is the average temperature of the *i*th month. The yearly amount of respiratory loss for 1986 would therefore be:

$$\sum_{i=\text{Jan}}^{i=\text{Dec}} 0.0373.q_i \qquad [10]$$

The calculated maintenance respiration rate for the whole of 1986 was $0.343\ \text{mol kg}^{-1}$.

The growth respiration rate of above-ground biomass over a period of time could be estimated from the increase in wood density and the difference in respiratory rate of new shoots together with maintenance respiration. Taking 1985 as a typical year, the average densities of new and 1-year-old (1984) shoots were 0.400 and $0.533\ \text{g cm}^{-3}$ of fresh biomass respectively, with an increment of $0.133\ \text{g cm}^{-3}$ in 1 year. The mean shoot water content is 50%, so the dry matter increase on a dry weight basis for shoots over 1 year is equal to 0.133/0.5, equivalent to $266\ \text{g kg}^{-1}$.

The yearly loss through growth respiration can then be calculated from the ratio of growth respiration to maintenance respiration, multiplied by the yearly loss due to maintenance respiration:

$$\frac{0.46 - 0.11}{0.11} \times 0.343 \times 30 = 32.7\ \text{g kg}^{-1}$$

Therefore, over 1 year's growth, the percentage of photosynthate lost in biomass conversion is given by:

$$\frac{32.7}{32.7 + 266} = 11.0\%$$

Total respiratory losses of forest in 1986

Total forest biomass was 11.268 kg m^{-2} in 1986. Total maintenance respiration was:

$$11.268 \times 0.343 \text{ mol kg}^{-1} \times 30 \text{ g mol}^{-1} = 116 \text{ g m}^{-2}$$

The total biomass increment in 1986 was the sum of increases in above- and below-ground biomass, fallen leaves, dead young shoots and fallen sheaths of new shoots:

$$0.877 + 0.600 + 0.232 + 0.027 + 0.020 = 1.756 \text{ kg m}^{-2}$$

The respiratory cost of this biomass increment was calculated to be:

$$1.756 \times \frac{11}{89} = 0.217 \text{ kg m}^{-2}$$

Gross primary production (P_g) could then be calculated by summing P_n with respiratory losses due to maintenance and growth:

$$P_g = 2.06 + 0.116 + 0.217 = 2.383 \text{ kg m}^{-2}$$

6.5 DISCUSSION

The above-ground biomass of 5.58 kg m^{-2} in 1986 is similar to the higher estimate of Suwannapinunt (1983) for *T. siamensis* in Thailand. However, Veblen *et al.* (1980) quote a much higher above-ground biomass for *C. culeou* in Chile. Nevertheless, our estimate of P_n for *P. pubescens* is higher than any of those quoted in the Introduction. It is of the same order of magnitude as the average of yearly yields of local crops. For example, 2.8 kg m^{-2} y^{-1} is typical for land cropped each year with barley or wheat, followed by rice. However, local crops are subject to intensive cultivation all year round. The forest under study is subject to little management, and its productivity could be increased significantly by the application of fertilizer. Our estimate of P_n is also comparable to grasslands throughout the tropics reported elsewhere in this book, and is on a par with the higher estimates of tropical and temperate forests quoted by Lieth and Whittaker (1975).

Previous estimates of the P_n of *P. pubescens* at Miao Shan Valley have been made from the harvest records of stems. Harvest weights were multiplied by their dry matter content and divided by the percentage of total biomass comprised by the stem. Losses to dead material were not included. For the period 1980–84, these estimates ranged from 1.55 to 1.98 kg m^{-2}, comparable to the result of this study. However, since these previous estimates did not include any measure of dead material or decomposition, they are probably underestimates of true P_n. If so, then our estimate of P_n is low. This is most probably due to the fact that 1986 was an arid year, with less precipitation and more evaporation than on average for the site. Previous studies have identified precipitation as one of the most important factors affecting the productivity of bamboo forests. Ma (personal communication), using harvest records of *P. pubescens* at Miao Shan from 1967 to 1981, found the following relationship between stem production and annual precipitation (r = 0.93):

$$y = 35.\log x - 90$$

where y is stem production and x is annual precipitation (mm). Chen (1982) found a close relationship between precipitation in August, September and October and the number of new shoots emerging in the following spring but one.

In 1986, the total biomass of *P. pubescens* increased by 1.5 kg m^{-2}, while the total dead material rose by 0.4 kg m^{-2}. This indicates that *P. pubescens* dominated ecosystems are net sinks for global carbon. Taking the annual increment of material at Miao Shan as an average, approximately 40×10^6 kg of organic matter may be accumulated yearly over the total area covered by *P. pubescens* in China.

Most previous studies of bamboo productivity have concentrated solely on the above-ground compartment. At Miao Shan, the below-ground systems contributed roughly half of the total biomass and were a major component of total P_n. Thus studies which fail to take below-ground production into account can, at best, only provide minimal estimates of net primary production.

The ratio of annual increase in biomass to total biomass is given by 1.76/11.27 = 0.156 for 1986. Similar ratios have been found for the above-ground biomass of other bamboo species, e.g. 0.15, 0.147 and 0.137 for *T. siamensis* (Suwannapinunt, 1983); 0.152 for *F. spathacea* and 0.20 for *S. fangiana* (Taylor and Qin, 1987).

A loss of 11% of photosynthate in growth respiration is somewhat lower than the figures suggested elsewhere (Charles-Edwards, 1982). There is little with which to compare the results from the photosynthesis study, due to the previous lack of portable gas exchange systems suitable for use at remote sites. However, it may be concluded that *P. pubescens* has photosynthetic characteristics similar to those of evergreen C_3 trees (Jarvis and Sandford, 1986), and is very sensitive to low humidity, high temperature and soil water deficit. Thus it often shows severe midday depression of photosynthesis during summer. In addition, the highest photosynthetic rates are found in the autumn. *P. pubescens* seems to be well adapted to the mild climate and plentiful rainfall of the middle to lower reaches of the Yangtze River.

6.6 CONCLUSIONS

This study has demonstrated that the productivity of *P. pubescens* at Miao Shan Valley is comparable to the higher estimates of P_n for tropical and temperate forests. A large proportion of both total biomass and P_n is provided by the below-ground systems of *P. pubescens*, confirming the importance of below-ground sampling in productivity studies of bamboo. The accumulation of both above- and below-ground biomass, and below-ground dead vegetation, observed during the study period, suggests that these bamboo forest systems are acting as net sinks for carbon. Since they cover large areas they may have considerable significance to current global cycling of carbon.

6.7 SUMMARY

1. Changes in the above-ground biomass of a *Phyllostachys pubescens* forest in Sichuan, China were monitored non-destructively from derived relationships between the circumference, volume, age and density of shoots. Below-ground biomass was monitored by excavating pits.
2. Above-ground biomass increased from 4.7 kg m^{-2} at the end of 1985 to 5.6 kg m^{-2} at the end of 1987. Over the same period, below-ground biomass increased by 0.6 kg m^{-2} to 5.7 kg m^{-2}. Below-ground biomass constituted almost 50% of the total. Dead material increased by about 0.4 kg m^{-2} in 1986.
3. Mean relative rates of decomposition varied from 0.04 to 0.1 g g^{-1} $month^{-1}$, both at 50 cm depth.
4. Total net primary production (P_n) for 1986 was calculated to be 1.9 kg m^{-2}, while total gross primary production (P_g) for the same year was 2.23 kg m^{-2}. The canopy intercepted over 3300 MJ m^{-2} of incident solar radiation, at a photosynthetic conversion efficiency of approximately 0.01.
5. The value of P_n determined in this study is higher than for previous

estimates on other species of bamboo. The increase in living and dead organic matter indicates that there is a net loss of carbon from the global carbon cycle to this ecosystem.

REFERENCES

Anon. (1986) Study of management techniques in even-year of *Ph. pubescens* forests. *Journal of Bamboo Research*, **5**(2), 64–77.

Bahadur, K.N. (1979) Taxonomy of bamboos. *Indian Journal of Forestry*, **2**, 222–41.

Beadle, C.L. (1986) *Photosynthesis and Related Measurements*. Report to UNEP from the Chinese Regional Centre.

Chang, P.-X. (1982) An investigation on the distribution of bamboo species in Zhejiang, Fujian and Jiangxi Provinces. *Journal of Bamboo Research*, **1**(2),77–90.

Charles-Edwards, D.A. (1982) *Physiological Determinants of Plant Growth*. Academic Press, Sydney.

Chen, G.-X. (1982) Studies on the etiology of the bamboo basal stalk (shoot) rot. *Journal of Bamboo Research*, **1**(2), 54–61.

Cheng, Y.-L. (1983) Potentialities of high yield and managerial techniques of annual-working bamboo groves of *Phyllostachys pubescens*. *Journal of Bamboo Research*, **2**(2), 71–81.

Hong, X. (1988) On present situation of bamboo artificial board manufacture in Zhejiang Province. *Journal of Bamboo Research*, **7**(3), 66–72.

Hsiung, W. (1987) Bamboo in China: new prospects for an ancient resource. *Unasylva*, **39**, 42–9.

Huang, Q.M. (1982) Dry matter distribution of bamboo seedlings, in *Cultivation Techniques of Bamboo* (*Phyllostachys pubescens*), Ministry of Forestry, Peoples Republic of China.

Janzen, D.H. (1976) Why bamboos wait so long to flower. *Annual Review of Ecology and Systematics*, **7**, 347-91.

Jarvis, P.G. and Sandford, A.P. (1986) Coniferous forests, in *Photosynthesis in Contrasting Environments* (eds N.R. Baker and S.P. Long), Elsevier, Amsterdam, pp. 167–219

Keng, P.C. (1984) A revision of the genera of bamboos from the world (VI). *Journal of Bamboo Research*, **3**(2), 1–22.

Keng, P.C. (1987) A systematic key of the tribes and genera of subfam. Bambusoideae (Gramineae) occurrence in China and its neighbours. *Journal of Bamboo Research*, **6**(3), 13–34.

Liao, G. (1988) Analysis and investigation of the structure of rhizome system of bamboo forests. *Journal of Bamboo Research*, **7**(3), 35–44.

Liese, W. (1989) Progress in bamboo research. *Journal of Bamboo Research*, **8**(2), 2–16.

Lieth, H. and Whittaker, R.H. (1975) *The Primary Production of the Biosphere*, Springer, New York.

Liu, J. (1987) A study on climatic zoning in *Phyllostachys pubescens* distribution range. *Journal of Bamboo Research*, **6**(3), 1–12.

Ma, N. (1989) Bamboo study in China. *Journal of Bamboo Research*, **8**(1), 75–82.

Qiu, F.-G. (1982) Glaze on *Phyllostachys pubescens* and its control. *Journal of Bamboo Research*, **1**(2), 72–4.

Roberts, M.J., Long, S.P., Tieszen, L.L. and Beadle, C.L. (1985) Measurement of plant biomass and net primary production, in *Techniques in Photosynthesis and Bioproductivity*, 2nd edn (eds. J. Coombs, D.O. Hall, S.P. Long and J.M.O. Scurlock). Pergamon Press, Oxford, pp. 1–19.

Suwannapinunt, W. (1983) A study on the biomass of *Thyrostachys siamensis* (Gamble) forest at Hin-Lap, Kanchanaburi. *Journal of Bamboo Research*, **2**(2), 227–37.

Taylor, A.H. and Qin, Z.-S. (1987) Culm dynamics and dry matter production of bamboos in Wolong and Tanajiahe Giant Panda Reserves, Sichuan, China. *Journal of Applied Ecology*, **24**, 419 - 33.

Veblen, T.T., Schlegel, F. and Escobar, B. (1980) Dry matter production of two species of bamboo (*Chusquea culeou*) and (*C. tenuiflora*) in south-central Chile. *Journal of Ecology*, **68**, 397–404.

Yi, T. (1988) A study of the genus *Fargesia* Fr. from China. *Journal of Bamboo Research*, **7**(2), 1–118.

Yuan, Z. and Xu, J. (1989) A preliminary study on the climate- ecology of *Phyllostachys pubescens* in Anhui Province. *Journal of Bamboo Research*, **8**(2), 37–44.

Zhang, J.X. (1988) Modelling the response of leaf photosynthetic rate in wheat and bamboo. *Photosynthetica*, **22**(4), 526–34.

7

Remote sensing of grassland primary production

J.M.O. SCURLOCK

7.1 INTRODUCTION

The measurements of primary productivity described in this volume provided an unrivalled opportunity to examine the potential of remote sensing for determination of biomass, leaf area and primary production at diverse natural sites within the tropics. Data on vegetation were already being collected, and the facilities to carry out additional remote sensing experiments were already present.

The work described here examined the possibility of using a rugged, inexpensive red/near-infrared sensor to monitor changes in above-ground biomass and leaf area index (L) throughout the year, for four natural tropical grass ecosystems of contrasting type. In addition, controlled environment studies were used to test its application to 'model canopies' of single and mixed species, under conditions where frequent sampling was possible during canopy development. The application of remote sensing to estimate light interception by plant canopies and ecosystem productivity was also attempted, using the comprehensive data on light interception and productivity (Scurlock *et al.*, 1989).

At the time of this study, there were few reports on the use of remote sensing for estimation of ground cover, light interception and net primary production. The subsequent sections therefore include exploration of instrument sensitivity and testing the validity of the approach (section 7.2), and application of the technique at the tropical study sites (section 7.3).

7.1.1 Remote sensing measurements

Canopy reflectance was measured using the Skye SKR100/110 spectral ratio meter (Skye Instruments, Llandrindod Wells, UK), which registers photon flux in 10 nm bandwidths centred on 660 nm (red) and 730 nm (near-infrared). These correspond quite well to the maxima for absorption of red radiation and reflectance of near-infrared radiation (NIR) by green plant tissue, although Tucker (1977) suggested that a slightly longer wavelength was optimal in the NIR. The ratio of red to near-infrared photon flux may also be read directly with this instrument. It was originally designed to measure light quality in glasshouses, but has already been used for remote sensing of temperate grassland (King *et al.*, 1986). The Skye spectral ratio meter was chosen for these studies under tropical conditions because it meets the requirements of cheapness, portability, reliability and ease of maintenance.

Red and near-infrared reflectance was measured as the ratio of photon flux reflected at 660 nm to that reflected at 730 nm, divided by the same ratio of reflected wavebands from a Lambertian reflector. This was accomplished by inserting a grey card (Grey card, Kodak Ltd.) between the sensor and the canopy, normal to the sensor axis and at a set distance from the sensor. The ratio of red to near-infrared reflectance (*R/NIR*) was calculated as follows:

$$R/NIR = \frac{\text{(canopy reflectance/grey card reflectance at 660 nm)}}{\text{(canopy reflectance/grey card reflectance at 730 nm)}}$$

To reduce the 180° field of view of the sensor head, a field stop was fitted, consisting of a matt black tube. The length of the stop tube was calculated to give a field of view of 40° (20° either side of vertical), i.e. a tube, 6.0 cm in length. Operating height was calculated so that the vegetation canopy sample was viewed as a circle of the desired radius (Figure 7.1); thus for a circle of 50 cm radius, the operating height was 139 cm. A small spirit level was mounted on the sensor head to check that it was vertical in use.

7.2 MODEL CANOPIES

These experiments were designed to evaluate the sensitivity of the Skye spectral ratio meter for remote sensing of vegetation canopies. In particular, different canopy structures were investigated to assess how they affected the usefulness of the technique.

7.2.1 Methods

The model canopies consisted of dense populations of seedlings, grown in square pots of vermiculite (expanded mica), packed together closely as squares of 10 × 10 or 14 × 14 pots to form a continuous

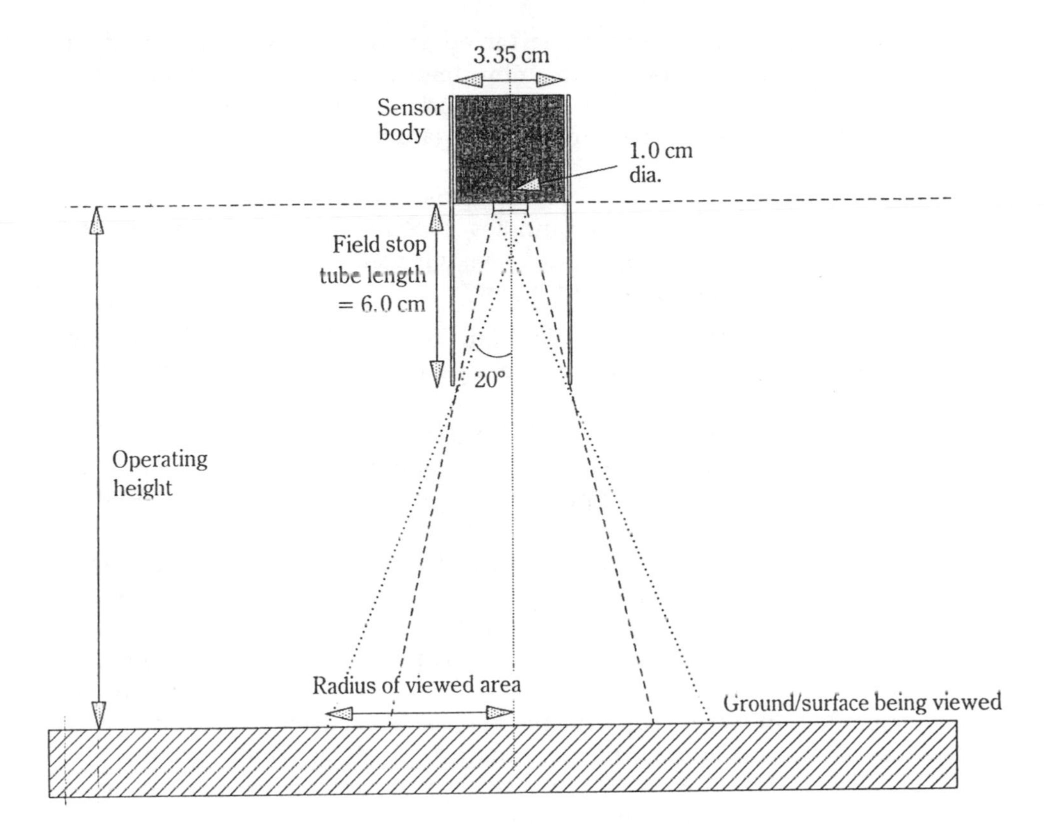

Figure 7.1. Illustration of arrangement of Skye Instruments red/near-infrared sensor with field stop tube, as used in the present study.

canopy. Studies were made on single-species canopies of maize and barley (*Zea mays*, *Hordeum vulgare*–graminaceous) and lettuce (*Lactuca sativa*–broad-leaf), grown in controlled environment chambers. Similar studies have been carried out on a simple controlled environment grown canopy of young cress (Curran and Milton, 1983), and on an outdoor canopy of potted fir trees (Ranson *et al.*, 1986). Measurements were also made outdoors on a mixed-species canopy of *H. vulgare* and *L. sativa*, and on a short mixed-grass lawn, containing *Lolium perenne*, *Festuca rubra* and *Agrostis* spp.

The controlled environment chambers were illuminated by either fluorescent tube lamps/incandescent bulbs or high-pressure sodium lamps/incandescent bulbs, with a photoperiod of 14 h. In both cases, the number of incandescent lamps was adjusted to give a *R/NIR* similar to that of natural daylight, and a photon flux of about 150 μmol m^{-2} s^{-1}. Temperatures were 25 and 10°C for the light and dark periods, respectively. The short mixed-grass lawn was measured in March 1986 (mean daily max./min. temperature 10/4°C) whilst the outdoor model canopy of *H. vulgare* and *L. sativa* was grown during July 1986 (mean daily max./min. temperature 22/15°C).

Experiments

The experiments on model canopies evolved as a series of tests to evaluate the technique of remote sensing using the Skye instrument. The experiments are described below.

1. Six destructive harvests were made over a 12-day period (i.e. one harvest every two days) for a canopy of young *Z. mays*, which has an erectophile canopy (vertically oriented leaves), grown in 5 × 5 cm square pots in a controlled-environment chamber.
2. In order to compare results for a planophile canopy (horizontal leaves), an identical experiment was performed over a 12-day period for a canopy of young *L. sativa*.
3. To obtain more frequent sampling during canopy development, a similar experiment involved daily destructive harvests over a 12-day period for a canopy of young *H.vulgare*.
4. Twelve destructive harvests were made over a 40-day period (i.e. every 3–4 days) for a mixture of *H. vulgare* and *L. sativa* grown outdoors in larger (10 × 10 cm) pots. This experiment was intended to allow development of more mature plants at wider spacing.
5. A mixed canopy of *H. vulgare* and *L. sativa* was grown in a controlled environment chamber over a 22-day period, for comparison with the outdoor experiment 4 above. No destructive harvests were made in this case, and measurements were restricted to *R/NIR*, light interception and ground cover.

Canopy reflectance, biomass and leaf area index

At each harvest, pots of seedlings were randomly sampled after measuring *R/NIR* at 5 grid points above the canopy. This was achieved by removing samples of 8 pots, selected by random numbers, and replacing them from a 'reserve bank' of extra pots, so that the canopy as a whole was kept intact. Leaf area was determined using an area meter (Delta-T Devices, Cambridge, UK) before drying all above-ground material to constant weight at 90°C.

Ground cover and light interception

Ground cover was measured with point quadrats. A 10 × 10 grid of points, 2 cm apart, was marked on a clear acetate sheet. This was held over the canopy at five locations and the number of 'hits' on green vegetation recorded to derive percentage ground cover.

Light interception was determined using miniature tube solarimeters (type TSM, Delta-T Devices, Cambridge, UK). Mean values of incident energy flux and energy flux at the base of the canopy were obtained over short (5 min) periods for 5 locations within the canopy. A continuous record of incident light energy was also kept, in order to calculate the apparent efficiency of light energy conversion to biomass.

Net primary production

Net primary production (P_n) was determined for the model canopies as the cumulative increase in biomass, above and below ground. Since all the plants were young and grown from seed, it was assumed that no net turnover occurred by death and decomposition. Below-ground biomass was determined by carefully separating roots from the vermiculite medium and drying to constant weight at 90°C. However, the quantity of vermiculite was too large for handsorting in experiment 4. Here, below-ground biomass was determined by growing plants in pre-weighed vermiculite, which was dried and weighed with the roots included.

Data analysis

R/NIR was plotted against both biomass and $\log_e$ (biomass), leaf area index (L) and $\log_e(L)$. The goodness of fit of data to a straight line was determined by preliminary inspection, by linear regression analysis and calculation of Pearson's correlation coefficient (r) (Snedecor and Cochran, 1956).

7.2.2 Results

Remote sensing of biomass and leaf area index

Inspection of graph plots for the model canopies suggested the correlations were stronger between *R/NIR* and $\log_e$ (biomass), and between *R/NIR* and $\log_e$ (L), than for untransformed data (Figures 7.2a, b).

Table 7.1 shows the coefficient of determination (r^2) and the standard error of the regression coefficient (expressed as a decimal fraction) of *R/NIR* on biomass, $\log_e$ (biomass), L and $\log_e$ (L). On the basis of maximum r^2 combined with minimum standard error, *R/NIR* shows a closer correlation with $\log_e$ (biomass) and $\log_e$ (L) than with untransformed data, for the model canopies.

Closer inspection of these data suggests that the *L. sativa* canopy (experiment 2) appears to form a regression distinct from the other model canopies' data. The regression coefficient of *R/NIR* on log (biomass) obtained for experiment 2 was compared statistically (t-test)

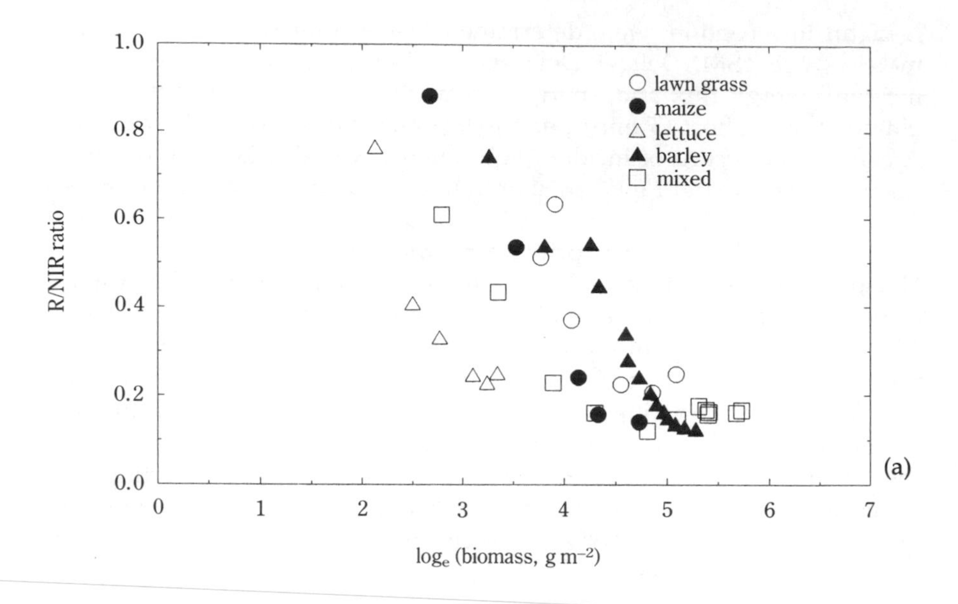

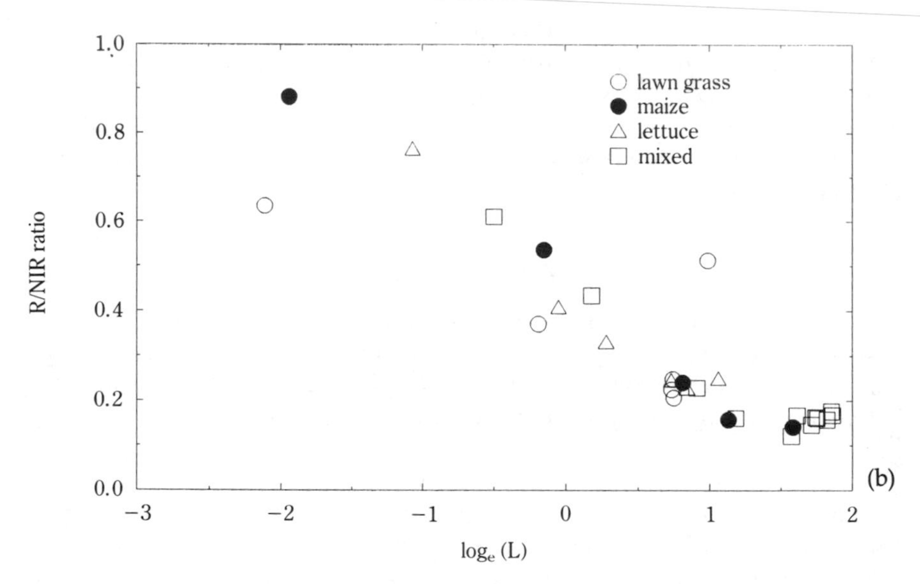

Figure 7.2. (a) The relationship between red/near-infrared reflectance ratio (*R/NIR* ratio) and $\log_e$ [biomass] for various types of model canopies. Symbols as indicated on the figure, where barley = *Hordeum vulgare*; lettuce = *Lactuca sativa*; maize = *Zea mays*; mixed = *H. vulgare*/*L. sativa*; lawn grass = mixed sward dominated by *Lolium perenne*, *Festuca rubra* and *Agrostis* spp; and (b) the relationship between *R/NIR* and $\log_e$[L] for various model canopies, as indicated on the figure and above.

Table 7.1. Remote sensing of biomass and leaf area index: linear regression analysis

	Linear regression of R/NIR ratio on			
	Biomass (g m^{-2})	$\log_e$ (*biomass*)	*Leaf area index* (L)	$\log_e$ (L)
All model canopies combined (n = 30)	r^2 = 0.417 se = 0.718*	r^2 = 0.535 se = 0.202	r^2 = 0.595 se = 0.494	r^2 = 0.873 se = 0.344
Kenyan grassland June to Dec. 1986 (n = 120)	r^2 = 0.233 se = 0.879*	r^2 = 0.182 se = 0.250	Only four mean values available	
Thailand grassland April 1986 to March 1987 (n = 139)	r^2 = 0.310 se = 0.904*	r^2 = 0.306 se = 0.227	r^2 = 0.263 se = 1.192*	r^2 = 0.217 se = 1.362*
Mexico grassland I. Jan. to Aug. 1986 (n = 120)	r^2 = 0.705 se = 0.519*	r^2 = 0.688 se = 0.374	r^2 = 0.799 se = 0.385	r^2 = 0.651 se = 0.505
II. Sept. to Dec. 1986 (n = 80)	r^2 = 0.659 se = 0.366	r^2 = 0.581 se = 0.113	r^2 = 0.742 se = 0.318	r^2 = 0.457 se = 0.874*
Brazil inundated grassland Jan. to Nov. 1986 (n = 110)	Only ten mean values available		r^2 = 0.261 se = 0.706*	r^2 = 0.367 se = 0.586*

r^2 = correlation coefficient squared (all correlations significant at $P < 0.05$).
se = standard error of regression coefficient, expressed as decimal fraction.
* Non-significant regression.

with that obtained for all other model canopies, and found to be significantly different (P<0.01). This might be explained by the planophile nature of the *L. sativa* canopy (as opposed to the other model canopies, which contained erectophile grasses).

Remote sensing of ground cover and light interception

Regression analysis of the data from the model canopies suggests a logarithmic relationship between *R/NIR* and percentage light interception (r^2=0.859, coefficient of variation 0.068) (Figure 7.3a) in preference to a linear relationship (r^2 = 0.714, coefficient of variation 0.272). For the purpose of direct comparison, the plot of *R/NIR* against

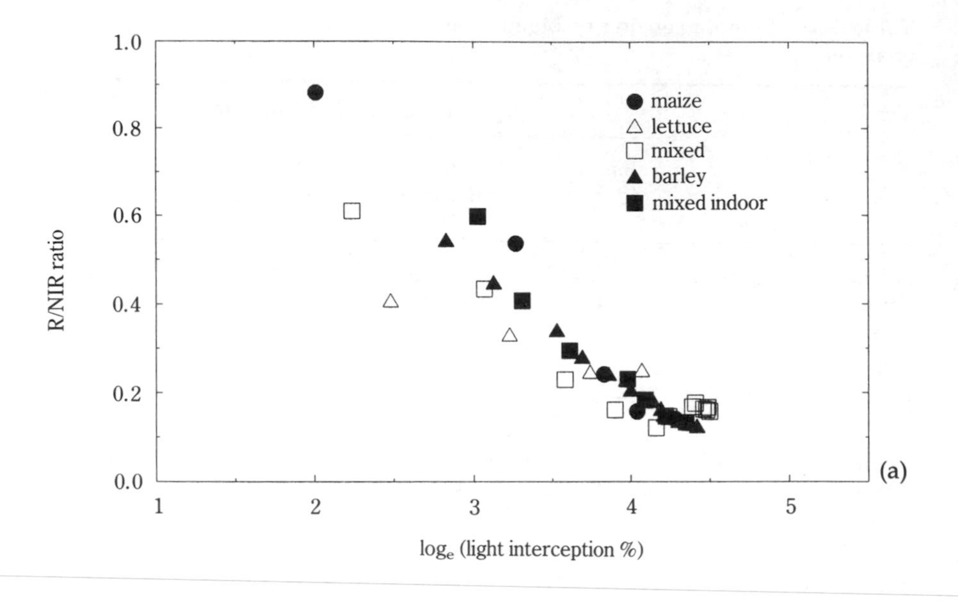

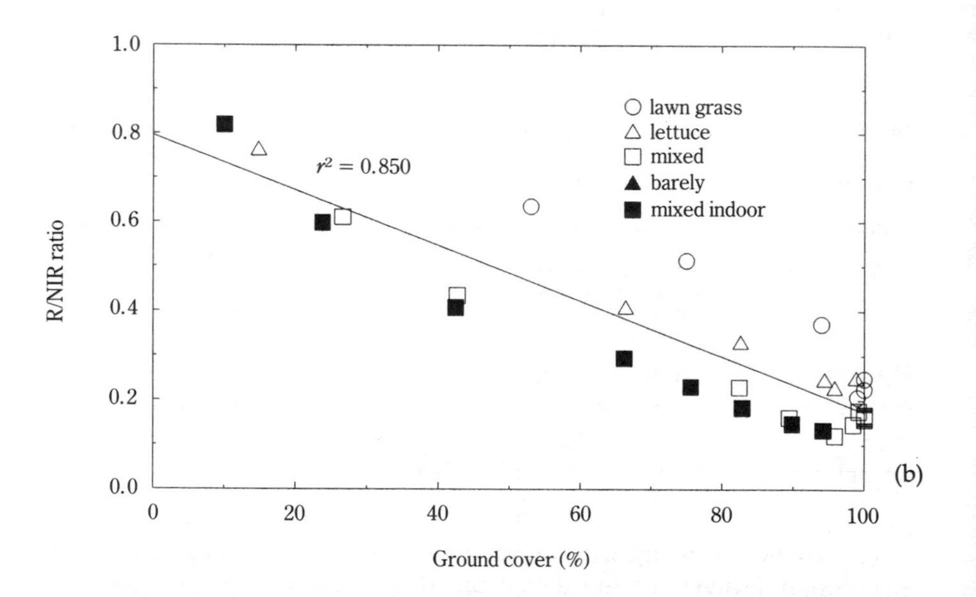

Figure 7.3. (a) The relationship between *R/NIR* and log$_e$ (light interception) for various types of model canopy. See marked symbols on figure and legend to Figure 7.2; (b) regression of *R/NIR* on percentage ground cover for various types of model canopy in combination.

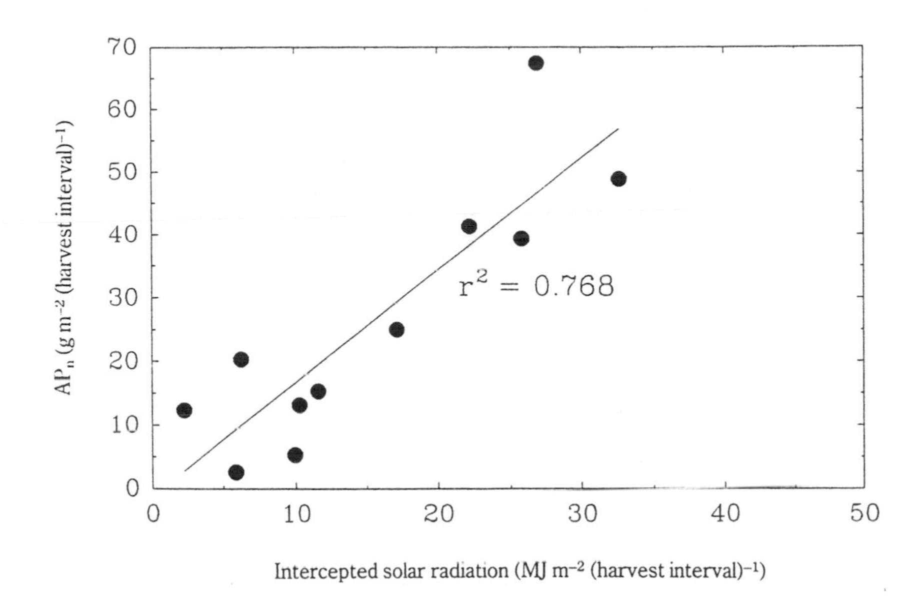

Figure 7.4. The relation between the proportion of incident light intercepted by different canopies and their *R/NIR* reflectance. (See Discussion for explanation of points marked '?'.)

light interception for the model canopies was drawn on the same axes as data from the tropical grass ecosystem sites (Figure 7.4).

The plot of *R/NIR* against percentage ground cover for the model canopies is shown in Figure 7.3b. Regression analysis of data suggests that this correlation is marginally more significant in linear form (r^2= 0.850, coefficient of variation 0.141) than in logarithmic form (r^2= 0.814, coefficient of variation 0.066).

Remote sensing of net primary production

The first four determinations of net primary production (P_n) showed a very good linear correlation between *R/NIR* and P_n (r^2 = 0.994) for the young *L. sativa* plants of experiment 2, performed under controlled incident radiation. However, experiment 4, conducted on *L. sativa* and *H. vulgare* plants outdoors, showed no such correlation between *R/NIR* and either P_n or AP_n (above-ground net primary production). Nevertheless, a good correlation was found between AP_n and intercepted radiation for this experiment (Figure 7.5), as was also found by Kumar and Monteith (1981) and Steven *et al.* (1983) for sugar-beet. Intercepted radiation was calculated as the product of mean light interception (%) and incident radiation (MJ m^{-2}), but the latter showed considerable variation with periods of sunshine and rain.

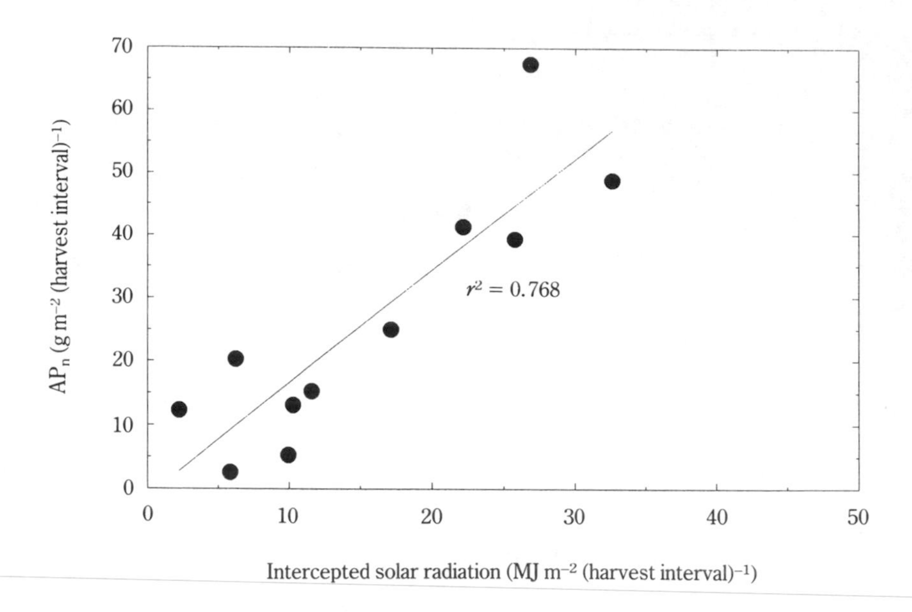

Figure 7.5. Relationship of the amount of new shoot (above-ground) biomass produced between harvests and the corresponding quantity of solar radiation intercepted for a mixed canopy (*H. vulgare*/*L. sativa*) grown in the field, in southern England. Solar radiation here is the total short-wave radiation (400–2000 nm). Harvest intervals were 3 or 4 days.

7.3 TROPICAL GRASSLAND SITES

7.3.1 Methods

The experiments on tropical grasslands were carried out at four sites of contrasting character between January 1986 and March 1987.

Biomass, leaf area index, light interception and net primary production were determined as described in the preceding chapters. Before clipping each quadrat at the terrestrial grassland sites, *R/NIR* was determined from a height of 1.39 m; the area viewed on the ground was a circle of radius 0.5 m, with the same centre as the clipped quadrat. For the purposes of this study, the circular scanned areas were assumed to be equivalent to the rectangular clipped areas

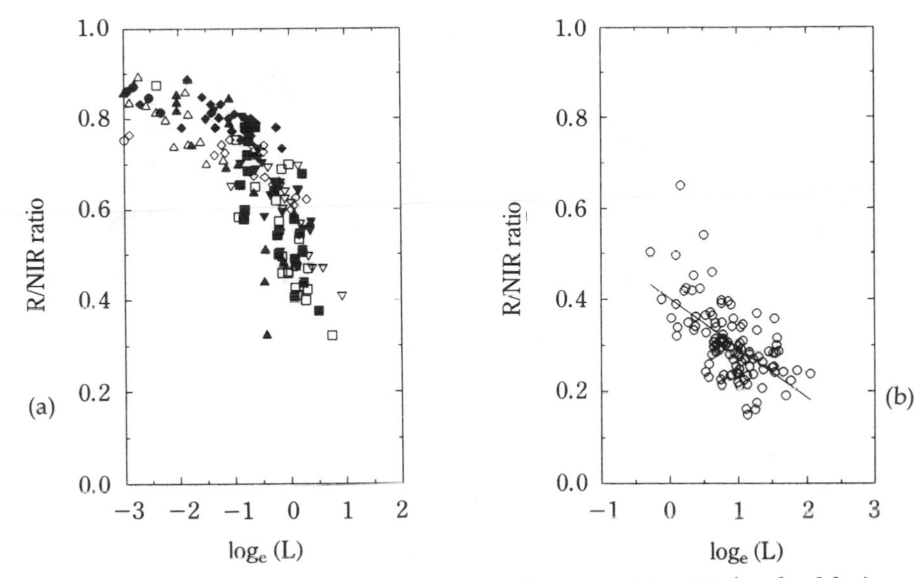

Figure 7.6. (a) The relationship between *R/NIR* and $\log_e(L)$ for the Mexican grassland site for January to December 1986. Symbols as given in Figure 7.8(c). (b) The relationship between *R/NIR* and $\log_e(L)$ for the Brazilian inundated grassland site. (M.T.F. Piedade, INPA, Brazil, unpublished data.)

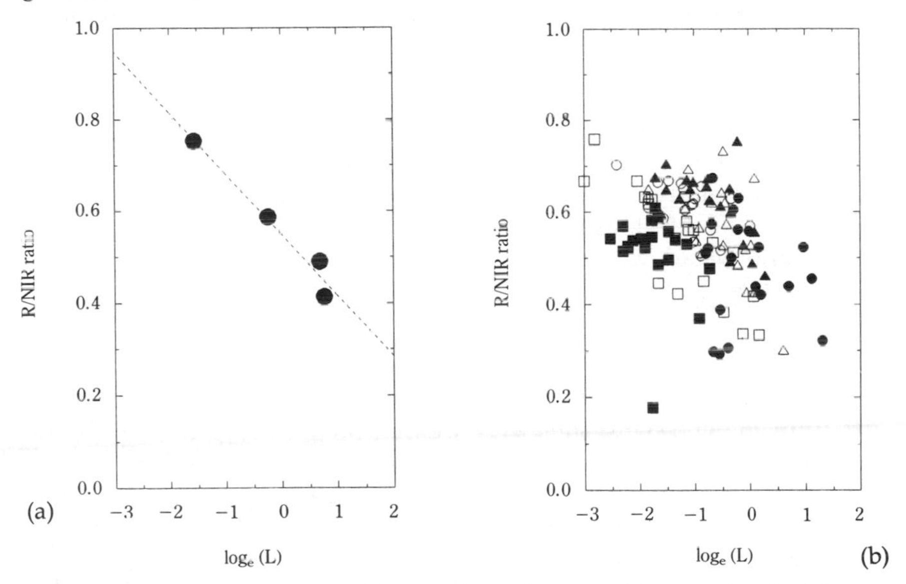

Figure 7.7. (a) The relationship between *R/NIR* and $\log_e(L)$ for the Kenyan grassland site. Each point is the mean of 20 quadrats from four different months. (J.I. Kinyamario, Kenya, unpublished data.) (b) The relationship between *R/NIR* and $\log_e(L)$ for the Thailand grassland site, for five different months in 1986, and one in 1987. Symbols as given in Figure 7.8b.

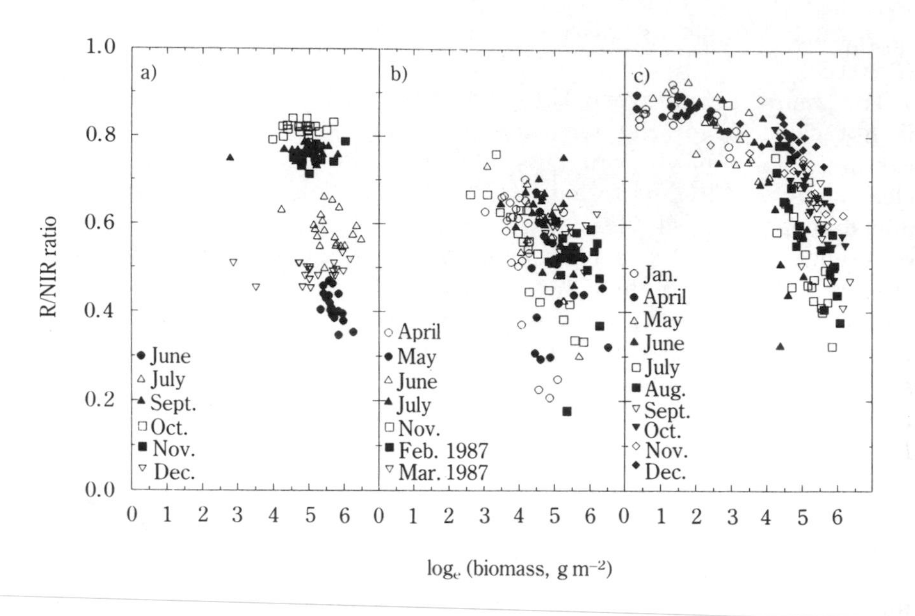

Figure 7.8. The relationship between *R/NIR* and $\log_e$(biomass) for the grassland site in (a) Kenya; (b) Thailand; and (c) Mexico. Different symbols indicate measurements in different months. All measurements were made in 1986, unless stated otherwise on the figure.

(1.0 m×0.25 m). For the inundated site in Brazil, *R/NIR* was determined by mounting the sensor at the end of a pole held about 1.0 m above the canopy in the areas to be sampled.

R/NIR was plotted against both biomass and $\log_e$ (biomass), *L* and $\log_e$ (*L*), for all the grassland sites. The goodness of fit of data to a straight line was determined as for the model canopies.

7.3.2 Results

Remote sensing of biomass and leaf area index

Data from the grassland sites were plotted on axes similar to those for the model canopies (Figures 7.6–7.8), enabling comparison of data from both sets of experiments. It is apparent by inspection that variation is much greater for the natural grassland sites than for the model canopies.

Table 7.1 shows the coefficient of determination (r^2) and the standard error of the regression coefficient (expressed as a decimal fraction) of *R/NIR* on biomass, $\log_e$(biomass), *L* and $\log_e$(*L*). All these correlations were significant ($P<0.001$), but in several cases the regression coefficient was not significantly different from zero ($P>0.05$).

The distinction between $\log_e$-transformed and untransformed data is not as clear for the terrestrial grassland sites as for the model canopies. At the Mexican site, the association of *R/NIR* with *L* is clearly stronger than with $\log_e (L)$, but the relationship between *R/NIR* and biomass is not clear; there is also no statistically significant difference between data collected during the growing and senescing seasons. Data for Kenya and Thailand show high standard errors compared with Mexico. On visual inspection, the regressions obtained for the different terrestrial grassland sites appear to have different slopes and intercepts, but statistical comparison of the regression coefficients using a *t*-test found them not significantly different at $P=0.05$, due to the considerable variation in the data for the Kenya and Thailand sites. At the inundated site in Brazil, there was no relationship between biomass and *R/NIR* ratio, but *R/NIR* was more closely correlated with $\log_e (L)$ than with *L* (Figure 7.6b).

Each plot from the grassland sites is composed of data from individual months, displayed as different symbols in Figures 7.6–7.8. Although the spread of data is fairly good at the Mexican and Thailand sites, the Kenyan data (Figure 7.8a) form a series of clusters with little or no overlap. From initial values obtained at the end of the long wet season (June), a considerable increase in *R/NIR* is observed for only a modest decrease in biomass as the dry season progresses into October and November. In response to the short wet season in December, *R/NIR* returns to near its June value, again for little change in biomass. This suggests that the relationship between *R/NIR* and canopy biomass is not consistent throughout the year.

Remote sensing of ground cover and light interception

Insufficient data were available to allow plotting of *R/NIR* against light interception for the individual sites. A plot of *R/NIR* against percentage light interception, using data from all four sites together with data from the model canopies, is shown in Figure 7.4. Although a significant relationship was apparent for model canopies, this was not seen for the tropical grasslands.

Remote sensing of net primary production

Mean *R/NIR* was plotted against P_n estimated on a monthly basis for all four tropical grassland sites. Regression analysis showed the correlation was non-significant for each site (*r* not significantly different from null hypothesis at $P=0.05$). The failure to find any

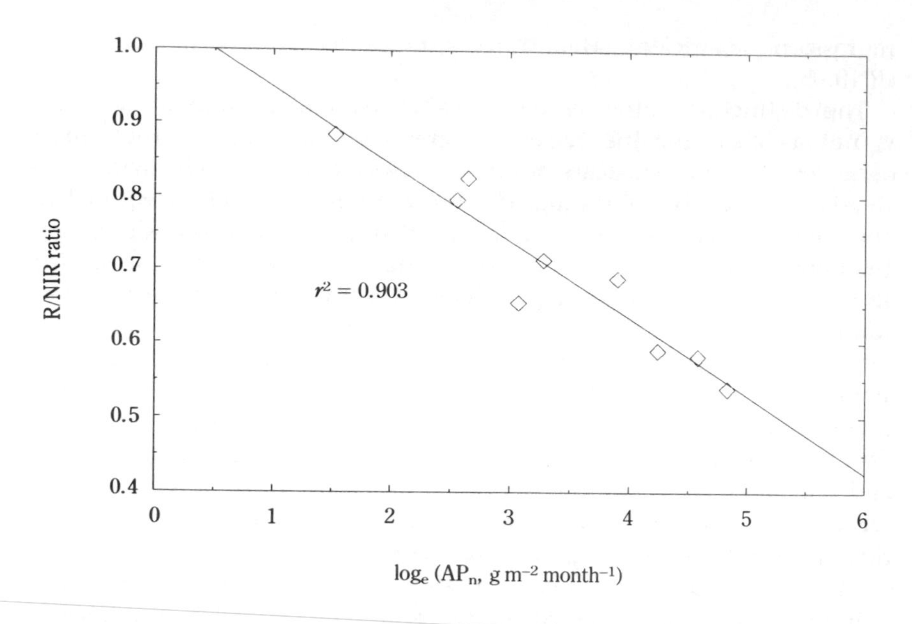

Figure 7.9. Regression of *R/NIR* on $\log_e(AP_n)$ for the Mexican grassland site, where AP_n = above-ground net primary production. Measurements made between April and December 1986, following burning of the site.

correlation between *R/NIR* and P_n might have been due to variability in the estimates of below-ground production, which showed considerable monthly fluctuations. In order to test this hypothesis, the association of *R/NIR* with AP_n was investigated. In the case of the Mexican grassland, which was burned in February 1986, a strong correlation (r^2=0.903) was found between *R/NIR* and $\log_e(AP_n)$ for 9 months following burning (Figure 7.9). However, in all other cases, the correlations were found to be non-significant.

7.4 DISCUSSION

7.4.1 Remote sensing of biomass and leaf area index

All the model canopies showed similar strong correlations between *R/NIR* and $\log_e$ (biomass), and between *R/NIR* and $\log_e(L)$. This suggests that the Skye spectral ratio meter is a suitable inexpensive instrument for remote sensing of canopy parameters (King *et al.*, 1986). Together with the good correlations obtained for light interception and ground cover in model canopies, these results confirm the validity of the experimental approach.

Although the planophile canopy structure of *L. sativa* (experiment (2)) may have caused its regression coefficient for $\log_e$ (biomass) to be significantly different from the other model canopies, it is interesting that this did not also produce a different regression coefficient for $\log_e$ (L). The data from the short lawn grasses showed no difference from the model canopies, suggesting that 'ideal' graminaceous canopies (in this case containing few dead leaves) may be similarly monitored by remote sensing.

The tropical grassland sites present a different picture. Although R/NIR is correlated with $\log_e$ (biomass), and also with either L or $\log_e$ (L), the type and consistency of these relationships varies from site to site. The relationship between R/NIR and canopy parameters, for both crops and natural ecosystems, has been shown to be linear by some workers (Tucker *et al.*, 1981; Boutton and Tieszen, 1983). However, others report a curvilinear or logarithmic relationship (Asrar *et al.*, 1984; Gross *et al.*, 1986).

In the present study, the Mexican site showed statistically consistent regressions during both the growing season and periods of senescence, suggesting that biomass and L could be estimated by remote sensing at any time of year with a reasonable degree of precision. However, data from the Thailand site show a high degree of variation, which would severely limit the accuracy of remote sensing estimates. Remote sensing at the Kenyan site could lead to serious underestimation of biomass during the dry season if data from the growing season were used as a guide to calibrate the remote sensing technique. Evidence from the three terrestrial grassland sites supports the notion that precision of remote sensing estimates is linked to the amount of dead matter in the canopy. In the present study, the proportion of dead matter present within the canopy or as litter varied as follows: Kenya 54–80% (mean 70%); Thailand 28–48% (mean 41%); Mexico 0–69% (mean 23%); Brazil 8–19% (mean 12%). Boutton and Tieszen (1983) demonstrated that for more than 70% dead matter, the standard error of biomass estimates became unacceptably high (>30% of the mean). Curran (1981) and Hardisky *et al.* (1986) acknowledge that dead matter in canopies tends to reduce the estimate of biomass and parameters by increasing red reflectance. The extent of this distortion of data can be difficult to assess, unless a great deal is known about the accumulation, decomposition and physical removal of dead matter. Thus careful 'ground truth' work is needed in this case before biomass and turnover of material can be estimated by remote sensing.

Although the canopy of the inundated grass in Brazil did not contain a high proportion of dead matter, the relatively poor correlation between *R/NIR* and *L* at this site may have been due to changing background reflectance from small floating plants and algae, since this can also be important where the canopy is not completely closed (Huete *et al.*, 1985).

The regression coefficient of *R/NIR* on both $\log_e$ (biomass) and $\log_e$ (*L*) is larger for the model canopies than for the grasslands. This might be caused by the higher proportion of non-photosynthetic tissue, such as stems and sheaths,in the grassland canopies, but it may also be a result of the high planting density used in the model canopy experiments giving rise to a more fully closed canopy structure (peak values of *L* were 9.7 and 6.5 for experiments 3 and 4, respectively). It follows that whilst *R/NIR* may be used as an indicator of both canopy biomass and *L* in grassland, the precision of estimation may be hard to assess without additional information on grassland type and dynamics. Canopy structure and composition must also be considered if the remotely sensed estimates are to be regarded as reliable.

7.4.2 Ground cover and light interception

The combined plot of *R/NIR* against light interception for the grassland sites does not show the consistent decline in *R/NIR* with increasing light interception seen in the model canopy systems (Figure 7.4). This is an important distinction. Remote sensing has been used successfully to estimate light interception in crop systems (Steven *et al.*, 1983; Gallo *et al.*, 1985), but the uniform structure of most crop canopies (with little or no dead matter present) facilitates the estimation of light interception, presenting a relatively homogeneous population of leaf surfaces to the remote sensing instrument. The complex nature of natural ecosystem canopies, with an irregular distribution of dead matter, tends to confound remote sensing studies (e.g. Drake, 1976; Gross *et al.*, 1986). It also makes it very difficult to accurately estimate light interception by photosynthetic tissues. The three outlying points marked '?' in Figure 7.4 were all obtained in periods when the canopy contained a high proportion of dead matter. True light interception by photosynthetic tissues may actually be much lower than suggested for these points, which would shift them all to the left and give a relationship closer to that for the model canopies. Advance in remote sensing of light interception efficiency of canopies requires the development of methods of distinguishing interception by live and dead matter. Only when good-quality 'ground truth' data are available, by the direct measurement of light interception by live photosynthetic tissues in complex natural ecosystem canopies, will it be possible to test indirect measurement of light interception by remote sensing.

7.4.3 Net primary production

As well as light interception in crops, crop productivity has been estimated successfully by remote sensing (Steven *et al.*, 1983; Gallo *et al.*, 1985). However, the distinction between the agricultural row crop and the natural plant community extends also into the area of primary production. Crop systems show little turnover of biomass and often show a much larger ratio of above- to below-ground biomass in comparison to natural systems. Measurement of productivity over monthly intervals or less is therefore subject to much greater experimental error for natural ecosystems than for crops. An additional problem, highlighted by the comparison of model canopy experiments 2 and 4, is that the relationship between productivity and canopy light interception depends also on incident radiation. Although remote sensing may be used to estimate the fraction of light intercepted by the canopy, it cannot yield any information about the incident radiation itself. Only over long time intervals, and for certain known environments, may incident radiation be considered constant.

Both these sources of error, i.e. failure to estimate P_n accurately and variation in incident radiation, are likely to have a greater effect the shorter the period of study. Although it may yet be possible to relate natural ecosystem productivity to remote sensing data, the experiments which attempt to show this will need to be much longer than those reported here (ideally 3–4 yearly cycles). Plant biomass production is a long-term process; if long-term remote sensing measurements are to be used for studying plant productivity, then long-term experiments to provide ground truth data throughout the seasons and between contrasting years will be needed to calibrate and later confirm the validity of the method.

One possibility would be to divide the yearly cycle of plant growth and decay into discrete periods, corresponding to the environmental constraints on productivity (rainfall, temperature, etc.). Remote sensing measurements and determinations of P_n would be timed to correspond to these periods, and would need to continue for several yearly cycles. However, such research work would require an intimate knowledge of the study areas and plant characteristics beforehand.

Plant productivity is a function of the amount of incident light energy, the fraction of light intercepted and the efficiency with which the intercepted light is utilized in the formation of new biomass. Claims that remote sensing can estimate global photosynthesis or global productivity should therefore be treated with caution. Even if we can overcome the problems of estimating the maximum conversion efficiency, i.e. potential photosynthetic productivity, of certain ecosystems by remote sensing, we cannot be certain how much of this capacity is realized. Periodic stresses probably limit those areas of the world where productivity does approach photosynthetic capacity to certain tropical regions with a relatively constant environment, or to semi-arid regions where a sharply defined growing season results in a short-lived plant canopy.

Such conclusions have wide implications for the use of remote sensing for global environmental monitoring. Remote sensing by aircraft and satellite has an enormous potential to gather data for modelling productivity and carbon cycling at both ecosystem and global level. However, if accurate remote sensing is limited to particular seasons or particular types of ecosystem, these limitations must be understood. Long-term verification of remote sensing studies with 'ground truth' data is the only way to do this (Scurlock *et al.*, 1989).

7.5 SATELLITE REMOTE SENSING

Satellites have been used for remote sensing of the Earth's surface since the early 1960s. Resolution is limited by the size of the smallest picture unit or pixel; in the case of the US NOAA satellites' Advanced Very High Resolution Radiometer (AVHRR) this is roughly 1000 m. Considerably higher spatial resolution, perhaps better suited to monitoring of vegetation in small sample areas, is available on the NASA Landsat series (80 and 30 m) and on the French SPOT series of satellites (20 m).

7.5.1 Satellite applications to vegetation monitoring

Satellite data have been used for vegetation mapping (Justice *et al.*, 1985; Tucker *et al.*, 1985) and detection of tropical forest clearance (Tucker *et al.*, 1984). However, primary rain forest and regrowing secondary forest may be spectrally indistinguishable, and difficulties have been experienced in discriminating between open forest and scrubland (Tucker *et al.*, 1984; Singh, 1987).

On a quantitative level, Prince and Astle (1986) conclude that predictive equations for estimating biomass and ground cover in southern African rangelands using *R/NIR* derived from the Landsat Multi-spectral Scanner (MSS) are applicable only to sites with similar vegetation. Tucker *et al.* (1986) demonstrate a correlation between atmospheric CO_2 variations and a satellite-derived vegetation index, arguing that this is an indicator of intercepted radiation and therefore photosynthetic carbon flux. However, as noted by Warrick (1986), the relationships between reflectance spectra and net primary productivity carry a high degree of uncertainty at present, so it may be too early to propose such applications of satellite data. Further 'ground truth' verification, and further testing of carbon-cycle models, is required to establish the validity of this approach. Milton (1980) has also recognized that demand for good-quality ground data collection will increase with the use of the latest high-resolution satellite sensors.

The need for an integrated approach to remote sensing is recognized by Badhwar *et al.* (1986). They recommend a workplan commencing with determination of ground truth from (necessarily) fairly small sample sites, followed by collection and analysis of data from aircraft and then satellite-borne sensors. In this way, it is possible to appreciate the inevitable losses of detailed data as the scale of the study increases. Ripple (1985) found differences between simulated 'turbid atmosphere' and 'clear atmosphere' satellite determinations of *R/NIR* and normalized difference vegetation index (NDVI) for a natural grassland canopy, demonstrating the effect of atmospheric conditions on data quality. Atmospheric correction, perhaps by detailed comparison of satellite data against ground reflectance data, would appear to be most important.

Thus, from the point of view of the plant ecophysiologist, the 'bottom-up' approach of Badhwar *et al.* (1986) is preferable to the 'top down' studies exemplified by Tucker *et al.* (1986). Although the construction of global models of CO_2 flux through natural ecosystems could be much helped by satellite remote sensing, this will only come about if information is available for understanding of canopy reflectance in each category of ecosystem considered. Hardisky *et al.* (1986) report that even within one ecosystem (coastal wetland), different canopy types exhibit distinct relationships between biomass and NDVI measured with Landsat Thematic Mapper (TM) bands 4 and 3 (Figure 7.10). This is probably due more to the orientation of reflecting surfaces within the canopy than to actual differences in the

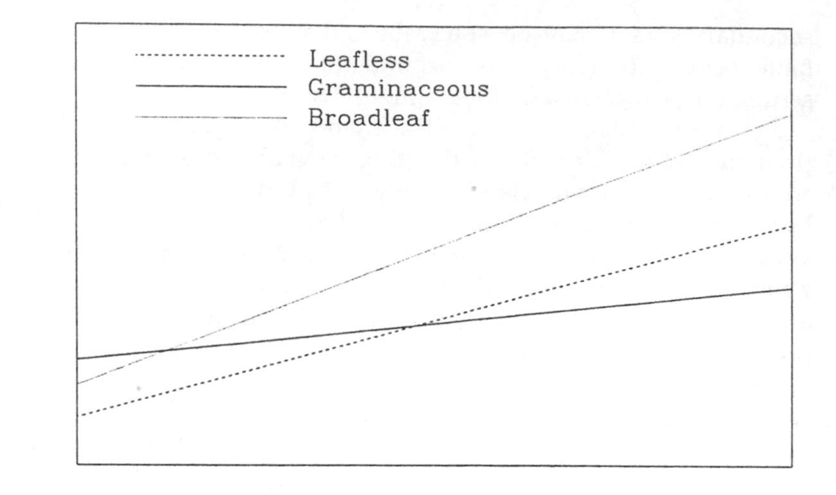

Figure 7.10. Different relationships between NDVI and biomass, exhibited by different canopy types in a coastal wetland ecosystem. After Hardisky *et al.* (1986)

reflectance of these surfaces. It might be concluded that accurate biomass prediction models should therefore incorporate details of canopy composition within an ecosystem. Seeking a generalized relationship for an ecosystem type may not be sufficient.

7.6 CONCLUSIONS

Experiments with model canopies show strong correlations between *R/NIR* measured with the Skye spectral ratio meter and various canopy parameters, confirming the validity of this experimental approach.

The complex nature of natural ecosystem canopies, with an irregular distribution of dead material, causes problems with otherwise valid remote sensing techniques. The precision of estimation of biomass and *L* may be low, and it is necessary to find a way of distinguishing light interception by live and dead matter. Only when good-quality 'ground truth' data are available, by the direct measurement of parameters such as light interception in complex natural ecosystem canopies, will it be possible to test their indirect measurement by remote sensing.

Although remote sensing by satellite and aircraft has an enormous potential for gathering data and application to modelling, its limitations must be understood. Remote sensing of plant productivity, in particular, requires long-term 'ground truth' validation.

7.7 SUMMARY

1. The high spatial variability within tropical grasslands over short distances requires a large number of destructive samples for any precise determination of net primary production. The method is therefore highly labour-intensive. There is a need for methods which are rapid and non-destructive to allow frequent monitoring of biomass dynamics and net primary production over larger areas. The field studies under the UNEP Project on Primary Productivity and Photosynthesis provided an unrivalled opportunity to examine the potential of remote sensing of biomass, leaf area and net primary production at diverse natural grassland sites within the tropics.
2. Detailed data on these vegetation parameters were already being collected at the UNEP study sites. Remote sensing measurements at these tropical study sites were complemented by further tests and studies on artificially manipulated canopies under controlled environment conditions.
3. Experiments with model canopies showed strong correlations between *R/NIR* and various canopy parameters, confirming the validity of this experimental approach. However, the complex nature of tropical grassland canopies, with an irregular distribution of dead material in both time and space, decreased the precision of otherwise valid remote sensing techniques.
4. The precision of estimation of biomass and leaf area index in natural tropical grasslands by remote sensing is low. The potential of remote sensing of primary productivity is severely limited in tropical grasslands by the difficulties of distinguishing light interception by live and by dead matter. These findings have important implications for the use of remote sensing in current studies of tropical grassland biomass and net primary productivity by aircraft and satellite.

REFERENCES

Asrar, G., Fuchs, M., Kanemasu, E.T. and Hatfield, J.L. (1984) Estimating absorbed photosynthetic active radiation and Leaf Area

Index from spectral reflectance in wheat. *Agronomy Journal*, **76**, 300–6.

Badhwar, G.D., MacDonald, R.D. and Mehta, N.C. (1986) Satellite-derived leaf area index and vegetation maps as input to global carbon cycle models – a hierarchical approach. *International Journal of Remote Sensing*, **7**, 265–81.

Boutton, T.W. and Tieszen, L.L. (1983) Estimation of plant biomass by spectral reflectance in an East African grassland. *Journal of Range Management*, **36**, 213–16.

Curran, P.J. (1981) Multispectral remote sensing for estimating biomass and productivity, in *Plants and the Daylight Spectrum* (ed. H. Smith), Academic Press, London, pp. 65-99.

Curran, P.J. and Milton, E.J. (1983) The relationship between the chlorophyll concentration, *L* and reflectance of a simple vegetation canopy. *International Journal of Remote Sensing*, **4**, 247–55.

Drake, B.G. (1976) Seasonal changes in reflectance and standing crop biomass in three salt marsh communities. *Plant Physiology*, **58**, 696–9.

Gallo, K.P., Daughtry, C.S.T. and Bauer, M.E. (1985) Spectral estimation of absorbed photosynthetically active radiation in corn canopies. *Remote Sensing of Environment*, **17**, 221–32.

Gross, M.F., Klemas, V. and Levasseur, J.E. (1986) Remote sensing of *Spartina anglica* biomass in five French salt marshes. *International Journal of Remote Sensing*, **7**, 657–64.

Hardisky, M.A., Gross, M.F. and Klemas, V. (1986) Remote sensing of coastal wetlands. *Bioscience*, **36**, 453–60.

Huete, A.R., Jackson, R.D. and Post, D.F. (1985) Spectral response of a plant canopy with different soil backgrounds. *Remote Sensing of Environment*, **17**, 37–53.

Justice, C.O., Townshend, J.R.G. and Holben, B.N. and Tucker, C.J. (1985) Analysis of the phenology of global vegetation using meteorological satellite data. *International Journal of Remote Sensing*, **6**, 1271–318.

King, J., Sim, E.M. and Barthram, G.T. (1986) A comparison of spectral reflectance and sward surface height measurements to estimate herbage mass and leaf area index in continuously stocked ryegrass pastures. *Grass and Forage Science*, **41**, 251–58.

Kumar, N. and Monteith, J.L. (1981) Remote sensing of crop growth, in *Plants and the Daylight Spectrum*. (ed. H.Smith), Academic Press, London, pp. 133–44.

Milton, E.J. (1980) A portable multiband radiometer for ground data collection in remote sensing. *International Journal of Remote Sensing*, **1**, 153–65.

Prince, S.D. and Astle, W.L. (1986) Satellite remote sensing of rangelands in Botswana. I. Landsat MSS and herbaceous vegetation. *International Journal of Remote Sensing*, **7**, 1533–53.

Ranson, K.J., Daughtry, C.S.T., Biehl, L.L. and Bauer, M.E. (1986) Sun-view angle effects on reflectance factors of corn canopies. *Remote Sensing of Environment*, **18**, 147–61.

Ripple, W.J. (1985) Asymptotic reflectance characteristics of grass vegetation. *Photogrammetric Engineering and Remote Sensing*, **51**, 1915–21.

Scurlock, J.M.O., Long, S.P., Imbamba, S.K., Garcia, E., Kamnalrut, A. and Hall, D.O. (1989) Remote sensing of primary production in natural tropical grasslands and artificial mixed species canopies, in *Proceedings of the 11th International Congress of Biometeorology* (eds. Driscoll, D.M. and Box, E.O.). SPB Academic Publishers, The Hague, pp. 379–97.

Singh, A. (1987) Spectral separability of tropical forest cover classes. *International Journal of Remote Sensing*, **8**, 971–9.

Snedecor, G.W. and Cochran, W.G. (1956) *Statistical Methods Applied to Experiments in Agriculture and Biology*, Iowa State University Press, Ames, Iowa.

Steven, M.D., Biscoe, P.V. and Jaggard, K.W. (1983) Estimation of sugar beet productivity from reflection in the red and infra-red spectral bands. *International Journal of Remote Sensing*, **6**, 1335–72.

Tucker, C.J. (1977) Asymptotic nature of grass canopy spectral reflectance. *Applied Optics*, **16**, 1151–7.

Tucker, C.J., Fung, I.Y., Keeling, C.D. and Gammon, R.H. (1986) Relationship between atmospheric CO_2 variations and a satellite-derived vegetation index. *Nature*, **319**, 195–9.

Tucker, C.J., Holben, B.N., Elgin, J.H. and McMurtrey, J.E. (1981) Remote sensing of total dry-matter accumulation in winter wheat. *Remote Sensing of Environment*, **11**, 171–89.

Tucker, C.J., Holben, B.N. and Goff, T.E. (1984) Intensive forest clearing in Rondonia, Brazil, as detected by satellite remote sensing. *Remote Sensing of Environment*, **15**, 255–61.

Tucker, C.J., Townshend, J.R.G. and Goff, T.E. (1985) African land-cover classification using satellite data. *Science*, **227**, 369–75.

Warrick, R.A. (1986) Photosynthesis seen from above. *Nature*, **319**, 181.

8

Synthesis and conclusions

M.B. JONES, S.P. LONG and M.J. ROBERTS

8.1 DEFINING THE SCOPE OF TROPICAL GRASSLANDS

Tropical grasslands were defined in Chapter 1 as any ecosystem within the tropics in which graminaceous species are a dominant or co-dominant character of the vegetation. In this sense, tropical grasslands are virtually synonymous with the term savanna which Bourlière and Hadley (1970) used to refer to a tropical formation where the grass stratum is continuous and with individual or discontinuous clumps of trees and shrubs. Two important characteristics of the grass stratum are that it is burnt from time to time, and that the main patterns of growth are closely associated with alternating wet and dry seasons.

The definition of the word 'savanna' has generated much, often fruitless, discussion (see Pratt *et al.* 1966; Huntley and Walker, 1982; Bourlière and Hadley, 1983) and some workers have even suggested that its use should be avoided altogether.

However, it is probably too important a term to jettison when used to identify in broad terms the vast, and poorly researched, wooded grasslands and grassy woodlands of the tropics. Tropical grasslands include different types of vegetation in a broad transition zone between 'closed' tropical forests and 'open' deserts but they share a number of structural and functional characteristics that set them apart from other terrestrial ecosystems of the tropics. Some of the more important characteristics of tropical grasslands are listed in Table 8.1, which has been adapted from Bourlière and Hadley (1983). The grasslands investigated during this project reflect the very diverse nature of tropical grasslands (see Table 1.1). They range from the 'typical' savanna grassland in Nairobi National Park, Kenya (Chapter 3) through the 'savanna forest' community at Hat Yai in southern Thailand (Chapter 2) to the saline grasslands in Mexico (Chapter 4) and the floodplain grasslands in the central Amazon (Chapter 5). Their diversity highlights the danger of making generalizations about 'typical tropical grasslands' as net primary production and the factors

Table 8.1. Some characteristics of tropical grasslands and related biological consequences (adapted from Bourlière and Hadley, 1983)

Characteristics	*Consequences*
(i) Transitional zone between closed tropical forests and open deserts	(a) complex tree–grass interactions (b) physiognomic gradient
(ii) Marked seasonality in climate – alternating wet and dry seasons	(a) seasonal growth and development (b) large proportion of underground biomass
(iii) Relatively simple but dynamic systems with few species	(a) fluctuation in dominant species (b) considerable and frequent disturbance
(iv) Changing 'quality' of primary production	(a) immediate availability of most primary production for use by consumers (b) low content of secondary compounds (c) storage of some primary production
(v) Relatively high net production	(a) high production/biomass ratio (b) rapid plant turnover
(iv) Very high proportion of species with C_4 photosynthesis	(a) high radiation conversion efficiency (b) high water use efficiency (c) high nitrogen use efficiency (d) low nutritive value

controlling it vary considerably from one location to another. Also, the inclusion of measurements of productivity of bamboo in China (Chapter 6) illustrates the differences between relatively closely related species of grasses but with very different life forms.

8.2 INTRODUCTION AND THE DATABASE

A primary goal of the United Nations Environment Programme (UNEP) project described in this book has been to provide reliable estimates of net primary production (P_n) for grass ecosystems in the tropics and sub-tropics. Because of apparent shortcomings in the methodology used in previous measurements it was important to examine the extent of prior underestimation of production and C-flow within these systems and to examine the significance of spatial and temporal variation in production and carbon flow. The results of measurements made in five contrasting ecosystems have been

Table 8.2. Summary of information available on the Technical Co-ordination Centre Database and time in months over which the measurements were made

Site	*Nairobi Kenya*	*Chapingo Mexico*	*Hat Yai Thailand*	*Manaus Brazil*	*Shanghai China*
Climate					
Air temp.	48	54	48	24	36
Precip.	48	54	48	20	36
Pan evap.	48	54	48	20	36
Solar rad.	34	40	36	18	36
Canopy light					
Interception	36	36	36	18	36
R/NIR	24	40	24	18	—
Above-ground					
Biomass	36(×3)	54	40(×4)	18	24
Litter	36	54	40	18	24
Decomp.	36	54	40	18	24
L	36(×3)	54	40(×4)	18	24
AP_n	36	54	40	18	24
Below-ground					
Biomass	36	54	40	18	8(3 mo)
Dead	36	54	40	18	
Decomp.	36	54	40	18	24
BP_n	36	54	40	18	8(3 mo)
Diurnal curves photosynthesis/ transpiration	24	36	—	18	12

Generally sample numbers were 20 for above-ground vegetation and 10–20 for below-ground vegetation. Diurnal curves are from 1 to 2 days in each month with hourly measurements of *ca* 8–12 leaves.
(×3), (×4) = species categories.
(3 mo) = at 3 monthly intervals.

described in preceding chapters. In each case they have estimated P_n over 18–54 months (Table 8.2) with measurements made at monthly intervals. From these measurements it has been possible to quantify the flow of carbon through these ecosystems and to account for losses to different trophic levels. In addition, the relationships between photosynthesis and the pattern of production throughout the annual cycle have been studied and these have been related to the prevailing climate and environmental stresses. This has generated upwards of 10 000 data items per site.

These data have been transferred at regular intervals from the regional centres to a central database at the Technical Co-ordination

Centre, at the University of Essex (Figure 8.1). The data were validated and cross-checked against the original recording sheets. The database was constructed via commercially available software (Lotus 1-2-3 v. 2.1, Lotus Inc., Cambridge, MA; Figure 8.2). This software was chosen because of its wide availability, its general purpose potential, and its wide compatibility with other more specialized packages. Figures 8.1 and 8.2 illustrate the organization of hardware and software used in the database. Table 8.2 outlines the headings of data held within the database. Chapters 2–6 highlight some of the key findings which have emerged at each location, but represent only a small selection of the total data available for each site. This chapter uses this database further to examine more broadly specific goals laid down in Chapter 1 and then to place these in the wider context of other productivity studies and the possible significance of tropical grasslands and temperate bamboo forests to the global carbon-cycle.

8.3 NET PRIMARY PRODUCTION OF TROPICAL GRASS ECOSYSTEMS – HAS IT BEEN UNDERESTIMATED IN THE PAST?

The rationale for the studies contained within this book was that net primary production (P_n) may only be ascertained by taking account of losses of vegetation simultaneously with measurements of biomass change. It was obvious at the time of the International Biological Programme (IBP) studies, which used the method of Milner and Hughes (1968), that the IBP methods would underestimate true net production by grassland ecosystems (Coupland, 1979). Nevertheless, at many sites, production estimates were substantially increased over pre-IBP values which were largely based on the simple measurement of peak biomass or the difference between peak and minimum biomass (e.g. Bourlière and Hadley, 1970). Using the IBP methods it was possible to separate data for periods when additions to the live portion of the canopy were rapid in relation to losses, from those when losses exceeded gains. The effect was to avoid the tendency from earlier measurements for losses to simply cancel gains. With these methods, the more frequent the harvests the closer estimates of net productivity come to true net production. Consequently, knowledge of harvest intervals is required before any attempt can be made to assess the actual underestimation of production in previous studies using the IBP methods.

The methods chosen for the studies described in this book were specifically designed to allow direct estimation of quantities of

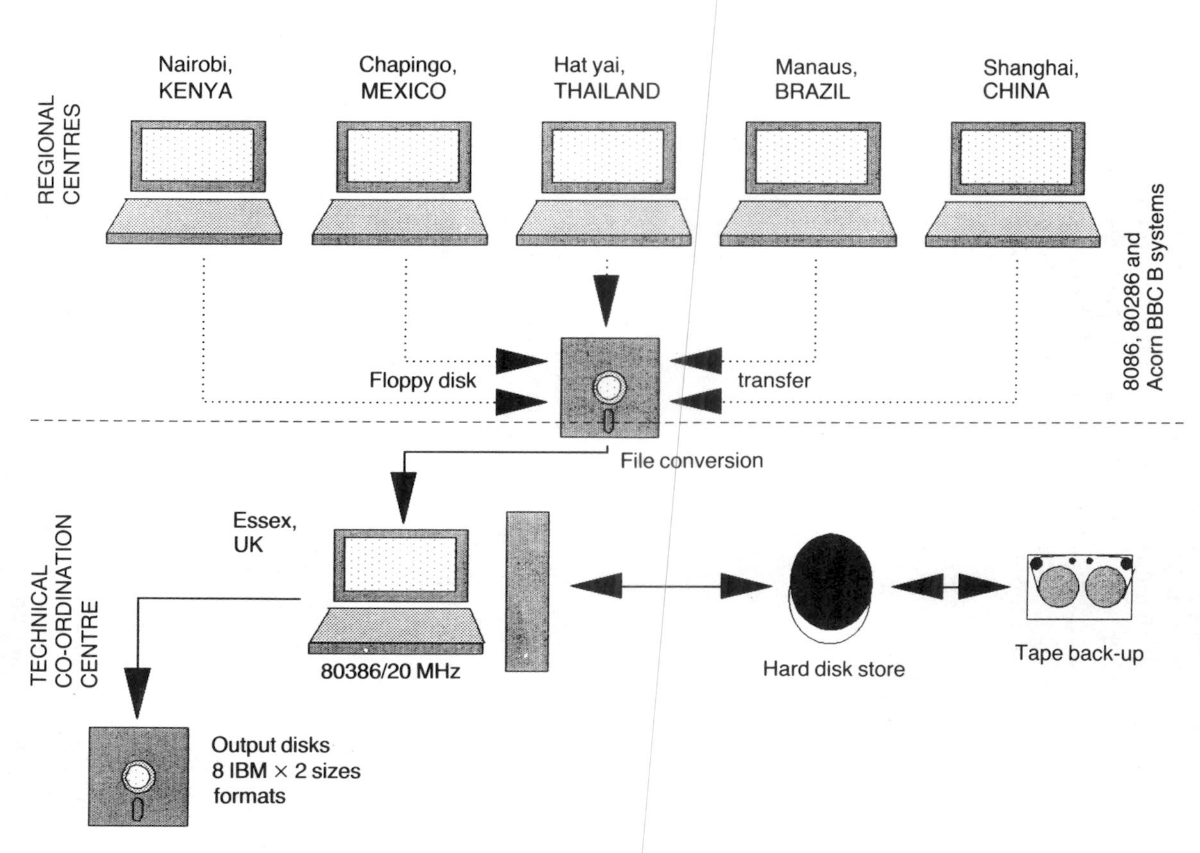

Figure 8.1. Flow diagram, illustrating the relationship of hardware used in establishing the UNEP Project database on the five study sites.

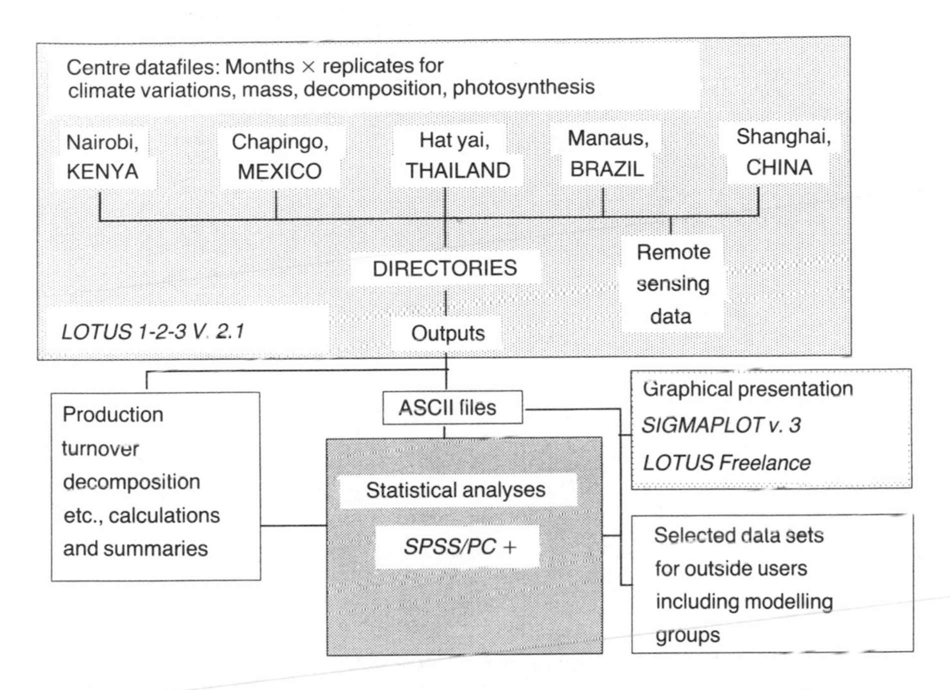

Figure 8.2. The arrangement of software used in the storage and analysis of the UNEP Project database. Lotus 1-2-3 and Lotus Freelance are registered trademarks of Lotus Development Corp., Sigmaplot is a registered trademark of Jandel Scientific Inc., and SPSS/PC+ is from SPSS U.K. Ltd.

vegetation lost each month through mortality so that the flow of material could be directly assessed. However, since this involved monthly measurements of changes in biomass and litter, it was also possible to estimate P_n via the recommended IBP grasslands method (Milner and Hughes, 1968) of summing the positive increments of biomass. Thus, direct comparisons could be made for each site of production estimated by equations which take direct account of death (pp. 18–20) and those which simply infer production from the direction of biomass change , as used in the IBP studies (Milner and Hughes, 1968; Singh *et al.*, 1979, 1980). Figure 8.3 compares these estimates for the four grassland sites. Use of the IBP method always underestimated P_n, but by varying amounts. In the floodplain *Echinochloa polystachya* grassland at Marchantaria, the estimate based on the IBP method was 83% of that estimated when account was taken of losses through mortality. In the monsoonal grassland of Hat Yai in southern Thailand the IBP method estimated production at just 26% of that determined when account was taken of losses through mortality, i.e. a 4-fold underestimation (Figure 8.3). In a comparison of these

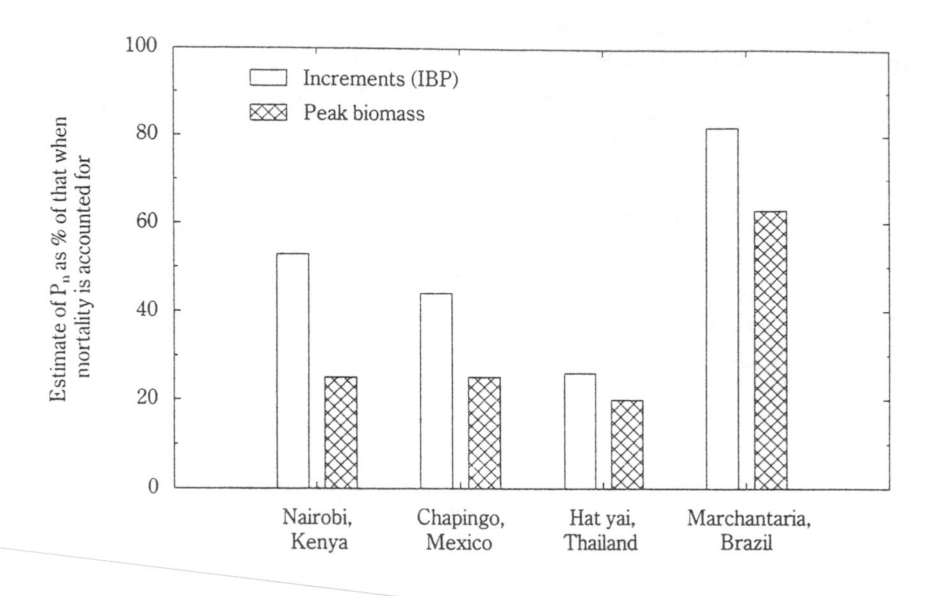

Figure 8.3. Estimates of net primary production (P_n) at the four grassland sites in 1984/85 (1985/86 at Marchantaria) based on the IBP positive increments method and on peak biomass, as a proportion of P_n estimated by taking account of losses through mortality; i.e. equation 1 (Chapter 1).

studies with IBP it should be noted that for a majority of IBP sites there was inadequate information to estimate production from increments in biomass and production was instead estimated from peak biomass. Figure 8.3 illustrates that for the sites examined here, this procedure results in an even greater underestimation of P_n. For the three terra firma grassland sites, peak biomass and summation of the positive increments in biomass suggest P_n values *ca* 20% and *ca* 35% of P_n determined when full account is taken of losses through mortality.

It was well recognized within the framework of IBP that the procedures used would underestimate P_n (Coupland, 1979); however this has not prevented the values generated from being widely used in estimation of global production (e.g. Hall, 1989). The present studies show that for three terra firma tropical grasslands of strongly contrasting character, underestimation resulting from the use of the IBP methods is between 2- to 5-fold, and that estimates of production for this biome should be increased accordingly. Lieth (1973) and Jordan (1981) suggested a mean production of 800 g m^{-2} y^{-1} for tropical grasslands. Assuming an average underestimation of 3.5-fold in the present study then this figure should now be revised upwards to 2800 g m^{-2} y^{-1}, giving 56 Pg across the suggested 20×10^6 km^2 of tropical grasslands as a whole.

Differences in underestimation between sites may be related to the seasonality of growth and death. Underestimation was least on the annually inundated floodplain site at Marchantaria. Here the vegetation shows a near annual growth cycle with almost 11 months of continuous growth, followed by a brief and distinct period of death prior to the recruitment of a new generation of plants. Since there is relatively little overlap of production and death, it could be expected, as observed, that production estimated by IBP mass increments method would differ little from estimates which take direct account of losses through death (see Chapter 5). At the other extreme the grassland at Hat Yai, where there is only moderate seasonality and a mixture of species which have different phases of production and senescence, shows almost continuous production which overlaps with significant losses through death (see Chapter 4). Estimates based on increments cannot take account of this overlap between production and death, and thus would be expected to be particularly in error in this type of community. This proves to be the case (Figure 8.3). The failure of peak biomass to provide realistic estimates of net production in this community is illustrated by the contrasting magnitudes of production and biomass from year to year at Hat Yai (Figure 8.4). The grasslands at Chapingo and Nairobi National Park show more seasonality with very distinct dry periods and in consequence slightly less underestimation of P_n.

In an earlier assessment of techniques for determining P_n (Singh *et al.*, 1975), it was suggested that correction factors could be applied to relate P_n determined by one technique to P_n determined by another. Thus P_n determined from the peak biomass method could be scaled up to P_n as it would be determined by addition of positive increments, simply by multiplication with a single constant. However, the techniques examined all assumed that net production could be adequately estimated from measurement of biomass and litter dynamics. Might production determined by combining biomass dynamics with direct estimation of losses through mortality, as used here, be calculated simply by scaling-up from estimates based on biomass and litter dynamics? Figure 8.3 shows that underestimation by the standard IBP is far from a constant proportion, and thus a single constant could not be used for scaling up. Could nevertheless a single value be used within specific grassland types? Figure 8.4 shows that at Hat Yai in 1984 the standard IBP method estimated P_n as only

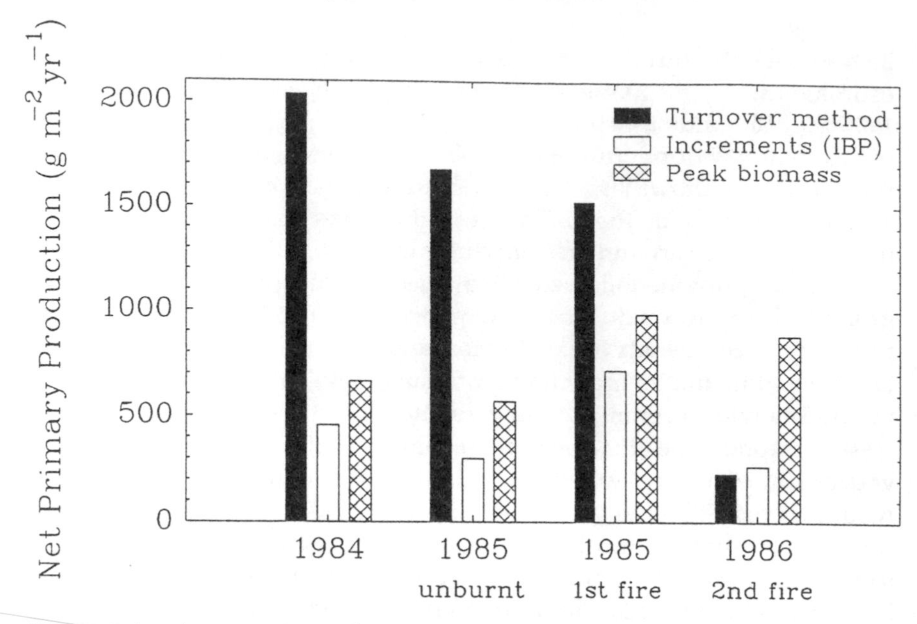

Figure 8.4. A comparison of estimates of net primary production (P_n) for the grassland at Hat Yai, Thailand (Chapter 4) based on: taking account of losses through mortality, i.e. equation 1 (Chapter 1); the 'IBP positive increments method'; and peak biomass. Comparisons are based on monthly measurements of biomass and litter dynamics, and decomposition, in 1984, for burnt and unburnt areas of the study site in 1985, and for areas burnt a second time in 1986.

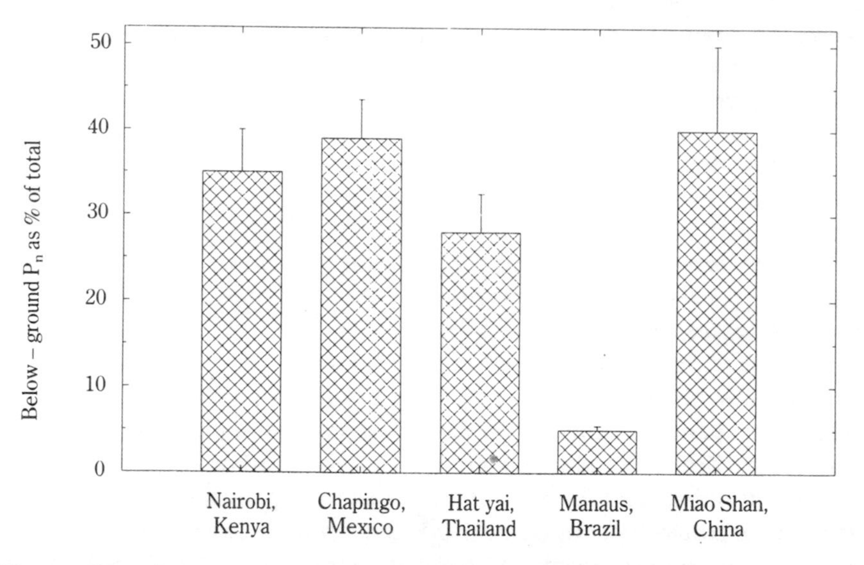

Figure 8.5. A comparison of root and rhizome production (BP_n) as a proportion of total net primary production (P_n) for the five sites in 1985/86.

20% of that measured by the methods recommended here. In 1985 the estimate by the IBP methods was even lower at just 15%. However, in 1986 the IBP method actually overestimated P_n giving a value of 110% of P_n estimated in this study. Thus, even within one site the relationship with the IBP method is not constant.

Less than 10% of the IBP grassland studies made direct measurements of below-ground biomass (Long *et al.*, 1989), in most cases biomass and production was estimated as a fixed proportion of above-ground biomass and production (Murphy, 1975). Below-ground production (BP_n), by comparison to above-ground production (AP_n), is far more difficult to measure. Analysing amounts of biomass, dead vegetation and decomposition below ground in these studies required 10-fold more time than the equivalent analysis of above-ground vegetation. If a simple factor could be used to relate BP_n to AP_n this would save time and resources, perhaps allowing examination of further sites. Figure 8.5 shows that the proportion of production in roots and rhizomes varied greatly between sites, from under 10% at Marchantaria, Manaus, Brazil to 40% on the saline grassland at Chapingo. Such variation is not unexpected. The Marchantaria site has water and nutrients in abundance for most of the year (Chapter 5) and thus strong investment in root production would not be expected. Similarly, at the Chapingo site where water is limited for much of the year a strong investment in roots and rhizomes would be necessary for surviving the periods of drought. This might suggest that BP_n could be estimated from AP_n only after calibration in each environment. Figures 8.6 and 8.7 show by reference to the Chapingo and Hat Yai sites that this approach would also be in serious error. BP_n as a proportion of total P_n at the Chapingo site, varied markedly over the 3 years illustrated, from *ca* 40% in 1984/5 to over 70% in 1986/7, which followed a fire on the site. BP_n as a proportion of P_n at the Hat Yai site varies from 12% on the unburnt site in 1985 to 32% on the burnt site in 1985. However, on the sites which were burnt a second time at the beginning of 1986, BP_n was negative, i.e. there was a net loss of material either through an excess of respiration over translocation to the rhizomes or through a net loss of material through translocation from the rhizomes. The increment and peak biomass methods used in the IBP studies do not allow for the possibility of negative production. This study however, shows that a tropical grassland stressed either through too frequent burning and/or through drought, can show a strongly negative BP_n, representing a loss of carbon from the biomass and soil. Given the recognized global increases in the frequency of firing of tropical grasslands (Hammond, 1990) and the predicted

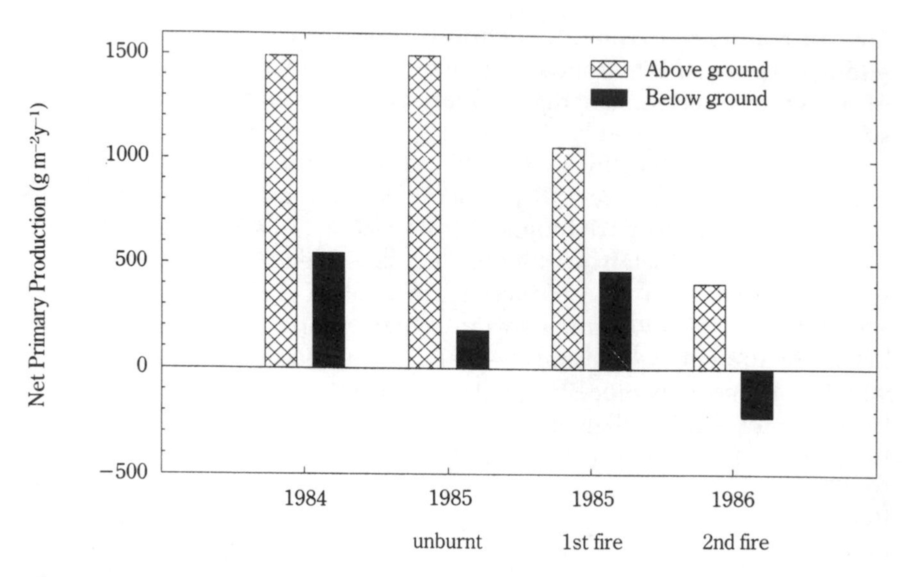

Figure 8.6. A comparison of above- and below-ground annual net primary production for the grassland at Hat Yai, Thailand. Comparisons are based on monthly measurements of biomass and litter dynamics, and decomposition, in 1984, for burnt and unburnt areas of the study site in 1985, and for areas burnt a second time in 1986.

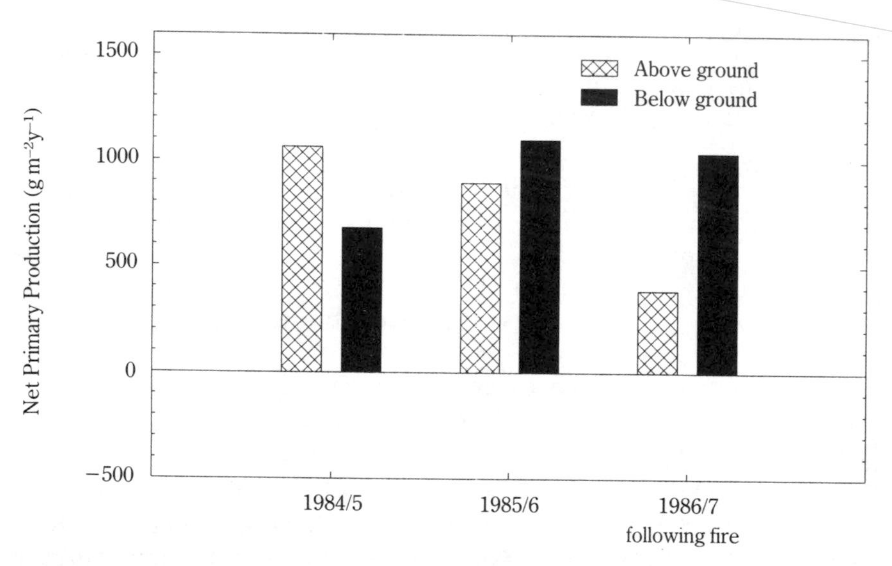

Figure 8.7. A comparison of above- and below-ground annual net primary production for the grassland at Chapingo, Mexico. Comparisons are based on monthly measurements of biomass and litter dynamics, and decomposition, in 1984/85, 1985/86, and in 1986/87, following a single fire.

increase in the frequency of droughts in the semi-arid tropics (Houghton *et al.*, 1990) such negative productivities and hence losses of below-ground biomass have important implications for the levels of soil carbon.

8.4 COMPARISON OF PAST AND PRESENT MEASUREMENTS OF PRODUCTIVITY

The values of annual P_n reported from the five locations in Chapters 2–6 show large differences, particularly between the semi-aquatic Amazonian floodplain where annual P_n was almost 10 000 g m^{-2} and the other locations where P_n varied between 136 and 2040 g m^{-2}, depending on year and location (Figures 8.8–8.10). The productivity of the *Echinochloa polystachya* stand in the Amazonian floodplain is possibly the highest P_n ever recorded for a natural higher plant community. This occurs in an environment which is extremely unstable with an annual cycle from terrestrial to aquatic as water levels rise. During most years the area is inundated with nutrient-rich waters from the Amazon River for >10 months. Growth then occurs without limitation by water and perhaps without nutrient limitation. When the area dries, at low water of the annual cycle of river level, net production ceases and the vegetation dies back rapidly (Chapter 5). The three terra firma grassland ecosystems in this study experience a very different set of environmental conditions with marked dry and rainy seasons. During the dry seasons, primary productivity declines markedly. However, the above- and below-ground compartments show rather different responses to the hydrological regime. Below-ground net primary production (BP_n) appears to continue after above-ground production has ceased during a period of water stress. Conversely, when rainfall reduces the level of stress, AP_n appears to be stimulated at the expense of below-ground production (Chapters 2–4). BP_n showed marked fluctuations at all the terrestrial grassland sites. During some months BP_n was negative, indicating a period of translocation of reserves to the above-ground component while during other months it reached levels which in the case of the saline grassland in Mexico were twice as high as the maximum recorded monthly AP_n. During these periods there was clearly a very active translocation of assimilates to the below-ground component. The pattern of these changes is complex but appears to be a response to a combination of changing climatic conditions and the cyclic growth pattern of the grass plant (Robson *et al.*, 1988).

The occurrence of fire also has dramatic effects. Recovery after the removal of all the above-ground vegetation by fire depends on the translocation of reserves from below-ground. Some grasses are better adapted than others for this recovery. In the saline grasslands in Mexico, recovery was rapid and complete while in the moist savanna in Thailand there was considerable variability between species in the level of recovery and this was also strongly influenced by the availability of water (cf. Figures 8.8a and 8.8d). Recovery was markedly reduced if fire occurred when soil moisture levels were low (cf. Figures 8.8c and 8.8d).

There have been a number of previous measurements of tropical grassland productivity, largely typical savanna grasslands, dating mainly from the beginning of the 1960s. Most of these measurements were made either as part of the IBP or were stimulated by it. Table 8.3 summarizes many of these measurements and because of the wide range of techniques used, it also attempts to specify the methods used in making the measurements. All these methods are 'conventional' in that they involve destructive harvesting of plant biomass. Alternative methods of estimating primary productivity are discussed in section 8.5.

8.4.1 The importance of decomposition in estimating grassland net primary production.

There has been a good deal of discussion already concerning the earlier methods of estimating primary productivity. Most of them measured production above ground as the sum of living biomass increments over a particular period of time, which was either the growing season or a whole year (Milner and Hughes, 1968). As has been pointed out, most of them failed to take into account root production and many of them did not measure losses of live biomass between harvests. These both led to an underestimation of primary productivity. We have shown that if complete account is taken of mortality above and below ground, the primary productivity is from 2 to 5 times higher than if mortality is unaccounted for (Figure 8.3) and 3 to 10 times higher if below-ground biomass is also ignored (Figure 8.3; Long *et al.*, 1989). However, our estimates in this study still almost certainly underestimate production since they do not take account of carbon losses due to root exudation or to pest attack between the sampling intervals. The magnitude of these potential losses is not quantified. These results illustrate a very important characteristic of grasses in general; that is, the very rapid turnover of plant parts due to

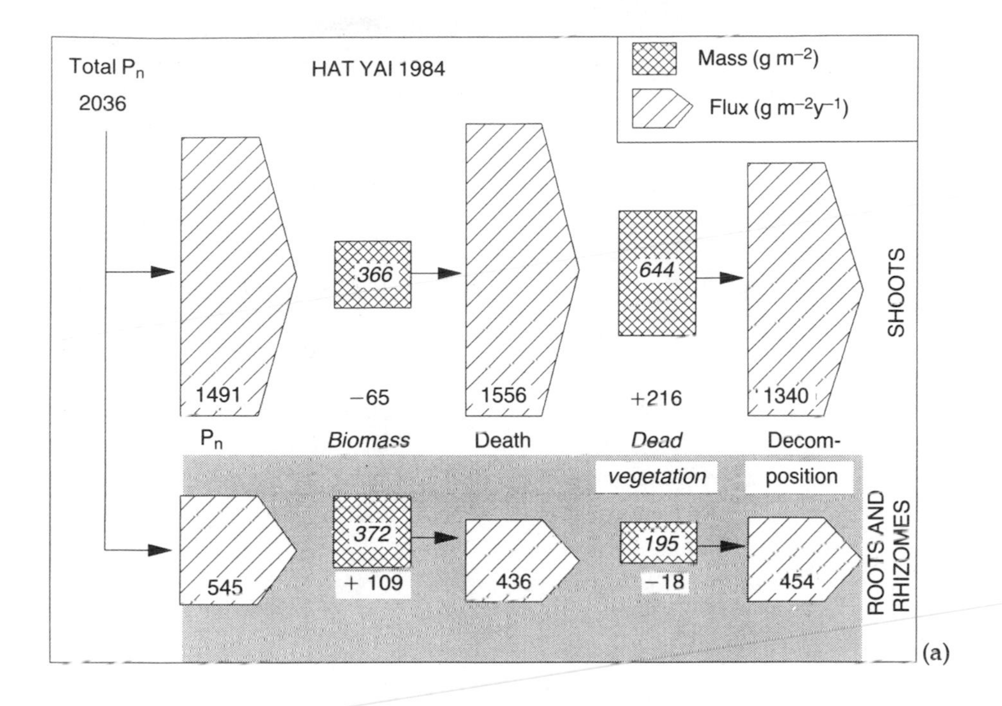

(a)

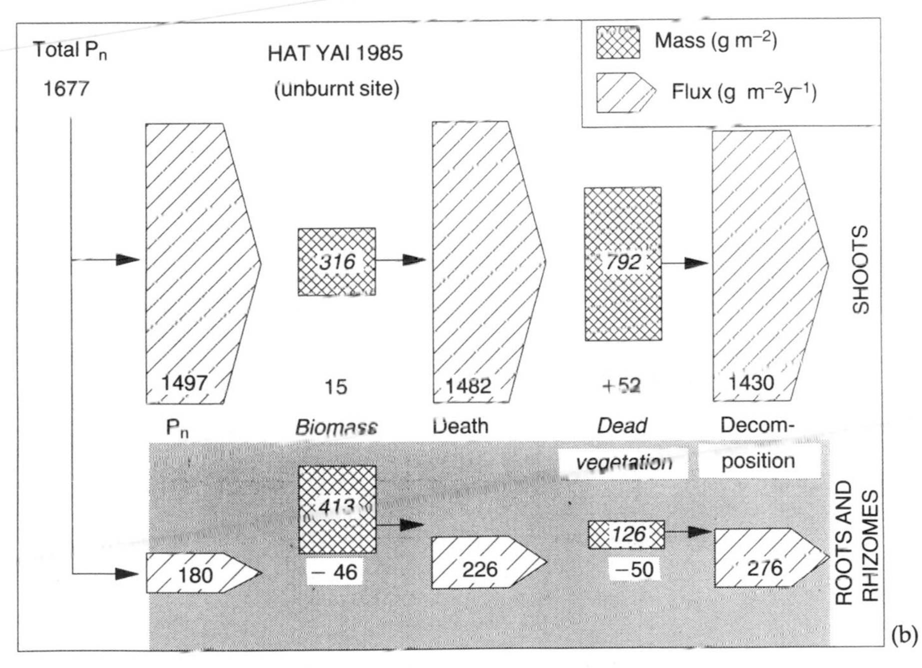

(b)

Figure 8.8. (caption overleaf)

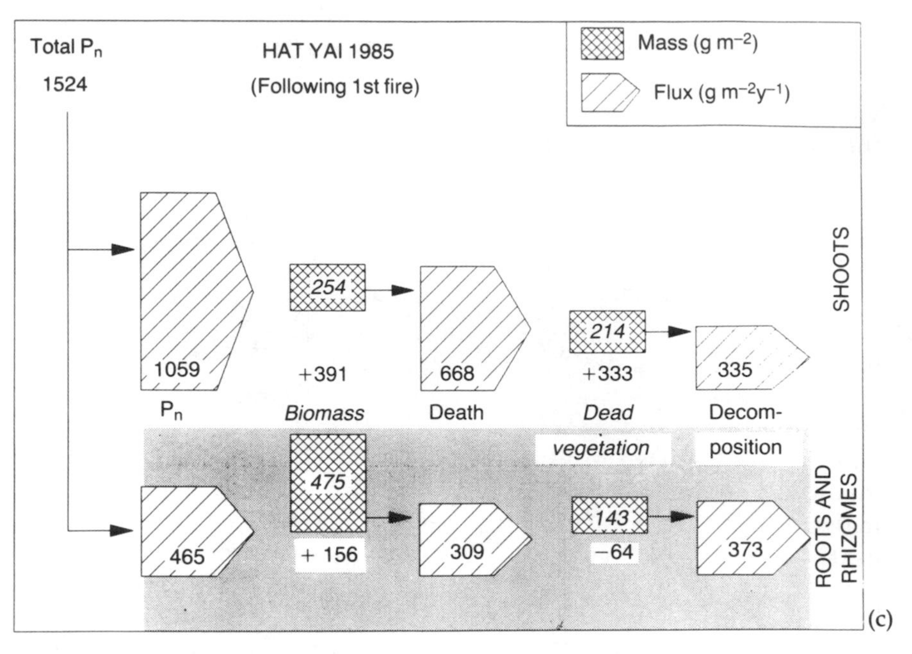

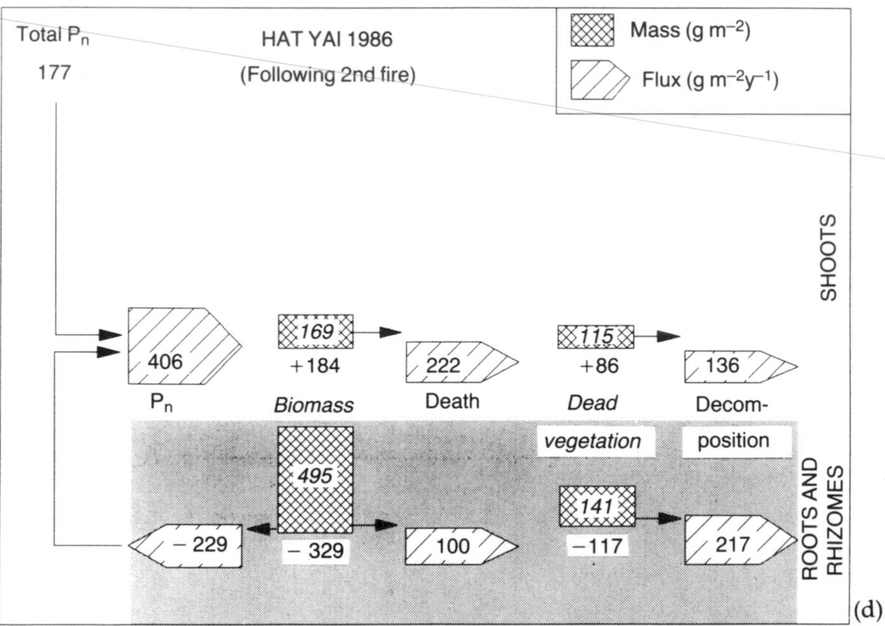

Figure 8.8. Fluxes and mean quantities of dry mass in the grassland at Hat Yai, Thailand. The arrow shaped polygons, representing the rate variables of annual fluxes (g (dry mass) m^{-2} y^{-1}) and their direction, and rectangles representing the state variables of mean mass of vegetation (g (dry mass) m^{-2}). Numbers below the rectangles indicate the net change over the year in that quantity, thus '−65' below the box labelled '366', representing the amount of shoot biomass, indicates a net decrease in the dry matter in shoots over the year of 65 g m^{-2} y^{-1}. Net flows are illustrated for the site (a) in 1984; (b) for unburnt areas of the study site in 1985; (c) for burnt areas in 1985; and (d) for areas burnt for a second time at the beginning of 1986.

virtually continuous death and detachment. This can be seen here as the main reason for the previous underestimations of grassland P_n.

Other recent studies of tropical grasslands have suggested a similar degree of underestimation by IBP methods. Cox and Waithaka (1989) in Nairobi National Park, Kenya, estimated that if the maximum difference in living biomass during a 60-day period in the dry season and 71 days in the short rainy season were taken as a measure of net production and extrapolated to a full year, production estimates would have been only 56 and 75% of the values actually obtained for mounds and inter-mounds, respectively, using the Wiegert and Evans (1964) technique to account for mortality. At the same location, Deshmukh and Baig (1983) used a different method to estimate mortality which involved marking clusters of shoots with cotton tags, removing dead material and then measuring the production of dead material at the next harvest. Underestimation of net primary production when mortality was ignored was 46% for *T. triandra* and 48% for *Sorghum phleoides* during the long rainy season. During the dry season, estimates based on harvest samples only would have indicated that P_n was zero whereas when mortality was taken into account there was a small amount of production equivalent to about 1 g m^{-2} day^{-1} (Deshmukh and Baig, 1983).

It is clear that harvest data alone cannot be used to estimate the true value of P_n, but the amount by which previous studies that have not accounted for mortality have underestimated production cannot be predicted (cf. Figures 8.3 and 8.4). Even if account is taken of differences in harvest intervals, the rate at which mortality occurs will be influenced by a range of factors such as climate, soil and grazing pressure. This varies with season but also from one location to another. For example, the results from the three terrestrial grassland sites reported here indicate that the above-ground rate of mortality on an annual basis is almost four times as high as at Hat Yai, Thailand and more than twice as high in Chapingo, Mexico as it was in Nairobi National Park, Kenya.

Techniques used to estimate the mortality component of the net primary productivity measurement frequently differ between studies and have seldom given comparable results. The most commonly used methods have been the 'paired-plots' method of Wiegert and Evans (1964), the 'litter-bag' technique used in this study (Roberts *et al.*, 1985) and the marking of individual grass shoots (Deshmukh and Baig, 1983). Wiegert and Evans (1964), using both the litter-bag and paired-plot methods, obtained substantially lower rates of decomposition from the former than from the latter and this was also found to be the case by Abouguendia and Whitman (1979) in a mixed grass prairie.

Table 8.3. Net annual primary production (g m^{-2}) above ground (AP_n), below ground (BP_n) and total (P_n) and annual rainfall values for certain selected tropical grasslands

Site	*Grassland type*	AP_n	BP_n	P_n	Rainfall (*mm*)	*Source of data*	*Measurement**
Fete Ole (Senegal)	*Aristida* dominated	82	—	—	209	Singh and Joshi (1979)	(a)
	Panicum and *Chloris* dominated	256	—	—	209	(data of J.C. Bille)	(a)
	Panicum and *Zornia* dominated	476	—	—	209	(data of J.C. Bille)	(a)
Pilani (India)	Mixed grass	217	61	278	388	Kumar and Joshi (1972)	(a)
Welgevonden (South Africa)	Bushveld open savanna	710	—	—	436	Singh and Joshi (1979) (data of J. Van Wyk)	(a)
Kurukshetra (India)	Mixed grass and forbs	2407	1131	3538	790	Singh and Yadava (1974)	(d)
Rwenzori National Park (Uganda)	*Hyparrhenia-Themeda*	730	1572	2302	900	Strugnell and Piggott (1978)	(a)
Jhansi (India)	*Sehima-Heteropogon*	1014	524	1538	*ca* 700	Shankar *et al.* (1973)	(b)
Lamto (Ivory Coast)	Palm savanna	498	—	—	1158	Singh and Joshi (1979) (data of M. Lamotte)	(a)

Lamto (Ivory Coast)	*Loudiata simplex* grass savanna	830	1320	2150	1300	Menaut and Cesar (1979)	(c)
Lamto (Ivory Coast)	Andropogonae grass savanna	1540	2040	3580	1300	Menaut and Cesar (1979)	(c)
Mokwa (Nigeria)	*Hyparrhenia-Andropogon*	614	—	—	1115	Ohiagu and Wood (1979)	(c)
N. Bengal (India)	*Axonopus compressus* dominated	269 354	108 179	377 533	— —	Nandi and Pal (1983)	(a) (b)
Nairobi National Park (Kenya)	*Themeda-Setaria*	1071	—	—	460	Desmukh (1986)	(c)
Nairobi National Park (Kenya)	*Themeda* grassland	3228	—	—	850	Cox and Waithaka (1989)	(c)
Olokemeji (Nigeria)	Derived savanna	680	—	—	1168	Hopkins (1968)	(a)
Serengeti (Tanzania)	Tall grass savanna	520	—	—	*ca* 700	Bourlière and Hadley (1970)	(a)

* Methods of measurement: (a) difference between maximum and minimum biomass; (b) sum of positive increments on successive sampling dates; (c) the sum of all changes in biomass and losses due to mortality, above-ground only; (d) as in (c), but accounting for changes in below-ground biomass and losses in addition to those above-ground, but decomposition not accounted for.

Direct comparisons of the three different techniques at one site at the same time have not been made. However, three different measurements of productivity have recently been made in Nairobi National Park and they have used three different techniques. Litter-bags were used by Kinyamario and Imbamba (Chapter 3), paired-plots were used by Cox and Waithaka (1989) and the marking of individual shoots was used by Deshmukh and Baig (1983). Unfortunately the measurements were not made at the same time but the marked differences in estimates of above-ground annual net primary productivity of 811 g m^{-2} y^{-1} by Kinyamario and Imbamba, 3228 g m^{-2} y^{-1} by Cox and Waithaka (1989) and 1071 g m^{-2} y^{-1} by Deshmukh and Baig (1983) may not be totally attributable to spatial differences or year-to-year variation.

The incorporation of decomposition estimates into the calculation of P_n in the present studies has, in all cases, substantially elevated the value of P_n above previous estimates using the IBP method. The importance of the decomposition component in the tropics is a reflection of the high rates of decomposition in these ecosystems. Decomposition rates are strongly influenced by temperature and moisture. In the tropics temperature is unlikely to be limiting to rates of decomposition, soil moisture being the dominant influence on rates.

Decomposition in these ecosystems will influence the availability of nutrients required for growth. When water is in plentiful supply, growth is most likely to be nutrient limited. Under these conditions the rapid decomposition of plants and release of mineral elements is necessary to achieve the high productivities which are observed. The accumulation of dead vegetation in grasslands which are undisturbed by grazing or fire has the effect of sequestering minerals as well as carbon. Conversely, it is likely that the effect of grazing and fire is a stimulation of productivity brought about largely by increasing the turnover rate of the vegetation.

8.4.2 Heterogeneity within grasslands

The heterogeneity of many grassland areas means that many replicate harvests need to be taken in order to reduce confidence intervals and establish statistically significant differences. In this study it was established for the three terra firma grasslands that 20 replicate quadrats, of an optimized size, should be harvested on each sampling date to obtain a standard error of 10% of the mean. Examination of the studies in Table 8.3 shows that very few have used these replicate numbers, thus the error in estimates of biomass alone are likely to exceed ±10%. However, this should not detract attention from the fact that heterogeneity in the distribution of biomass in tropical grasslands is high and in need of quantitative assessment. Furthermore, Singh and Misra (1969) suggest that it may be an important factor in maintaining these ecosystems.

The scale of heterogeneity can vary considerably. Small-scale heterogeneity over a few metres can be accounted for by many replicate harvests, as in the present studies, but when the scale becomes larger then measurements in the different microhabitats should probably be made. For example, Cox and Waithaka (1989) working in Nairobi National Park, Kenya made a distinction between large earth mounds, probably formed by the combined activities of rhizomyid mole rats (*Tachvoryctes splendens*) and fungus-gardening termites (*Macrotermes* spp.), and areas between these mounds. The inter-mound areas were dominated by *Themeda triandra*, similar to the areas investigated in the same National Park by Deshmukh (1986) and in Chapter 3. The mounds were about 5–32 m in diameter and 0.4–1.2 m in height and were dominated by different grasses from those in the inter-mound areas. Cox and Waithaka (1989) found that the seasonal pattern of AP_n differed markedly between mound and inter-mound sites. During the dry season, production was greater on the mounds than on the inter-mounds but during the wet season the opposite was true. Annual estimates of AP_n for the two microhabitats were similar at 2905 g m^{-2} for mounds and 3228 g m^{-2} for inter-mounds, respectively.

8.4.3 The effect of grazing

An important component of grassland ecosystems are the grazers, which have been intimately associated with the development of the grassland structure. Although the largest proportion of grass may be consumed by large ungulates there are a number of other primary consumers including insects. Also, dead and decomposing plant material may also be consumed by earthworms, termites, fungi and bacteria. In this study the estimates of primary production were based on the assumption that no consumption by grazers took place between harvests. Large herbivores were excluded from the sampling area by fences to prevent grazing. As a result however, the effect of grazing on primary production is not known. In the absence of such exclosures, losses to grazers can be very large. McNaughton (1985) found that 17–94% of the AP_n is harvested by grazing animals in the Serengeti Park in Tanzania. Cox and Waithaka (1989) estimated that grazers removed 74.5% and 21.6% of AP_n from mounds and inter-mounds respectively in Nairobi National Park, Kenya. It is likely that the effect of grazing is a stimulation of productivity brought about largely by an increase in the turnover rate of mineral nutrients and carbon. However, this will be dependent on grazing intensity and the magnitude of the effect in natural grasslands is not known.

8.5 REMOTE SENSING AS A VALUABLE SHORT CUT FOR ESTIMATING PRODUCTION AND CARBON FLUXES?

Conventional measurements of primary productivity in grassland communities require frequent, labour-intensive harvesting of large numbers of quadrats, sorting of the harvested material and the monitoring of losses through death and decomposition. Although providing some of the most detailed estimates of biomass dynamics and production in tropical grasslands, this study has only provided information of 4×1 ha sites in a biome of over 20×10^6 km^2, and then over no more than 4 years. Assessing the present status of tropical grasslands requires far more extensive sampling over longer periods. In the present study it was only practicable to monitor a few small sites without very large inputs of manpower. Because of the heterogeneous nature of grasslands this poses a particularly difficult problem in attempting to make global estimates of primary productivity. Also, unlike tropical forests for instance, tropical grasslands occur largely in areas of marked year to year variation in rainfall, and therefore marked variation in productivity is likely, as illustrated in these studies. As a consequence, reliable assessment of the productivity of these communities requires that monitoring should be carried out over a number of years.

One way of overcoming this problem may be to use remote sensing of biomass change and productivity via aircraft or satellite-mounted sensors. Satellite remote sensing of spectral reflectance is a very attractive method of studying wide areas at regular intervals.

Previous studies have shown that the reflectance ratio for red to near infra-red light (*R/NIR*) is closely correlated with the biomass of herbaceous communities (Chapter 7). Further, in temperate communities *R/NIR* has been shown to relate closely with leaf area and the capacity of vegetation to intercept radiation or light interception efficiency (Chapter 7). Given that productivity is mechanistically related to the quantity of radiation intercepted by vegetation (Monteith, 1978), it should be possible to predict the actual productivity of grasslands by combining *R/NIR* measurements with weather station records of radiation receipt.

However, a major limitation to the application of remote sensing techniques to determining the productivity of tropical grasslands is the lack of ground truth data. This development of relationships from ground truth measurements is of use only when reliable estimates of primary productivity are made. Consequently, there is a real need at present to combine these two areas in order to establish reliable relationships between reflectance ratios and productivity. There are major difficulties associated with this undertaking because in lightly grazed grasslands, particularly during dry periods, there is a rapid accumulation of dead vegetation. This may screen live vegetation from the view of the remote sensor and therefore alter the relationship between *R/NIR* and both biomass and light interception efficiency (Chapter 7). However, given the different spectral properties of dead vegetation it may be possible to use reflectance ratios other than *R/NIR* for correcting for interference by dead vegetation.

8.5.1 Remote sensing of biomass

A correlation of biomass with (*R/NIR*) is well recognized, as explained in Chapter 7. However, whilst extensive use of false colour images of NDVI developed from this ratio have been used to make semi-quantitative estimates of seasonal dynamics in plant biomass, this has rarely been accompanied by intensive ground truth measurements over consecutive months. The results of Chapter 7 show that whilst *R/NIR* may provide a precise prediction of biomass in simple crop communities, the relationship is far more complex for natural tropical grasslands, with the relationship of *R/NIR* to above-ground biomass and leaf area changing monthly and by a factor of more than 10 over the course of a year. The results show that at all three terra firma tropical grasslands examined here extensive and regular ground truth measurement would be essential to any meaningful estimation of plant biomass, and that calibrations based on ground truth measurements at a single point in the year would be meaningless in estimating biomass dynamics over the whole year.

8.5.2. Remote sensing of primary production

Monteith and Unsworth (1990) show that *R/NIR* reflectance is mechanistically linked to the capacity of a plant community to absorb photosynthetically active radiation, i.e. *R/NIR* reflectance is directly related to light interception efficiency (ε_i). Indeed from a theoretical standpoint the *R/NIR* reflectance should be a better predictor of ε_i than of leaf area index or live plant biomass (Monteith and Unsworth, 1990). For a range of crops, production has been shown to be directly proportional to the integrated quantity of light intercepted by the canopy over the growing season. Further, crops of the same photosynthetic type show remarkably similar efficiencies of conversion (ε_c) of intercepted light into dry matter production (Beadle *et al.*, 1985). Thus, if ε_i and the incident quantity of radiation are known, then production can be estimated. Warrick (1986) has suggested the extension of this technique to estimating net production from satellites. However, to be generally applicable ε_c must be constant or at least similar, within a vegetation type.

A goal of these studies was to examine the constancy and magnitude of ε_c between and within tropical grasslands (Chapter 1). ε_c has been estimated by plotting accumulated net production against the measured accumulated quantity of solar radiation intercepted by the plant communities, all dominated by C_4 grasses (Figure 8.11). The conversion efficiency suggested for healthy C_4 crops is 2.3 g (dry matter) MJ^{-1} (Piedade *et al.*, 1991). This was achieved by the *Echinochloa polystachya* dominated community of the Amazon floodplains. However, similar efficiencies were not observed in the other grasslands. The *Distichlis spicata* saline grassland at Chapingo showed less than half of this efficiency, whilst an even lower value was recorded in the *Eulalia trispicatum/Lophopogon intermedius* co-dominated community at Hat Yai. This however, still leaves the possibility that ε_c may be constant within a community type. However, results for the Hat Yai site show that ε_c can show marked year-to-year variation. In 1984 ε_c averaged over the year was 0.66 g MJ^{-1}, but in 1986, following two fires at the site and below-average precipitation in the early part of the year, ε_c averaged over the year was just 0.08 g MJ^{-1}. This is only one-thirtieth of the expected value of a C_4 community. Following the fires very little dead vegetation was present or was formed in the canopy that could interfere with the measurements of light absorption by the photosynthetically active elements of the canopy. So in this case a low ε_c cannot be attributed to overestimation of ε_i through interception of light by dead vegetation. The result shows that even within a site ε_c can vary by almost an order of magnitude invalidating any attempt to sense production remotely via an assumption of constant ε_c. Considering the biology of the species involved these results are perhaps not unexpected. On the burnt site at Hat Yai in 1986 an average of over 75% of the biomass was in non-photosynthetic tissues below-ground, thus respiratory losses through maintenance of this increased proportion of non-photosynthetic tissues would be expected to be considerably higher in 1986. Similarly, it can be seen that the community with the highest ε_c, *E. polystachya*, had the lowest proportion of biomass present as roots. The results suggest that whilst remote sensing of biomass and production has proved successful with crop stands, the same techniques are not easily or necessarily meaningfully transferred to tropical grasslands, and a greater understanding of the bases of variation in ε_c and in biomass to *R/NIR* reflectance ratios will be essential if these measures are to have any meaningful application in the remote sensing of biomass and production of the tropical grasslands biome.

8.6 MODELLING GRASSLAND PRODUCTIVITY

A major objective of the IBP Grassland Biome Programme was the development of ecosystem level models to simulate the intra-seasonal carbon, nitrogen and water dynamics of grassland ecosystems (Innis *et al.*, 1980). One product of this was, for instance, the primary producer sub-model described by Detling (1979). This is a model of production based upon the photosynthetic and respiratory characteristics of six component plant parts, viz. three layers of the root system, the crowns, expanding leaves and expanded leaves. Information gathered in the present project could be used in such a model to predict the productivity of natural grasslands, as has been done for managed temperate grasslands (Johnson and Thornley, 1983). The major physiological inputs to such a model are the photosynthetic characteristics of leaves in the canopy. In the present study, these measurements were made at all the locations using a portable infra-red gas analysis system. Information on other model inputs such as leaf area index and ambient temperatures and light levels were also collected. This is stored in the central database at the Technical Co-ordination Centre, at the University of Essex. However the universal applicability of such a model depends on knowledge of nutrient and water limitations which are not modelled in the managed grasslands (Johnson and Thornley, 1983). The supply of water will limit growth of tropical grasslands on many occasions (cf. Chapters 2–4). Saugier *et al.* (1974) constructed a mechanistic model of grass growth based upon physiological properties, soil physical characteristics and meteorological driving variables (the Matador Model). Model parameters were derived largely from laboratory data, and validity was tested against field measurements of such parameters as CO_2 flux, leaf temperature, soil water and plant biomass. The authors concluded that the sections of the model dealing with physical problems such as radiation penetration, micro-climate and soil heat flux and water were relatively straightforward. However, the sections dealing with physiological parameters such as stomatal activity and the influence of water stress on photosynthesis and growth were less exact. It was suggested that the model could be used to examine production processes, including water movement, in species for which adequate physiological data are known, and in environments for which information of physical properties is also available. The lack of this relatively detailed information clearly imposes severe limitations on the widespread use of such physiological models.

Models which are perhaps more widely used at the ecosystem level are those similar to the 'Chikugo model' of Uchijima and Seino (1985). This is an empirical model and uses a non-linear regression equation to describe the dependence of P_n on annual net radiation (R_n) and a radiative dryness index (RDI) which is the ratio of net radiation to the product of annual precipitation and the latent heat of evaporation. The equation used by Uchijima and Seino (1988) is as follows:

$$P_n = 0.29 (\exp(-0.216\ RDI))\ R_n$$

and it is used to estimate P_n above ground from weather parameters such as global solar radiation outside the atmosphere, percentage potential sunshine duration, albedo, temperature, air humidity and precipitation. These empirical models have been used to construct P_n distribution maps for Japan (Uchijima and Seino, 1985) and the continents (Uchijima and Seino, 1987) using climatic data from these areas. Estimates of productivity made using these models have been verified using published plant production data and in most cases there appears to be good agreement. They apparently give better predictions of P_n than those given by the 'Miami' model of Lieth (1973) which was even simpler in form and based on precipitation and temperature data alone. Obviously the weakness of this empirical approach is the fact that P_n values used to verify the predictions may have been underestimated using conventional methods. However, although the spatial resolution of these maps is poor they do produce estimates which are in most cases within about 20% of the actual measurements of P_n if stresses such as fires are discounted.

8.7 SPECIAL CHARACTERISTICS OF TROPICAL GRASSLANDS WHICH INFLUENCE P_n AND CARBON FLOW

8.7.1 Seasonal variation in primary production

One of the most striking characteristics of tropical grasslands in general and savanna ecosystems in particular is the marked seasonal variation in primary production. The main driving force for this seasonal variation in the tropics is rainfall. The seasonality in P_n is most marked in the more arid grasslands, where growth can be restricted to less than three months, and seasonal changes in production are far less pronounced in more humid savannas. This contrast is apparent when seasonal variation in leaf area indices (L) at Chapingo (Chapter 3), where there was an annual and predictable period of pronounced drought, are compared to L at Hat Yai (Chapter 4), where periods of drought were far shorter, less pronounced and less predictable.

All grassland sites in the present study showed marked seasonal variations in productivity above ground. Where there is more than one dominant species there is little evidence that there are differences in their seasonal growth pattern. During the peak growth periods the above-ground growth rates were 132 g m^{-2} $month^{-1}$ in the saline grassland, 309 g m^{-2} $month^{-1}$ in the dry savanna, 425 g m^{-2} $month^{-1}$ in the moist savanna and 900 g m^{-2} $month^{-1}$ in the periodically inundated grasslands in the Amazon. The latter are amongst the highest growth rates recorded for natural communities.

Root growth is more difficult to correlate with seasonal climatic changes. This may be partly because of the difficulty of measuring root growth accurately but also because there appear to be shorter term changes in the pattern of distribution of assimilates to roots which are associated with developmental changes in the grasses, particularly those associated with initiation of reproductive growth. In general, however, root growth is apparently synchronous with shoot growth in the drier areas, whereas it takes place mostly at the end of the rains in moist humid environments, the underground reserves being actively used up at the beginning of the rains (Lamotte and Bourlière, 1983), and following fires (Chapters 3 and 4).

There appears to be a great deal of variation in the amount of primary production which is translocated below-ground to roots and rhizomes (Figures 8.6–8.10). Roots accounted for only about 5% of biomass in the semi-aquatic grass *E. polystachya* in the Amazon floodplain, approximately half the total biomass in the moist savanna in Thailand, dry savanna in Kenya and bamboo forest in China, and about 75% of the total biomass in saline grassland in Mexico. In addition, the present study has shown that the proportion of annual total P_n which occurs below ground varies considerably from one location to another (Figures 8.6 and 8.7). In North American grasslands, the tall-grass prairie below-ground production has been estimated at 48–64% of total P_n, at 61–80% in mixed grass prairie and at 70–78% in short-grass prairie. However, the proportion of biomass below ground is usually much higher here than in tropical grasslands, with root:shoot ratios varying from 2:1 to 13:1 (Stanton, 1988), suggesting far slower rates of organ turnover. Thus root production may well be higher in the tropics than in temperate regions but the faster turnover rates yield lower values for live biomass.

8.7.2 Biomass turnover

Biomass turnover represents the fraction of biomass which is exchanged or replaced during a particular time interval. It is calculated as the P_n divided by the mean standing biomass produced within a given time interval, normally a year. The reciprocal of the biomass turnover rate is the turnover time. Turnover rates can be calculated for different components of the vegetation. Normally this is done for above and below-ground live biomass but also the turnover of litter and dead vegetation gives information on transfer of material between components (Figures 8.8–8.10).

Turnover values in different communities vary widely (Table 8.4). This may be due to variation in climatic, edaphic and biotic factors. When all the above-ground standing vegetation is burnt once a year in a tropical savanna, the yearly rate of above-ground turnover obviously exceeds unity. At Lamto, on the Ivory Coast the values range from 130 to 200% depending on the vegetation type (Lamotte and Bourlière, 1983). In the present studies, annual turnover values for 1984/85 were 500, 430 and 420% at Hat Yai, Nairobi and Chapingo, respectively. This compares to 160% in 1985/86 for the floodplain vegetation at Marchantaria and 15% in the bamboo forest of Maio Shan. In all ecosystems apart from the bamboo forest, the annual value significantly exceeded 100%, thus illustrating the importance of continuous turnover and recycling within these communities. These values are generally higher than those previously reported because of the higher values of P_n. Nandi and Pal (1983) found that turnover rates of grassland communities were dependent on the method of calculating annual increments in biomass, even though they did not account for death and decomposition in their study. The turnover of above-ground biomass varies on a seasonal cycle as does primary production. Below-ground turnover is generally less than above ground, although fine roots obviously turn over faster than rhizomes.

This set of studies show that standing biomass is a poor guide to the actual flow of carbon through the primary producers of tropical grasslands. A graphic illustration of this is provided from the studies at Hat Yai, where the amounts of carbon passing through the herbaceous vegetation over the course of a year exceed the mean standing quantities by a factor of four (Figure 8.8a). This is in sharp contrast to the conclusions of major earlier syntheses of tropical

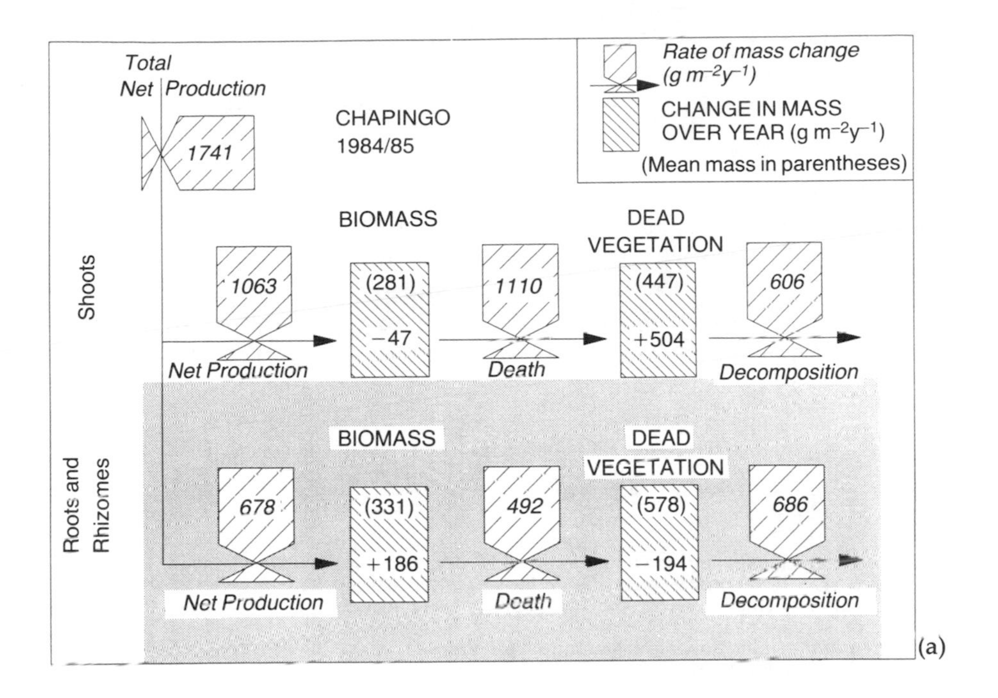

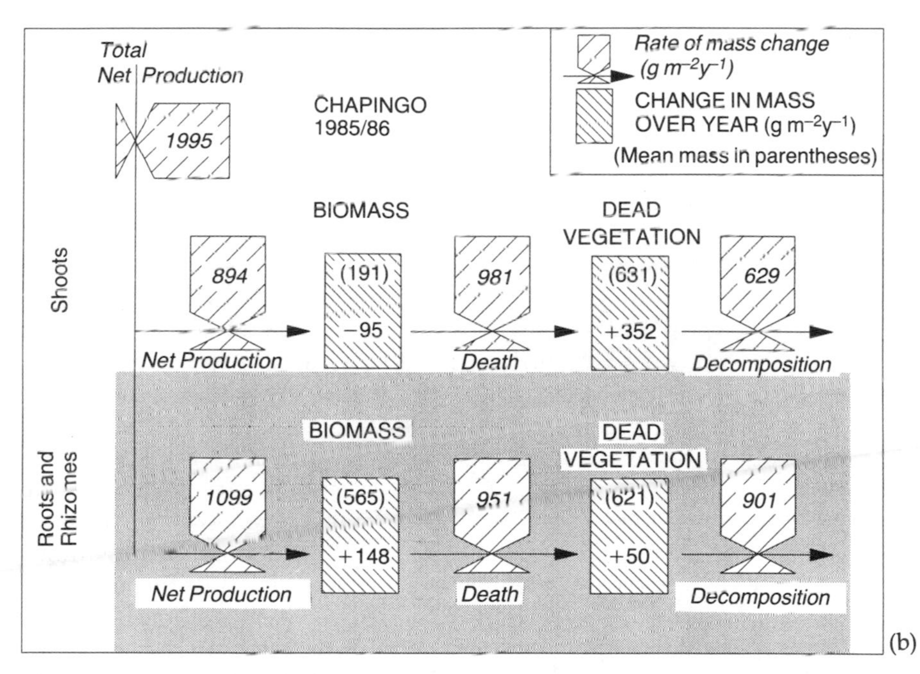

Figure 8.9. (caption overleaf)

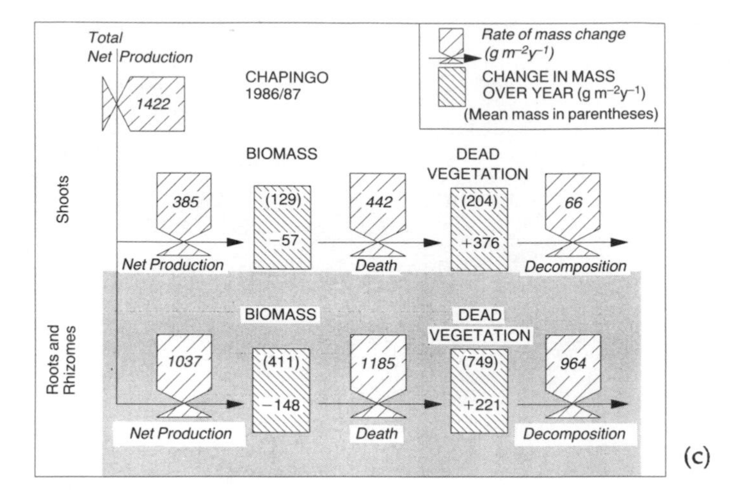

Figure 8.9. Fluxes and mean quantities of dry mass in the grassland at Chapingo, Mexico. The lower numbers in the rectangles indicate the net change over the year in that quantity, thus '−47' in the lower part of the rectangle labelled representing the biomass of shoots, indicates a net decrease in the dry matter in shoots over the year of 47 g m^{-2} y^{-1}. Net flows are illustrated for the site in (a) 1984/85; (b) 1985/86; and (c) 1986/87, following a fire in 1986.

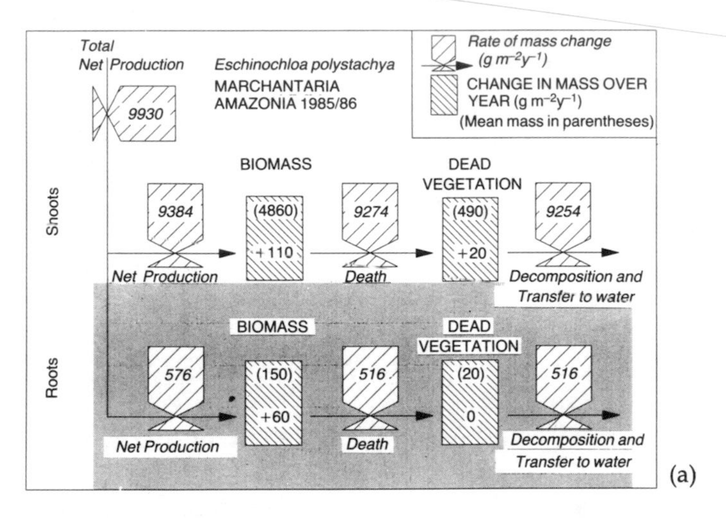

Figure 8.10. Fluxes and mean quantities of dry mass in (a) the *Echinochloa polystachya* dominated floodplain grassland on Ilha da Marchantaria, Brazil; and (b) the *Phyllostachys pubescens* bamboo forest at Miao Shan, China. The arrowed boxes, representing the rate variables of annual fluxes (g (dry mass) m^{-2} y^{-1}), and rectangles representing the state variables of mean mass of vegetation (g (dry mass) m^{-2}) are vertically scaled to their magnitude. The lower numbers in the rectangles indicate the net change over the year in that quantity, thus '+110' in the lower part of the rectangle representing the biomass of shoots, indicates a net increase in the dry matter in shoots over the year of 110 g m^{-2} t^{-1}. Net flows are illustrated for both sites in 1986.

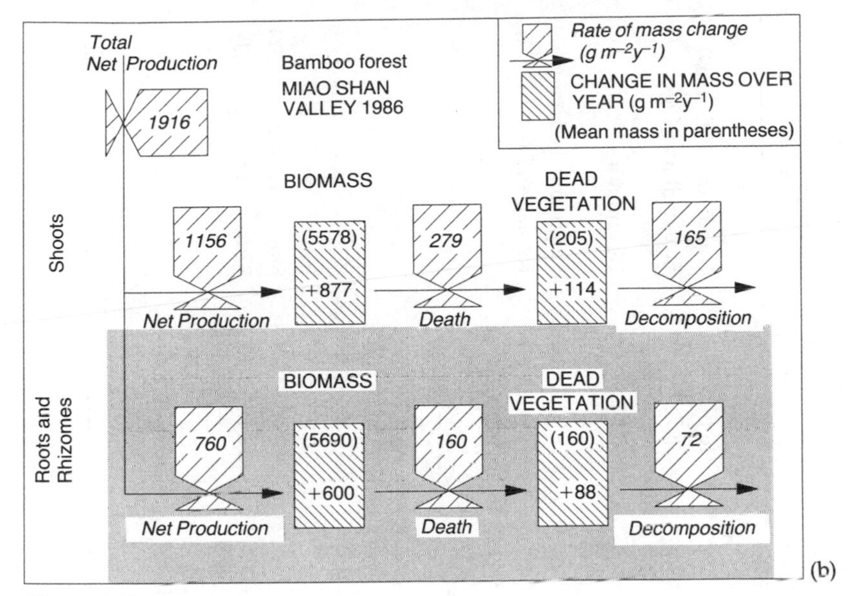

Figure 8.10.

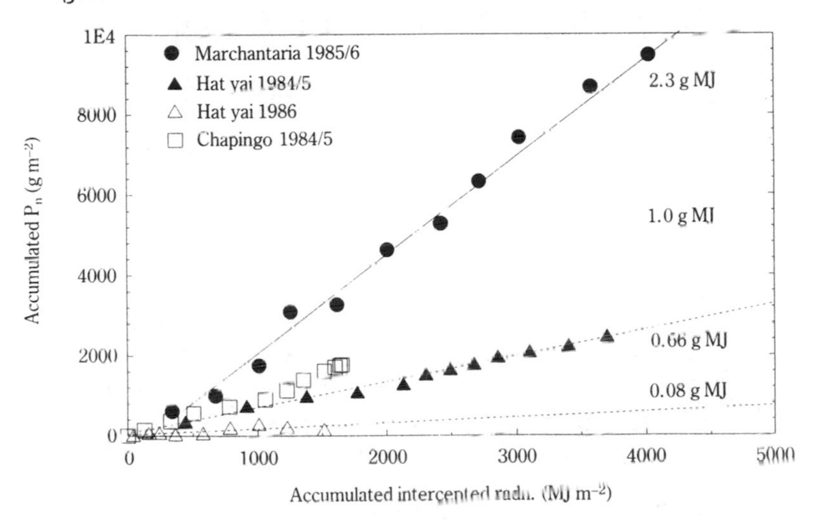

Figure 8.11. Accumulated monthly mean values of net primary production plotted against the accumulated quantity of solar radiation intercepted by the same communities. Lines are the best-fitting straight lines to the illustrated data points, as determined by the least squares method. Values, indicate the slope of these lines. Mean P_n is illustrated for *Echinochloa polystachya* at Marchantaria, Brazil; the *Distichlis spicata* dominated community at Chapingo, Mexico; the mixed grassland at Hat Yai, Thailand, for both unburnt sites in 1984/5 and for measurements made in 1986 on sites which had been burnt twice.

Table 8.4. Turnover of material in tropical grasslands

Site	*Grassland type*	*Turnover rate* (%)	*Turnover time* (*months*)	*Source of data*
(a) Above-ground				
N. Bengal (India)	*Axonopus compressus*	75	16.0	Nandi and Pall (1983)
Rwenzori National Park (Uganda)	*Hyparrhena-Themeda*	180	6.7	Strugnell and Piggott (1978)
Nairobi National Park (Kenya)	*Themeda-Setaria*	218	5.5	Deshmukh (1986)
Kurukshetra (India)	Mixed grass and forbs	122	9.8	Singh and Yadava (1974)
Pilani (India)	Mixed grass	167	7.2	Kumar and Joshi (1972)
Thailand (This study)	Moist savanna	240–473	2.5–5.0	
Kenya (This study)	Dry savanna	104*	12	
Mexico (This study)	Saline grassland	430–520	2.2–2.8	
Brazil (This study)	Semi-aquatic grassland	157	7.6	
China (This study)	Bamboo forest	23	52	
(b) Below-ground				
N. Bengal (India)	*Axonopus compressus* dominated	46	26.0	Nandi and Pal (1983)
Kurukshetra (India)	Mixed grass and forbs	97	12.4	Singh and Yadava (1974)
Pilani (India)	Mixed grass	56	21.4	Kumar and Joshi (1972)
Thailand (This study)	Moist savanna	44–146	8.2–27.2	
Kenya (This study)	Dry savanna	93	12.9	
Mexico (This study)	Saline grassland	200–230	5.2–6.0	
Brazil (This study)	Semi-aquatic grassland	n.d.	n.d.	
China (This study)	Bamboo forest	10	120	

* Two year average.
n.d. not determined.

grassland studies. Singh *et al.* (1979) concluded that annual root turnover in tropical grasslands ranged from 0.45 to 0.97, calculated as the ratio of below-ground production to peak below-ground biomass (cf. Table 8.4). Averaged over the 3 years (Figure 8.9), root turnover at Chapingo calculated on the same basis was 2.1. The difference, however, is almost certainly methodological. The studies by Singh *et al.* (1979, 1980) inferred production from positive increments in mass, as discussed earlier. If production is calculated here by this method, then the data for Chapingo similarly suggest a turnover of less than 1. This illustrates the major underestimation not simply of production, but of input of carbon to the ecosystem.

8.7.3 Photosynthesis and water use efficiency

All the ecosystems investigated here, apart from the bamboo forest, are dominated by C_4 grasses, i.e. species which utilize the C_4 pathway of photosynthetic carbon metabolism. C_4 photosynthesis is normally associated with high rates of photosynthesis at high light levels, a lack of light saturation at high irradiance, and higher cardinal temperatures for photosynthesis and growth than those of herbaceous C_3 plants (Ludlow, 1985). The C_4 pathway therefore confers the potential for high growth rates and high P_n, although these are only achieved in the absence of biological limitations to growth and in a favourable environment. Nevertheless there is a clear link between the maximum leaf net photosynthetic rates and P_n (Ludlow, 1985).

Measurements of leaf photosynthesis were made at monthly intervals throughout for at least one year at all the sites used in this study, except at Hat Yai. In general, the highest rates of photosynthesis reached about 40 $\mu mol(CO_2)\ m^{-2}\ s^{-1}$, which is comparable with many C_4 pasture grasses (Ludlow, 1985). However, they also showed marked inhibition under conditions of stress. The light saturated rates (A_{sat}) obtained in the saline grassland at Chapingo (Chapter 3) were also lower than at other grassland sites (Long and Baker, 1986). In both Nairobi National Park and Marchantaria, Brazil, decrease in A_{sat} was closely associated with availability of water. In the canopy of bamboo (*Phyllostachys pubescens*), a C_3 species, marked depressions of A_{sat} developed during the course of hot sunny days, suggesting photoinhibition.

Another feature of plants with C_4 photosynthesis is their higher water use efficiency relative to C_3 plants of similar form in the same environment. At the terrestrial grassland sites it is clear that the seasonal variation in rainfall has a dominant effect on the annual cycle of growth. Under these circumstances it is likely that C_4 grasses will have a competitive advantage over plants with C_3 photosynthesis if they combine high growth rates with an efficient use of available water. C_4 species formed over 99% of the biomass of the ground layer vegetation at the Nairobi National Park site and at Chapingo. At Hat Yai a C_3 dicotyledon formed a significant part of the cover (*ca* 10%). Following the second fire, C_4 grasses diminished as a proportion of the total cover. One explanation of this may be that fire, in creating gaps in the vegetation, decreases the significance of competition as a factor determining survival and hence the significance of the increased water use efficiency associated with C_4 photosynthesis.

There is much evidence that the productivity of tropical grasslands is largely limited by water availability and in some cases a direct relation exists between rainfall and productivity. For example, Walter (1984) found a linear correlation between net primary productivity and rainfall for a number of locations in Namibia. However when rainfall exceeds 600–800 mm the relationship no longer appears to hold. The difficulty with establishing these relationships once again revolves around the methodological shortcomings in the measurement of P_n. Recognizing these difficulties, Deshmukh (1984) used peak above-ground biomass instead of P_n for grasslands in east and southern Africa to establish a linear relationship with rainfall. However, even this gross over-simplification appears to have little application in other geographical locations. This also applies to the relationship established by Sala *et al.* (1988) for sites in the central United States. In Chapter 3 it is shown that this relationship grossly underestimates the measured P_n at the site in Mexico. Although rainfall clearly influences productivity there appears to be no simple relationship which links them in any mechanistic way. As there is no information on actual water use by the grasslands in this or any other study it is only possible to determine a crude index of water use efficiency calculated as a ratio of P_n to annual rainfall. When Singh and Joshi (1979) attempted this they found no clear relationship between annual rainfall and efficiency of water use from above-ground production. There was, however, some indication that water use efficiency declined in areas of high rainfall.

8.8 A SOURCE OR A SINK FOR CARBON DIOXIDE?

8.8.1 Carbon pools and climate change

Patterns of climate change are predicted from the enhanced greenhouse effect being produced by rising levels of hetero-atomic gases in the troposphere, in particular CO_2. Assuming the IPCC 'business-as-usual' scenario and an approximate doubling of CO_2 concentrations towards the end of the next century, substantial changes in global climate are expected (Houghton *et al.*, 1990). In particular, a warming of much of the area currently occupied by tropical grasslands and a decrease in soil moisture. However, climate change is not simply a one-way process of temperature and soil moisture change affecting ecosystems. Terrestrial ecosystems contain considerably more carbon than that present in the atmosphere as CO_2. If climate change and rising CO_2 affect the size of these pools then there is great potential for positive and negative feedback effects. Clearly, meaningful models projecting future climates must link this feedback effect with climate. The initial step though must be to determine the present situation, particularly the quantities and fluxes of carbon in natural vegetation.

Current quantities of carbon as CO_2 in the atmosphere are approximately equal to those in plant biomass, whilst litter and soil carbon is estimated to contain twice as much carbon. Thus any change in the size of these pools could and may have a profound influence on the magnitude of global rise in CO_2 concentrations. Only about half the annual industrial emissions of CO_2 are currently accounted for by atmospheric concentration increase. Recently, it has become apparent that terrestrial ecosystems must account for up to half of this so-called 'missing sink' for CO_2, although other ecosystems notably tropical forests are, through their destruction, obvious contributors of CO_2 (Tans *et al.*, 1990). Gifford (1980) suggested that tropical grasslands were in fact the largest terrestrial sink for atmospheric CO_2, accounting for 27% of the 'missing sink'.

The consequences of making incorrect estimates of net primary production are becoming increasingly important at present, particularly in relation to the greenhouse effect and climate change. For example, the rate of change in the atmospheric CO_2 concentration depends on the cycling of carbon globally. The cycling of carbon between the atmosphere and the biosphere depends on the sequestration of carbon in plant material and it is only when we have accurate measurements of the primary productivity of all ecosystems that this level of sequestration can be determined. This has particular relevance to the 'missing sink' for CO_2, with the geochemical models of CO_2 cycling suggesting that the net release of CO_2 at present should be in the order of 1.4 Pg (1Pg = 1 Gigatonne = 10^{12} kg) carbon higher than is actually observed through measured atmospheric CO_2 increase (Hall, 1989).

Also, the global distribution of vegetation is predicted to change with response to the increasing 'greenhouse effect' (Houghton *et al.*, 1990). This is influencing carbon cycling; but the effect on this cycle will not be quantifiable until we can, for instance, accurately predict the effect of removal of tropical rain forests and their probable replacement by grasslands. Furthermore, there are also uncertainties about the net effect of increasing atmospheric CO_2 levels (CO_2 fertilization) on ecosystems. This is largely because of the poor or incomplete baseline data on primary productivity, particularly in the tropics and the environmental factors that limit it. Raising the atmospheric CO_2 concentration increases the rates of photosynthesis, production of dry matter and water use efficiency (Long and Hutchins, 1991). It has been suggested that in natural ecosystems other factors may well be limiting for much of the time and CO_2 stimulation will not occur. The effects of elevated CO_2 on the production of natural tropical vegetation has not been investigated, leaving a critical gap in our understanding of how the biosphere will respond to rising CO_2 levels. In arctic tundra, CO_2 elevation produced only a transient stimulation of community carbon gain (Oechel and Reichers, 1986). However, in a warm-temperate coastal marsh, elevation of CO_2, continuously over 4 years, has consistently led to increases in both primary production and biomass (Curtis *et al.*, 1989). Measurements of primary productivity, and the environmental factors which influence it, will be essential if we are to identify where increasing CO_2 levels may have a stimulatory effect on growth and productivity and in particular if they will lead to increased sequestration of carbon. However there are also more difficult problems to unravel such as the likely changes in competitive ability of individual species within a community as CO_2 levels increase and the possibility that species acclimate differentially when CO_2 levels change (Lemon, 1983). The ability to answer many of the questions outlined briefly here will be improved considerably if we have more reliable estimates of the primary productivity of the biosphere's major ecosystems.

The most important of these ecosystems are in the developing world with a limited research base. While measurements of primary productivity have been extensively made in developed areas of the world, similar measurements have been few and far between in the less developed areas in the past and this must give an increased impetus for this type of work to be done now. The work described in this book has attempted to provide this information for some important tropical and subtropical ecosystems, particularly tropical grasslands.

A consistent feature of all three terra firma grasslands investigated here was that in the absence of disturbance the sites accumulated organic matter. For example, at Chapingo in 1984/85 and 1985/86 input (P_n) exceeded output (decomposition) by 449 and 465 g m^{-2}. Over the 2 years, this was reflected in an increase in below-ground biomass of *ca* 330 g m^{-2} and an increase in above-ground litter of 850 g m^{-2} (Figure 8.9). Similar patterns of accumulation were apparent at both Nairobi National Park and Hat Yai (Figure 8.8). On average, and in the absence of fires, the three terra firma grasslands accumulated dry matter at *ca* 360 g m^{-2} y^{-1}. Assuming a carbon content of 0.40, this would translate to 144 g (C) m^{-2} y^{-1} and an accumulation across the 20 $\times$ 10^6 km^2 of tropical grassland of 2.9 Pg (C) y^{-1}, which is more than 50% of the amount of carbon in current fossil fuel emissions and twice the size of the 'missing sink' for CO_2. Fire which will remove much of this accumulated carbon, is a feature of tropical grasslands which will limit if not nullify this sequestering potential.

The results of this study suggest that even following single fires, accumulation of carbon below-ground will continue. At Chapingo at the beginning of 1986 a fire removed most of the accumulated litter above ground and in contrast to the preceding year, the biomass in roots and rhizomes decreased by 148 g m^{-2} over the year (Figure 8.9c), probably reflecting its utilization in the resynthesis of the above-ground canopy. Production, however, was only decreased slightly, relative to the preceding years (Figures 8.9a and 8.9b), and the quantity of dead vegetation below-ground continued to increase by *ca* 100 g m^{-2}, as in the preceding year. Even if this lower rate of sequestration by tropical grasslands were sustained it could account for 0.4 Pg (C) y^{-1}. In 1984/85 inputs at the Hat Yai site exceeded outputs by 242 g m^{-2}, resulting in accumulation of biomass in roots and rhizomes, and of litter above ground. A single fire at the Hat Yai site appeared to have relatively little effect on net production and appeared to stimulate accumulation of carbon in root and rhizome biomass. However, a second fire in 1986, coupled with a moderate drought in the first part of the year, decreased net production by *ca* 90%. Below-ground production was negative and *ca* 66% of the mass in below-ground biomass and dead vegetation was lost over the course of the year. Over this year outputs exceeded inputs by 176 g m^{-2}. These results illustrate the delicate balance which determines whether these grasslands are net sinks or sources of atmospheric CO_2. Increasing frequency of fire and decreasing soil moisture, both predicted changes in this ecosystem with climate change (Houghton *et al.*, 1990), may tip the balance in favour of carbon loss to the atmosphere. Although fire is considered essential to the maintenance of many tropical grasslands, frequency of burning has increased (Hammond, 1990). These results suggest that an increased frequency of burning is not simply removing the above-ground vegetation, but indirectly also results in a loss of below-ground biomass and dead vegetation. Thus emission of carbon to the atmosphere here is not simply the direct result of combustion, but in the longer term results from the remobilization of below-ground reserves in the resynthesis of the destroyed above-ground vegetation.

It is well established that developing forests can be net sinks for carbon via the accumulation of wood (Houghton *et al.*, 1990). The *Phyllostachys pubescens* bamboo forest however, showed (Figure 8.10b) that this is not simply due to accumulation of new wood above ground. In 1986, accumulation of stem material accounted for 5580 g m^{-2}; however, roots and rhizomes accounted for 5690 g m^{-2}. In total, inputs to this community during 1986, exceeded outputs by 1679 g m^{-2}. The investment in below-ground organs is of particular significance. If the stems are periodically harvested to provide a wood crop for construction then this accumulation represents a real sequestration of carbon from the atmospheric pool. For this one species of bamboo, occupying an estimated 50×10^3 km^2 (Chapter 6), this could represent a removal of 0.034 Pg (C) y^{-1}.

The most productive ecosystem of those examined here was the floodplain community dominated by the C_4 grass *Echinochloa polystachya*. However, losses of vegetation in decomposition and transfer of fragments to the water and sediment roughly equalled inputs in production. In this study decomposition and losses of vegetation by fragmentation could not be separated. It has however, been estimated for these communities that *ca* 3% of the production becomes incorporated into the accreting sediments (Hedges *et al.*, 1986), which would amount to 30 g m^{-2} y^{-1}. Floodplains occupy an estimated 10% of the total Amazon region (6×10^6 km^2). If it is assumed that grasses like *E. polystachya* occupy about 10% of the floodplains, then sequestration into the sediments would amount to 0.040 Pg (C) y^{-1}. Thus, these two communities alone could account for sequestration of 0.074 Pg (C) annually, i.e. about 5% of the 1.4 Pg (C) unaccounted for in the global carbon balance.

These estimates all assume that on decomposition, carbon from biomass is released to the atmosphere. Decomposition was estimated by weight loss from 2 mm mesh litter bags. Weight loss could and would result, not simply from microbial respiration, but also from the production of small particulate matter and soluble organics from the decomposing litter. A striking feature of all of the grasslands studied, is that the annual flux of organic matter through the vegetation far exceeds the biomass. Thus although the average quantity of shoot biomass at the Hat Yai site in 1984/85 was 366 g m^{-2}, the net production was 1490 g m^{-2}. Examination of flow diagrams for all sites (Figures 8.8–8.10) reveals a similar pattern. Much of the carbon lost on decomposition is likely to enter the soil as soluble organics and must provide some input to the soil refractile pool. Even if only 5% of this carbon enters the refractile pool this would represent a major sequestration of carbon. Assuming an average annual decomposition of 1500 g m^{-2}, based on these studies, 5% across tropical grasslands as a whole would amount to 0.6 Pg (C) y^{-1}, suggesting that this could be a major terrestrial sink for carbon.

8.9 CONCLUSIONS

The diverse nature of tropical grasslands was highlighted at the beginning of this chapter and the measurements of primary production made at the five locations described in Chapters 2–6 have illustrated how this is reflected not only in the variability of the absolute annual values of P_n, but also in seasonal and year to year variations. The major driving forces determining the type of grassland formed and its productivity are seasonal rainfall patterns, grazing, and recurrence of fire. Differences in the intensity and frequency of these events lead to the development of a range of diverse communities which are nevertheless all dominated by graminaceous species. At one end of the range are the spatially important 'dry' grasslands represented by savanna grassland in Kenya and saline grassland in Mexico and at the other by the floodplain grassland of the Amazon.

Despite the differences in grassland communities encountered, the unifying theme which emerges from this study is one of extremely dynamic ecosystems which are able to adapt to rapidly changing conditions. Two important adaptive features are the rapid turnover of biomass (Table 8.4) and the changing allocation of this biomass to above- and below-ground components. Although these characteristics of grasslands have been appreciated in the past (Singh *et al.*, 1979; Singh and Joshi, 1979), they have not been taken sufficiently into account in measurements of primary productivity. As a consequence it appears that the primary productivity of grassland has been underestimated to varying degrees in the past. It is unlikely however that the same degree of underestimation of P_n has occurred in other biomes because rates of turnover of biomass are generally much slower and the allocation of biomass to roots in most cases is smaller than in grasslands. The results of the present study should now put the contribution of grasslands and bamboo forest to global terrestrial primary production in a more prominent position relative to other terrestrial biomes. As tropical grasslands and savannas now form the world's largest terrestrial biome this upward revision of P_n will have significant effects on estimates of the amount of carbon cycling through the biosphere. Perhaps more importantly it could also mean that the pool of soil carbon in tropical grasslands is considerably higher than the current estimates suggest. As the amount of carbon stored globally in soil organic matter exceeds by a factor of more than two the amount stored in living vegetation it is possible that tropical grassland soils act as a substantial sink for atmospheric CO_2. Particularly as it has recently been argued that most other soils have much lower current rates of carbon accumulation than originally predicted (Schlesinger, 1990).

8.10 SUMMARY

1. A synthesis of the production studies described in the previous chapters shows that biomass increments and peak biomass, as used for the earlier IBP syntheses, are poor indicators of net primary production in tropical grasslands. Method comparisons suggest that for the three terra firma sites, underestimation was 2- to 5-fold. This degree of underestimation was not constant. It varied substantially from site to site and from year to year. The present studies suggest that the primary productivity of the 20×10^6 km^2 of tropical grasslands should be revised upwards to 56 Pg(dry matter) y^{-1} and 22 Pg(C) y^{-1}.
2. Remote sensing of the reflectance ratio of near infra-red to red (R/NIR) radiation has been suggested for extensive and repeated estimation of the biomass and net productivity of the tropical grasslands biome. The relationship of biomass to R/NIR reflectance, varied from site to site and month to month, and was therefore only useful if supported by extensive monthly calibration at each site. The R/NIR reflectance is directly related to the efficiency with which vegetation absorbs light (ε_i). In a range of crops the efficiency of conversion of intercepted light into dry matter (ε_c) has been shown to be constant, thus if ε_i is estimated from the R/NIR reflectance and available light known, production can be estimated remotely. For the communities studied here ε_c varied by more than 50%, and at the Hat Yai site was found to vary 10-fold between years. Such variation would invalidate attempts to sense production remotely on the assumption of constant ε_c.
3. The results show the five sites studied to be potential sites of net carbon accumulation. In the absence of fires the terra firma sites accumulated carbon at 144 g(C) m^{-2} y^{-1} and accumulated 40 g(C) m^{-2} y^{-1} with occasional fires (0.5 y^{-1}). These are equivalent to 2.9 and 0.8 Pg y^{-1}, respectively, if they can be extrapolated across the biome.

With more frequent fires and drought a net loss of 70 g(C) m^{-2} y^{-1} occurred, suggesting a delicate balance between these sites acting as sources and as sinks for carbon. Accumulation of carbon in the subtropical bamboo forest type (50 $\times$ 10^3 km^2) was estimated at 0.034 Pg(C) y^{-1} and in the Amazonian floodplain grass community (0.6 $\times$ 10^6 km^2) at 0.040 Pg(C) y^{-1}. The results show that all these grass-dominated communities have the potential to act as significant sinks for carbon and contribute to the amelioration of rising CO_2 levels. Management and climate variation can reduce or reverse this sink capacity.

REFERENCES

Abouguendia, Z.M. and Whitman, W.C. (1979) Disappearance of dead plant material in a mixed grass prairie. *Oecologia*, **43**, 23–9.

Beadle, C.L., Long, S.P., Imbamba, S.K., Hall, D.O. and Olembo, R.J. (1985) *Photosynthesis in Relation to Plant Production in Terrestrial Environments*. UNEP/Tycooly International, Oxford.

Bourlière, F. and Hadley, M. (1970) The ecology of tropical savannas. *Annual Review of Ecology and Systematics*, **1**, 125–52.

Bourlière, F. and Hadley, M. (1983) Present-day savannas: an overview, in *Tropical Savannas. Ecosystems of the World*, Vol 13 (ed. F. Bourlière), Elsevier, Amsterdam, pp.1–17.

Coupland, R.T. (1979) Problems in studying grassland ecosystems, in *Grassland Ecosystems of the World. IBP* Vol 18 (ed. R.T. Coupland), Cambridge University Press, Cambridge, pp. 31–7.

Cox, G.W. and Waithaka, J.M. (1989) Estimating above-ground net production and grazing harvest by wildlife on tropical grassland range. *Oikos*, **54**, 60–6.

Curtis, P., Drake, B.G., Leadly, P.W., Arp, W. and Whigham, D.F. (1989) Growth and senescence of plant communities exposed to elevated CO_2 concentrations on an estuarine marsh. *Oecologia*, **78**, 20–6.

Deshmukh, I.K. (1984) A common relationship between precipitation and grassland peak biomass for east and southern Africa. *African Journal of Ecology*, **22**, 181–6.

Deshmukh, I.K. (1986) Primary production of a grassland in Nairobi National Park, Kenya. *Journal of Applied Ecology*, **23**, 115–23.

Deshmukh, I.K. and Baig, M.N. (1983) The significance of grass mortality in the estimation of primary production in African grasslands. *African Journal of Ecology*, **21**, 19–23.

Detling, J.K. (1979) Processes controlling blue grams production on the shortgrass prairie, in *Perspectives in Grassland Ecology, Ecological Studies 32* (ed. N.R. French) Springer-Verlag, New York, pp. 25–42.

Gifford, R.M. (1980) Carbon storage by the biosphere, in *Carbon Dioxide and Climate* (ed. G. Pearman), Australian Academy of Sciences, Canberra. pp. 167–81.

Hall, D.O. (1989) Carbon flows in the biosphere: present and future. *Journal of the Geological Society*, **146**, 175–81.

Hammond, A.L. (ed.) (1990) *World Resources 1990–91*. Oxford University Press, Oxford.

Hedges, J.I., Clark, W.A., Quay, P.D., Richey, J.E., Devol, A.H. and Santos, U.M. (1986) Compositions and fluxes of particulate organic material in the Amazon River. *Limnology and Oceanography*, **31**, 717–38.

Hopkins, D., (1968) Vegetation of the Olokemeji forest reserve, Nigeria V. The vegetation on the savanna site with special reference to its seasonal changes. *Journal of Ecology*, **56**, 97–115.

Houghton, J.T., Jenkins, G.J. and Ephramus, J.J. (eds)(1990) *Climate Change–The IPCC Scientific Assessment*, Cambridge University Press, Cambridge.

Huntley, B.J. and Walker, B.H. (eds)(1982) *Ecology of Tropical Savannas*, Springer-Verlag, Berlin.

Innis, G.S., Noy-Meir, I., Godron, M. and Van Dyne, G.M. (1980) Total-system simulation models, in *Grassland, Systems Analysis and Man*, IBP Vol. 19. (eds A.I. Breymmeyer and G.M. Van Dyne) Cambridge University Press, Cambridge, pp. 759–97.

Johnson, I.R., and Thornley, J.H.M. (1983) Vegetative crop growth model incorporating leaf area expansion and senescence, and applied to grass. *Plant, Cell and Environment*, **6**, 721–9.

Jordan, C.F. (ed.) (1981) *Tropical Ecology*. Hutchinson Ross, Stroudsberg.

Kumar, A. and Joshi, M.C. (1972) The effects of grazing on the structure and productivity of the vegetation near Pilani, Rajasthan, India. *Journal of Ecology*, **60**, 665–74.

Lamotte, M. and Bourlière, F. (1983) Energy flow and nutrient cycling in tropical savannas, in *Tropical Savannas. Ecosystems of the World, 13*. (ed. F. Bourlière), Elsevier, Amsterdam, pp. 583–603.

Lemon, E.R. (ed.) (1983) *CO_2 and Plants: the Response of Plants to Rising Levels of Atmospheric Carbon Dioxide*. Westview Press, Boulder, Colorado.

Lieth, H. (1973) Primary production: terrestrial ecosystems. *Journal of Human Ecology*, **1**, 303–32.

Lieth, H. and Whittaker, R.H. (eds) (1975) *Primary Productivity of the Biosphere*, Springer-Verlag, Berlin.

Long, S.P. and Baker, N.R. (1986) Saline terrestrial environments, in *Photosynthesis in Contrasting Environments*. (eds N.R. Baker and S.P. Long) Elsevier, Amsterdam, pp. 63–102.

Long, S.P., Garcia-Moya, E., Imbamba, S.K., Kamnalrut, A., Piedade, M.T.F., Scurlock, J.M.O., Shen, Y.K. and Hall, D.O. (1989) Primary productivity of natural grass ecosystems of the tropics: a reappraisal. *Plant and Soil*, **115**, 155–66.

Long, S.P. and Hutchins, P. (1991) Predicting the effects of climate change on grasslands – mechanisms. *Ecological Applications*, in press.

Ludlow, M. (1985) Photosynthesis and dry matter production in C_3 and C_4 pasture plants, with special emphasis on tropical C_3 legumes and C_4 grasses. *Australian Journal of Plant Physiology*, **12**, 557–72.

McNaughton, S.J. (1985) Ecology of a grazing ecosystem: the Serengeti. *Ecological Monographs*, **55**, 259–94.

Menaut, J.C. and Cesor, J. (1979) Structure and primary productivity of Lamto Savannas, Ivory Coast. *Ecology*, **60**, 1197–210.

Milner, C. and Hughes, R.E. (1968) *Methods for Measuring the Primary Production of Grasslands*. Blackwell Scientific Publications, Oxford.

Monteith, J.L. (1978) Reassessment of maximum growth rates for C_3 and C_4 crops. *Experimental Agriculture*, **14**, 1–5.

Monteith, J.L. and Unsworth, M.H. (1990) *Principles of Environmental Physics*. 2nd edn, Arnold, London.

Murphy, P.G. (1975) Net primary production in tropical terrestrial ecosystems, in *Primary Productivity of the Biosphere* (eds. H. Lieth and R.H. Whittaker), Springer-Verlag, Berlin, pp. 217–31.

Nandi, A.K. and Pal, B.C. (1983) Turnover rates of a grassland community. *Geobios*, **10**, 13–16.

Oechel, W.C. and Reichers, G.H. (1986) *Response of a Tundra Ecosystem to Elevated Atmospheric Carbon Dioxide*. US Department of Energy, Washington, DC.

Ohiagu, C.E. and Wood, T.G. (1979) Grass production and decomposition in Southern Guinea Savanna, Nigeria. *Oecologia*, **40**, 155–65.

Piedade, M.T.F., Junk, W. and Long, S.P. (1991) The primary productivity of the C_4 grass *Echinochloa polystachya* on the floodplains of the central Amazon. *Ecology*, in press.

Pratt, D.J., Greenway, P.J. and Gwynne, M.D. (1966) A classification of East African rangeland, with an appendix on terminology. *Journal of Applied Ecology* **3**, 369–82.

Roberts, M.J., Long, S.P., Tieszen, L.L. and Beadle, C.L. (1985) Measurements of plant biomass and net primary production, in *Techniques in Bioproductivity and Photosynthesis*, 2nd edn (eds J.

Coombs, D.O. Hall, S.P. Long and J.M.O. Scurlock), Pergamon Press, Oxford, pp. 1–9.

Robson, M.J., Ryle, G.J.A. and Woledge, J. (1988) The grass plant – its form and function, in *The Grass Crop*. (eds M.B. Jones and A. Lazenby), Chapman and Hall, London, pp. 25–84.

Sala, O.E., Parton, W.J., Joyce, L.A. and Laurenroth, W.K. (1988) Primary production of the central grassland region of the United States. *Ecology*, **69**, 40–5.

Saugier, B., Ripley, E.A. and Leuke, P. (1974) Modelling VIII. A mechanistic model of plant growth and water use for the Matador Grassland. Matador Project Technical Report 65. Canadian Committee for IBP, University of Saskatchewan, Saskatoon.

Schlesinger, W.H., (1990) Evidence from chronosequence studies for a low carbon storage potential of soils. *Nature*, **348**, 232–4.

Shankar, V., Shankarnavan, K.A. and Rai, P. (1973) Primary productivity, energetics and nutrient cycling in *Sehima-Heteropogon* grassland. I. Seasonal variations in composition, standing crop and net production. *Tropical Ecology*, **14**, 238–51.

Singh, J.S. and Joshi, M.C. (1979) Primary production, in *Grassland Ecosystems of the World. IBP* Vol 18 (ed. R.T. Coupland), Cambridge University Press, Cambridge, pp. 197–225.

Singh, J.S., Laurenroth, W.K. and Steinhurst, R.K. (1975) Review and assessment of various techniques for estimating net aerial primary production in grasslands from harvest data. *Botanical Reviews*, **41**, 181–232.

Singh, J.S. and Misra, R. (1969) Diversity, dominance, stability, and net production in the grasslands at Varanasi, India. *Canadian Journal of Botany*, **47**, 425–7.

Singh, J.S., Singh, K.P. and Yadava, P.S. (1979) Tropical grasslands ecosystems synthesis, in *Grassland Ecosystems of the World*, IBP Vol. 18 (ed. R.T. Coupland), Cambridge University Press, Cambridge, pp. 231–40.

Singh, J.S., Trlica, M.J., Risser, P.G., Redmann, R.E. and Marshall, J.K. (1980) Autotrophic subsystem, in *Grassland, Systems Analysis and Man*, IBP Vol. 19. (eds A.I. Breymmeyer and G.M. Van Dyne) Cambridge University Press, Cambridge, pp. 59–200.

Singh, J.S. and Yadava, P.S. (1974) Seasonal variation in composition, plant biomass, and net primary productivity of a tropical grassland at Kurukshetra, India. *Ecological Monographs*, **44**, 351–76.

Stanton, N.L. (1988) The underground in grasslands. *Annual Review of Ecology and Systematics*, **19**, 573–89.

Strugnell, R.G. and Piggott, C.D. (1978) Biomass, shoot-production and grazing of two grasslands in the Rwenzori National Park, Uganda. *Journal of Ecology*, **66**, 73–96.

Tans, P.P., Fung, I.Y. and Taro, T. (1990) Observational constraints on the global atmospheric CO_2 budget. *Science*, **247**, 1431–8.

Uchijima, Z. and Seino, H. (1985) Agroclimatic evaluation of net primary productivity of natural vegetations. (1). Chikugo model for evaluating net primary productivity. *Journal of Agricultural Meteorology*, **40**, 343–52.

Uchijima, Z and Seino, H. (1987) *Maps of net primary productivity of natural vegetation on continents*. National Institute of Agro-Environmental Sciences, Syushu National Agricultural Experiment Station, Chikugo, Japan.

Uchijima, Z. and Seino, H. (1988) An agroclimatic method of estimating net primary productivity of natural vegetation. *Japanese Agricultural Research Quarterly*, **21**, 244–50.

Walter, H. (1984) *Vegetation of the Earth and Ecological Systems of the Geo-biosphere*, 3rd edn, Springer-Verlag, Berlin.

Warrick, R.A. (1986) Photosynthesis seen from above. *Nature*, **319**, 181.

Wiegert, R.G. and Evans, F.C. (1964) Primary production and the disappearance of dead vegetation on an old field in southeastern Michigan. *Ecology*, **45**, 49–63.

Epilogue

This book summarizes data arising from the research conducted under UNEP Project FP1305-83-01(2405)(new). This research was initiated in 1983 with a review of methodology and the identification of regional centres to conduct the research. Field work commenced in 1984. The Project was initiated in response to the realization that understanding of primary production, carbon flow and photosynthetic resources within tropical grassland and sub-tropical bamboo forest ecosystems of the developing world, was lacking relative to understanding of many other terrestrial biomes. To-day tropical grasslands collectively represent the world's largest biome and their area continues to grow at the expense of tropical forest. Recent research on temperate grasslands and other herbaceous communities has suggested that methods applied to these communities in the IBP may have seriously underestimated production and carbon flow. It seemed likely that the same errors would apply to the IBP studies of tropical grasslands. The IBP studies had to date been the major source of information of production and carbon flow in tropical grasslands. Thus within the context of UNEP's former Outer Limits programme this lack of information on production and carbon flow within this major biome and the need to re-evaluate the earlier IBP estimates were seen as important gaps in understanding of the biosphere. The recognition of the impending global climate change that would be caused largely by rising levels of CO_2 and the possibility that tropical grasslands had the potential to act as either sources of or sinks for CO_2 underlined need to improve knowledge of this biome.

Many previous studies of tropical grasslands have taken the form of expeditions by, or temporary relocations of, experts from developed countries. This however has the disadvantage of a short-lived study in a biome notorious for its year-to-year and seasonal variations in climate and hence in system processes.The need identified, was not simply for more intensive and in-depth studies of primary production within this biome, but to catalyse the development of expertise and hence a capacity for long term studies within developing countries of the tropics. This would provide the capacity not only for describing the current situation, but also provide a basis for the longer-term monitoring and analysis of change in ecosystem function in relation to climate change over the coming years. Thus, training and support has been an integral part of this project. Indeed, some of the chapter authors attended our training courses and conducted their early research in this project as graduate students. For example, Dr Piedade conducted her graduate studies within the context of this project, and is now a permanent member of staff at the Instituto Nacional de Pesquisas da Amazônia and is the senior scientist co-ordinating all primary production studies within the joint Government of Brazil/IAEA Amazonia 1 project. She has published her work in major international journals and has been invited to present her work to a number of international meetings. Dr Kinyamario, also conducted his graduate studies within the early stages of this project and he is now a Lecturer at the University of Nairobi and has been invited to teach methods of Primary Production measurement on international training courses for other scientists. There are other similar examples within the ranks of the authors of this book, underlining the great success of the project in developing permanent expertise within the tropics.

A key factor in maintaining comparability of approach has been the provision of support through a network of centres. Five centres were established, each within the developing world, co-ordinated through Professor D.O. Hall at King's College, London, and provided with advice in methods, equipment and data analysis through Professor S.P. Long at the project's Technical Support Unit at the University of Essex. The success of the training and support afforded by the network of centres is emphasised by the contribution of a significant and unique, yet integrated, chapter in this book from each centre.

The Technical Support Unit also holds the database established through this research. The database now contains over 50 000 items of data and probably the longest running continuous data sets of monthly variation in biomass, production and vegetation turnover for any natural non-forest vegetation. This is a resource which has not escaped the modeller's attention and indeed has been chosen for an international comparison of ecosystem models of production and decomposition in the SCOPE Project on 'The Effects of Climate Change on Production and Decomposition in Forests and Grasslands'.

A major asset of this study is the unified approach used by researchers on different continents who have met regularly within the framework of the UNEP Study Group system to agree on methods and to discuss their results with both representatives of the centres and with selected international experts. However, there have also been many individual findings of significance. The project has identified the most productive vegetation yet known, a floodplain grassland of the Amazon (Chapter 5) which yielded 100 tons of dry matter per hectare per year. Studies of the semi-arid grasslands (Chapters 2–4) emphasize an initial hypothesis of the work, namely that the earlier IBP studies had grossly underestimated the productivity of these systems. In the humid grassland of Hat Yai in S. Thailand the error can be as great as 10-fold in one example. Of more concern though is the demonstration that this underestimation is not constant, but highly variable both within and between sites. Concurrent examination of production, biomass, storage of dead vegetation and decomposition shows that these systems have much potential to store carbon in the absence of fires and indeed could sequester very significant quantities of carbon from the atmosphere. However, increased frequency of burning in S. Thailand reverses this, making the grassland a net source of carbon.

Remote sensing has for some time now been recognized as a key tool in tracking vegetation change and even as a simple means of measuring production. There have however been few efforts as intense as those made here to establish ground truth for remote sensing of primary production in natural vegetation. For each site ground truth was examined at monthly intervals over at least one year for at least twenty different quadrats which were then destructively harvested. By combining their monthly measurements the chapter authors have been able to provide some of the most detailed information on the relationship of remote sensing signals with plant biomass, leaf area and production. Regrettably, but importantly, the results show the relationship to be far from constant. Similarly, solar radiation absorption by vegetation noted as such as a useful predictor of crop dry matter accumulation has also proved to be a poor predictor of production in these natural grasslands. The relationship varies not only between sites, but as is clearly demonstrated for S. Thailand also varies several fold between years within the same site. This finding must stimulate a detailed rethinking of the feasibility and approaches to remote sensing of the production of tropical grasslands.

The project, like all successful research, raises more questions than it answers. How applicable are these findings to a wider range of tropical grasslands? Of the very much increased quantities of carbon that this research shows to be entering the system through the turnover of the vegetation, how much is really sequestered into refractile carbon pools? What additional remote sensing signals need to be combined with estimates of canopy interception efficiency to be able to predict production with more repeatability, at least within a vegetation type? Finally, and most importantly, how will these systems respond to the possibility of decreased soil moisture levels and increased temperatures (both increasing the incidence of drought limitation) in parallel with rising CO_2 levels which will increase water use efficiency? The outcome of these two coupled effects of climate change will determine whether the world's largest terrestrial biome will become a net sink or source for CO_2 as the 'Greenhouse Effect' intensifies over the coming decades. A basis for answering some of these points is now being established through a current UNEP Project: 'Environmental changes and the productivity of tropical grasslands'; FP/6108-88-01 (2855). The results of this successor project, it is hoped, will provide decision-makers with scientifically sound data on how bioproductivity and certain types of vegetation would be affected by the predicted changes in climate of the earth.

Professor R.J. Olembo
Deputy Assistant Executive Director
United Nations Environment Programme
Nairobi.

Index